International Series in Operations Research & Management Science

Volume 296

More information about this series at http://www.springer.com/series/6161

Kurt Marti

Optimization Under Stochastic Uncertainty

Methods, Control and Random Search Methods

Springer

Kurt Marti
Institute for Mathematics
and Computer Science
University of Bundeswehr Munich
Munich, Germany

ISSN 0884-8289 ISSN 2214-7934 (electronic)
International Series in Operations Research & Management Science
ISBN 978-3-030-55661-7 ISBN 978-3-030-55662-4 (eBook)
https://doi.org/10.1007/978-3-030-55662-4

This Springer imprint is published by the registered company Springer Nature Switzerland AG
The registered company address is: Gewerbestrasse 11, 6330 Cham, Switzerland

Preface

Optimization problems in practice depend mostly on several model parameters, noise factors, uncontrollable parameters, etc., which are not given fixed quantities at the planning stage. Typical examples from engineering and economics/operations research are: material parameters, manufacturing errors, tolerances, noise terms, demand parameters, technological coefficients in input–output functions, cost factors, etc. Due to several types of stochastic uncertainties (physical uncertainty, economic uncertainty, statistical uncertainty, and model uncertainty), these parameters must be modeled by random variables having a certain probability distribution. In most cases, at least certain moments of this distribution are known. In order to cope with these uncertainties, a basic procedure in the engineering/economic practice is to replace first the unknown parameters by some chosen nominal values, e.g., estimates, guesses, of the parameters. Then, the resulting and mostly increasing deviation of the performance (output and behavior) of the structure/system from the prescribed performance (output and behavior), i.e., the "tracking error," is compensated by (online) input corrections. However, the online correction of a system/structure is often time-consuming and causes mostly increasing expenses (correction or recourse costs). This can be omitted to a large extent by taking into account already at the planning stage the consequences of the possible decisions and the known prior and sample information about the random data of the problem. Hence, instead of relying on ordinary deterministic parameter optimization methods—based on some nominal parameter values—and applying then just some correction actions, stochastic optimization methods should be applied: Incorporating stochastic parameter variations into the optimization process, expensive and increasing online correction expenses can be omitted or at least reduced to a large extent.

Consequently, for the computation of optimal decisions which are insensitive with respect to random parameter variations, hence, robust optimal decisions, appropriate deterministic substitute problems must be formulated first. Based on decision theoretical principles, these substitute problems depend on probabilities of failure/success and/or on more general expected cost/loss terms. Since probabilities and expectations are defined by multiple integrals in general, the resulting often nonlinear and also nonconvex deterministic substitute problems can be solved by approximate methods only. Two basic types of deterministic substitute problems occur mostly in practice cost minimization subject to expected recourse (correction) cost: Minimization of the expected primary costs subject to expected recourse cost constraints (reliability constraints) and remaining deterministic constraints, e.g., box constraints, and minimization of the expected total costs (costs of construction, design, recourse costs, etc.) subject to the remaining deterministic constraints.

After an introduction into the theory of dynamic control systems with random parameters, the major control laws are described, as open-loop control, closed-loop, feedback control, and open-loop feedback control, used for iterative construction of feedback controls. Taylor expansion methods are considered for solving stochastic optimization problems involving expected cost or loss functions. Applications of Taylor expansion methods to control problems with linear and sublinear cost functions and the computation of stochastic open-loop controls for tracking problems complete Chap. 1. A detailed presentation of the optimization of regulators under stochastic uncertainty is as follows in Chap. 2. Optimal stochastic open-loop control problems with more general dynamic equation are examined in Chap. 3. The construction of optimal feedback controls for control problems with random parameters by means of homotopy methods can be found in Chap. 4. Limit state functions for applying in reliability-based design problems arising in static and time-dependent stochastic optimization problems are constructed in Chap. 5.

For the numerical solution of the deterministic substitute problems as arising in the treatment of stochastic optimization and stochastic optimal control problems as considered in Part I, efficient numerical optimization routines are needed. Besides mathematical optimization techniques, one of the major methods for solving deterministic parameter optimization problems, as arising in the handling of optimization and optimal control problems with random parameters in Part I, are stochastic search methods. For many years, mostly (deterministic) mathematical programming (MP) procedures were used to solve (approximately) deterministic parameter optimization problems, hence, mathematical programs. However, due to the complexity of concrete optimization/control problems and their often lack of mathematical regularity as required of MP techniques, other optimization techniques like random search methods (RSM) became increasingly important. Hence, in the second Part II of this monograph basic results on the convergence and convergence rates of random search methods are presented. Moreover, methods, based on optimal stochastic decision processes, for the improvement of the—sometimes very low—convergence rates are presented: After the description of important classes of random search methods, such as random direction methods and procedures with absolutely continuous mutation sequences, conditions are considered which guarantee the convergence of random search methods. Moreover, convergence rates are determined. In order to improve the convergence behavior of RSM, the random search procedure is embedded into a stochastic decision process for an optimal control of the probability distributions of the mutation random variables. Basic theoretical results on the convergence and convergence rate of random search methods are taken from the doctoral thesis of G. Rappl. Several mathematical prerequisites are provided in the appendix.

Last but not least, I would like to thank Dipl. Math. Ina Stein for her excellent support in the LaTeX-typesetting as well as in the final proofreading. Moreover, I am indebted to Springer-Verlag for inclusion of the book into the Springer International Series in Operations Research & Management Science (ISOR). I would like to thank especially the Senior Editor for Business/Economics/Operations Research of Springer-Verlag Heidelberg, Germany, Christian Rauscher, and the Editor for Operations Research/Management Science and Information Systems of Springer-Verlag New York, USA, Neil Levine, for their advice until the completion of this book.

Munich, Germany
October 2020

Kurt Marti

Contents

Symbols and Abbreviations

$\mathbb{N}$	Set of natural numbers
$\mathbb{N}_0$	$\mathbb{N} \cup \{0\}$
$\mathbb{R}_+$	$\{x \in \mathbb{R}; x \geq 0\}$
$\mathbb{R}^d$	d-dimensional Euclidean space
S	Unit sphere of the $\mathbb{R}^d$
B^d	Borel-σ-algebra of the $\mathbb{R}^d$
$B^d \mid_M$	Restriction of B^d to $M \subset \mathbb{R}^d$
$\lvert A \rvert$	Cardinality of the set A
$A^{\complement}$	Complement of A
$\mathring{A}$	Interior of A
$\bar{A}$	Closure of A
rd A	Boundary of A
x^T	Transposed vector s (vectors are generally constructed as column vectors)
$\mid x \mid$	Euclidean norm
dist(x, A)	Distance from $x \in \mathbb{R}^d$ to $A \subset \mathbb{R}^d$
$K(x, r)$	Open sphere with center at x and radius r
det C resp. $\lvert C \rvert$	Determinant of C
$\lfloor r \rfloor$	Largest whole number $\leq r$
$\Gamma(x)$	Euler gamma function
$\sim$	$a_n \sim b_n \iff a_n/b_n \to l$
f	Function to be minimized
D	Set of admissible points
D_a	$\{x \in D : f(x) \leq a\}$ Level set for the level a
f^*	$\inf\{f(x) : x \in D\}$
D^*	$\{x \in D : f(x) \leq f(y)$ for all $y \in D\}$ Minimal set
x^*	Point from D^*
ε_x	Singular measure, concentrated on x
λ^d	d-dimensional Lebesgue measure
P_X	Distribution of x

$T(\mu)$	Pushforward measure of μ under T
$A \otimes B$	Product-σ-algebra
$\mu \otimes \nu$	Product measure
p_n	Mutation transition probability
q_n	Selection transition probability
R-S-M	Random search method
R-D-M	Random direction method
γ	Uniform distribution on S
λ_n	Step length distribution
ZV	Random variable
TP	Transition probability

Part I
Stochastic Optimization Methods

Chapter 1
Optimal Control Under Stochastic Uncertainty

1.1 Stochastic Control Systems

Optimal control and regulator problems arise in many concrete applications (mechanical, electrical, thermodynamical, chemical, etc.) are modeled [3, 29, 33] by dynamical control systems obtained from physical measurements and/or known physical (a priori) laws. The basic control system (input–output system) is mathematically represented [16, 34] by a system of first order differential equations with random parameters:

$$\dot{z}(t) = g\Big(t, \omega, z(t), u(t)\Big), t_0 \leq t \leq t_f, \ \omega \in \Omega \tag{1.1a}$$

$$z(t_0) = z_0(\omega). \tag{1.1b}$$

Here, ω is the basic random element taking values in a probability space $(\Omega, \mathcal{A}, P)$, and describing the random variations of model parameters or the influence of noise terms. The probability space $(\Omega, \mathcal{A}, P)$ consists of the sample space or set of elementary events Ω, the σ-algebra $\mathcal{A}$ of events and the probability measure P. The plant state vector $z = z(t, \omega)$ is an m-vector involving direct or indirect measurable/observable quantities like displacements, stresses, voltage, current, pressure, concentrations, etc., and their time derivatives (velocities), $z_0(\omega)$ is the random initial state. The plant control or control input $u(t)$ is a deterministic or stochastic n-vector denoting system inputs like external forces or moments, voltages, field current, thrust program, fuel consumption, production rate, etc. Furthermore, $\dot{z}$ denotes the derivative with respect to the time t. We assume that an input $u = u(t)$ is chosen such that $u(\cdot) \in U$, where U is a suitable linear space of input functions $u(\cdot) : [t_0, t_f] \rightarrow \mathbb{R}^n$ on the time interval $[t_0, t_f]$. Examples for U are subspaces of the space $PC_0^n[t_0, t_f]$ of piecewise continuous functions

K. Marti, *Optimization Under Stochastic Uncertainty*, International Series in Operations Research & Management Science 296,
https://doi.org/10.1007/978-3-030-55662-4_1

$u(\cdot) : [t_0, t_f] \to \mathbb{R}^n$ normed by the supremum norm

$$\left\|u(\cdot)\right\|_\infty = \sup\left\{\left\|u(t)\right\| : t_0 \le t \le t_f\right\}.$$

Since feedback control (FB) laws can be approximated very efficiently, cf. [2, 17, 29], by means of *open-loop feedback (OLF)* control laws, see Sect. 1.2, for practical purposes we may confine to the computation of stochastic optimal open-loop (OL) controls $u = u(\cdot; t_b)$, $t_b \le t \le t_f$, on arbitrary *remaining time intervals* $[t_b, t_f]$ of $[t_0, t_f]$. Here, $u = u(\cdot; t_b)$ is stochastic optimal with respect to the information $\mathcal{A}_{t_b}$ available at the corresponding *initial* time point t_b.

1.1.1 Differential and Integral Equations Under Stochastic Uncertainty

In many technical applications the random variations are not caused by an additive white noise term, but by means of possibly time-dependent random parameters. Hence, in the following the dynamics of the control system is represented by a differential equation under stochastic uncertainty, i.e. a *system of ordinary differential equations (1.1a,b) with random parameters*. Furthermore, solutions of differential equations under stochastic uncertainty are defined here in the parameter (ω-point)-wise sense, cf. [5].

In case of a *discrete* or *discretized* probability distribution of the random elements, model parameters, i.e.

$$\Omega = \{\omega_1, \omega_2, \ldots, \omega_\varrho\}, \quad P(\omega = \omega_j) = \alpha_j > 0, j = 1, \ldots, \varrho, \quad \sum_{j=1}^{\varrho} \alpha_j = 1,$$

we can *redefine* (1.1a,b) by

$$\dot{z}(t) = g\Big(t, z(t), u(t)\Big), \quad t_0 \le t \le t_f, \tag{1.1c}$$

$$z(t_0) = z_0. \tag{1.1d}$$

With the vectors and vector functions

$$z(t) := \Big(z(t, \omega_j)\Big)_{j=1,\ldots,\varrho}, \qquad z_0 := \Big(z_0(\omega_j)\Big)_{j=1,\ldots,\varrho}$$

$$g(t, z, u) := \Big(g(t, \omega_j, z_{(j)}, u)\Big)_{j=1,\ldots,\varrho}, \qquad z := (z_{(j)})_{j=1,\ldots,\varrho} \in \mathbb{R}^{\varrho m}.$$

Hence, for a discrete distribution P we have again an ordinary system of first order differential equations for the ϱm unknown functions

$$z_{ij} = z_i(t, \omega_j), \qquad i = 1, \dots, m, j = 1, \dots, \varrho.$$

Results on the existence and uniqueness of the systems (1.1a,b) and (1.1c,d) and their dependence on the inputs can be found in [8].

Also in the general case we consider a *solution in the point-wise sense*. This means that for each random element $\omega \in \Omega$, (1.1a,b) is interpreted as a system of ordinary first order differential equations with the initial values $z_0 = z_0(\omega)$ and control input $u = u(t)$. Hence, we assume that to each deterministic control $u(\cdot) \in U$ and each random element $\omega \in \Omega$ there exists a unique solution

$$z(\cdot, \omega) = S\Big(\omega, u(\cdot)\Big) = S\Big(\omega, u(\cdot)\Big), \tag{1.2a}$$

$z(\cdot, \omega) \in C_0^m[t_0, t_f]$, of the integral equation

$$z(t) = z_0(\omega) + \int_{t_0}^{t} g\Big(s, \omega, z(s)u(s)\Big)ds, \; t_0 \le t \le t_f, \tag{1.2b}$$

such that $(t, \omega) \to S\Big(\omega, u(\cdot)\Big)(t)$ is measurable. This solution is also denoted by

$$z(t, \omega) = z_u(t, \omega) = z(t, \omega, u(\cdot)), \; t_0 \le t \le t_f. \tag{1.2c}$$

Obviously, the integral equation (1.2a–c) is the integral version of the initial value problem (1.1a,b): Indeed, if, for given $\omega \in \Omega$, $z = z(t, \omega)$ is a solution of (1.1a,b), i.e. $z(\cdot, \omega)$ is absolutely continuous, satisfies (1.1a) for almost all $t \in [t_0, t_f]$ and fulfills (1.1b), then $z = z(t, \omega)$ is also a solution of (1.2a–c). Conversely, if, for given $\omega \in \Omega$, $z = z(t, \omega)$ is a solution of (1.2a–c), such that the integral on the right hand side exists in the Lebesgue-sense for each $t \in [t_0, t_f]$, then this integral as a function of the upper bound t and therefore also the function $z = z(t, \omega)$ is absolutely continuous. Hence, by taking $t = t_0$ and by differentiation of (1.2a–c) with respect to t, we have that $z = z(t, \omega)$ is also a solution of (1.1a,b).

1.1.1.1 Parametric Representation of the Differential/Integral Equation Under Stochastic Uncertainty

In the following we want to justify the above assumption that the initial value problem (1.1a,b), the equivalent integral equation (1.2a–c), resp., has a unique solution $z = z(t, \omega)$. For this purpose, let $\theta = \theta(t, \omega)$ be an r-dimensional stochastic process, as, e.g., time-varying disturbances, random parameters, etc., of the system, such that the sample functions $\theta(\cdot, \omega)$ are continuous with probability one. Furthermore, let

$$\tilde{g} : [t_0, t_f] \times \mathbb{R}^r \times \mathbb{R}^m \times \mathbb{R}^n \to \mathbb{R}^m$$

be a continuous function having continuous Jacobians $D_\theta \tilde{g}$, $D_z \tilde{g}$, $D_u \tilde{g}$ with respect to θ, z, u. Now consider the case that the function g of the process differential equation (1.1a,b) is given by

$$g(t, \omega, z, u) := \tilde{g}\Big(t, \theta(t, \omega), z, u\Big),$$

$(t, \omega, z, u) \in [t_0, t_f] \times \Omega \times \mathbb{R}^m \times \mathbb{R}^n$. The spaces U, Z of possible inputs, trajectories, resp., of the plant are chosen as follows: $U := Reg([t_0, t_f]; \mathbb{R}^n)$ is the Banach space of all regulated functions $u(\cdot) : [t_0, t_f] \to \mathbb{R}^n$, normed by the supremum norm $\| \cdot \|_\infty$.

Furthermore, we set $Z := C_0^m[t_0, t_f]$ and $\Theta := C_0^r[t_0, t_f]$. Here, for an integer ν, $C_0^\nu[t_0, t_f]$ denotes the Banach space of all continuous functions of $[t_0, t_f]$ into $\mathbb{R}^\nu$ normed by the supremum norm $\| \cdot \|_\infty$. By our assumption we have $\theta(\cdot, \omega) \in \Theta$ a.s. (almost sure). Define

$$\Xi = \mathbb{R}^m \times \Theta \times U;$$

Ξ is the space of possible *initial values, time-varying model/environmental parameters, and inputs* of the dynamic system. Hence, Ξ may be considered as the total space of inputs

$$\xi := \begin{pmatrix} z_0 \\ \theta(\cdot) \\ u(\cdot) \end{pmatrix}$$

into the plant, consisting of the random initial state z_0, the random input function $\theta = \theta(t, \omega)$, and the control function $u = u(t)$. Let now the mapping $\tau : \Xi \times Z \to Z$ related to the plant equation (1.1a,b) or (1.2a-c) be given by

$$\tau\Big(\xi, z(\cdot)\Big)(t) = z(t) - \left(z_0 + \int_{t_0}^{t} \tilde{g}\Big(s, \theta(s), z(s), u(s)\Big) ds \right), \; t_0 \le t \le t_f. \tag{1.3a}$$

Note that for each input vector $\xi \in \Xi$ and function $z(\cdot) \in Z$ the integrand in (1.3a) is piecewise continuous, bounded, or at least essentially bounded on $[t_0, t_f]$. Hence, the integral in (1.3a) as a function of its upper limit t yields again a continuous function on the interval $[t_0, t_f]$, and therefore an element of Z. This shows that τ maps $\Xi \times Z$ into Z.

Obviously, the initial value problem (1.1a,b) or its integral form (1.2a–c) can be represented by the operator equation

$$\tau(\xi, z(\cdot)) = 0. \tag{1.3b}$$

Operators of the type (1.3a) are well studied, see, e.g., [8, 18]: It is known that τ is continuously Fréchet (F)-differentiable [8, 18]. Note that the F-differential is a generalization of the derivatives (Jacobians) of mappings between finite-dimensional spaces to mappings between arbitrary normed spaces. Thus, the F-derivative $D\tau$ of τ at a certain point $\left(\bar{\xi}, \bar{z}(\cdot)\right)$ is given by

$$\begin{aligned}
&\left(D\tau\left(\bar{\xi}, \bar{z}(\cdot)\right) \cdot \left(\xi, z(\cdot)\right)\right)(t) \\
&= z(t) - \left(z_0 + \int_{t_0}^{t} D_z\tilde{g}\left(s, \bar{\theta}(s), \bar{z}(s), \bar{u}(s)\right) z(s) ds \right. \\
&+ \int_{t_0}^{t} D_\theta\tilde{g}\left(s, \bar{\theta}(s), \bar{z}(s), \bar{u}(s)\right) \theta(s) ds \\
&\left. + \int_{t_0}^{t} D_u\tilde{g}\left(s, \bar{\theta}(s), \bar{z}(s), \bar{u}(s)\right) u(s) ds \right), \; t_0 \le t \le t_f,
\end{aligned} \tag{1.3c}$$

where $\bar{\xi} = \left(\bar{z}_0, \bar{\theta}(\cdot), \bar{u}(\cdot)\right)$ and $\xi = \left(z_0, \theta(\cdot), u(\cdot)\right)$. Especially, for the derivative of τ with respect to $z(\cdot)$ we find

$$\left(D_z\tau\left(\bar{\xi}, \bar{z}(\cdot)\right) \cdot z(\cdot)\right)(t) = z(t) - \int_{t_0}^{t} D_z\tilde{g}\left(s, \bar{\theta}(s), \bar{z}(s), \bar{u}(s)\right) z(s) ds, \; t_0 \le t \le t_f. \tag{1.3d}$$

The related equation

$$D_z\tau\left(\bar{\xi}, \bar{z}(\cdot)\right) \cdot z(\cdot) = y(\cdot), \; y(\cdot) \in Z, \tag{1.3e}$$

is a linear vectorial Volterra integral equation. By our assumptions this equation has a unique solution $z(\cdot) \in Z$. Note that the corresponding result for scalar Volterra equations, see, e.g., [31], can be transferred to the present vectorial case. Therefore, $D_z\tau\left(\bar{\xi}, \bar{z}(\cdot)\right)$ is a linear, continuous one-to-one map from Z onto Z. Hence, its inverse $\left(D_z\tau\left(\bar{\xi}, \bar{z}(\cdot)\right)\right)^{-1}$ exists. Using the implicit function theorem [8, 18], we obtain now the following result:

Lemma 1.1 *For given $\bar{\xi} = \left(\bar{z}_0, \bar{\theta}(\cdot), \bar{u}(\cdot)\right)$, let $\left(\bar{\xi}, \bar{z}(\cdot)\right) \in \Xi \times Z$ be selected such that $\tau\left(\bar{\xi}, \bar{z}(\cdot)\right) = 0$, hence, $\bar{z}(\cdot) \in Z$ is supposed to be the solution of*

$$\dot{z}(t) = \tilde{g}\Big(t, \bar{\theta}(t), z(t), \bar{u}(t)\Big), t_0 \le t \le t_f, \tag{1.4a}$$

$$z(t_0) = \bar{z}_0 \tag{1.4b}$$

in the integral sense (1.2b). Then there is an open neighborhood of $\bar{\xi}$*, denoted by* $V^0(\bar{\xi})$*, such that for each open connected neighborhood* $V(\bar{\xi})$ *of* $\bar{\xi}$ *contained in* $V^0(\bar{\xi})$ *there exists a unique continuous mapping* $S : V(\bar{\xi}) \to Z$ *such that a)* $S(\bar{\xi}) = \bar{z}(\cdot)$*; b)* $\tau\Big(\xi, S(\xi)\Big) = 0$ *for each* $\xi \in V(\bar{\xi})$*, i.e.* $S(\xi) = S(\xi)(t), t_0 \le t \le t_f$*, is the solution of*

$$z(t) = z_0 + \int_{t_0}^{t} \tilde{g}\Big(s, \theta(s), z(s), u(s)\Big)ds, t_0 \le t \le t_f, \tag{1.4c}$$

where $\xi = \Big(z_0, \theta(\cdot), u(\cdot)\Big)$*; c) S is continuously differentiable on* $V(\bar{\xi})$*, and it holds*

$$D_u S(\xi) = -\Big(D_z \tau\big(\xi, S(\xi)\big)\Big)^{-1} D_u \tau\Big(\xi, S(\xi)\Big), \xi \in V(\bar{\xi}). \tag{1.4d}$$

An immediate consequence is given next:

Corollary 1.1 *The directional derivative* $\zeta(\cdot) = \zeta_{u,h}(\cdot) = D_u S(\xi)h(\cdot) \in Z, h(\cdot) \in U$*, satisfies the integral equation*

$$\zeta(t) - \int_{t_0}^{t} D_z \tilde{g}\Big(s, \theta(s), S(\xi)(s), u(s)\Big)\zeta(s)ds$$

$$= \int_{t_0}^{t} D_u \tilde{g}\Big(s, \theta(s), S(\xi)(s), u(s)\Big)h(s)ds, \tag{1.4e}$$

where $t_0 \le t \le t_f$ *and* $\xi = \Big(z_0, \theta(\cdot), u(\cdot)\Big)$.

Remark 1.1 Taking the time derivative of Eq. (1.4e) shows that this integral equation is equivalent to the so-called *perturbation equation*, see, e.g., [16].

For an arbitrary $h(\cdot) \in U$ the mappings

$$(t, \xi) \to S(\xi)(t),\ (t, \xi) \to \Big(D_u S(\xi)h(\cdot)\Big)(t), (t, \xi) \in [t_0, t_f] \times V(\bar{\xi}), \tag{1.4f}$$

are continuous and therefore also measurable.

The existence of a unique solution $\bar{z} = \bar{z}(t), t_0 \le t \le t_f$, of the reference differential equation (1.4a,b) can be guaranteed as follows, where *solution* is interpreted in the integral sense (1.2b), i.e. $\bar{z} = \bar{z}(t), t_0 \leq t \leq t_f$, is absolutely continuous, satisfies equation (1.4a) almost everywhere in the time interval $[t_0, t_f]$ and the initial condition (1.4b), cf. [6] and [36].

Lemma 1.2 *Consider an arbitrary input vector* $\bar{\xi} = \left(\bar{z}_0, \bar{\theta}(\cdot), \bar{u}(\cdot)\right) \in \Xi$, *and define, see (1.4a,b), the function* $\tilde{g}_{\bar{\theta}(\cdot),\bar{u}(\cdot)} = \tilde{g}_{\bar{\theta}(\cdot),\bar{u}(\cdot)}\left(t, z\right) := \tilde{g}\left(t, \bar{\theta}(t), z, \bar{u}(t)\right)$. *Suppose that*

(i) $z \to \tilde{g}_{\bar{\theta}(\cdot),\bar{u}(\cdot)}\left(t, z\right)$ *is continuous for each time* $t \in [t_0, t_f]$,
(ii) $t \to \tilde{g}_{\bar{\theta}(\cdot),\bar{u}(\cdot)}\left(t, z\right)$ *is measurable for each vector* z,
(iii) generalized Lipschitz-condition: For each closed sphere $K \subset \mathbb{R}^n$ *there exists a nonnegative, integrable function* $L_K(\cdot)$ *on* $[t_0, t_f]$ *such that*

(a) $\|\tilde{g}_{\bar{\theta}(\cdot),\bar{u}(\cdot)}\left(t, 0\right)\| \leq L_K(t)$, *and*
(b) $\|\tilde{g}_{\bar{\theta}(\cdot),\bar{u}(\cdot)}\left(t, z\right) - \tilde{g}_{\bar{\theta}(\cdot),\bar{u}(\cdot)}\left(t, w\right)\| \leq L_K(t)\|z - w\|$ *on* $[t_0, t_f] \times K$.

Then, the initial value problem (1.4a,b) has a unique solution $\bar{z} = \bar{z}(t; \bar{\xi})$.

Proof Proofs can be found in [6] and [36]. □

We observe that the controlled stochastic process $z = z(t, \omega)$ defined by the plant differential equation (1.1a,b) may be a non-Markovian stochastic process, see [3]. Moreover, note that the random input function $\theta = \theta(t, \omega)$ is not just an additive noise term, but may describe also a disturbance which is part of the nonlinear dynamics of the plant, random varying model parameters such as material, load or cost parameters, etc.

1.1.2 Objective Function

The aim is to obtain optimal controls being **robust**, i.e. most insensitive with respect to stochastic variations of the model/environmental parameters and initial values of the process. Hence, incorporating stochastic parameter variations into the optimization process, for a *deterministic* control function $u = u(t), t_0 \leq t \leq t_f$, the objective function $F = F\left(u(\cdot)\right)$ of the controlled process $z = z(t, \omega, u(\cdot))$ is defined, cf. [24], by the conditional expectation of the total costs arising along the whole control process:

$$F\left(u(\cdot)\right) := Ef\left(\omega, S\left(\omega, u(\cdot)\right), u(\cdot)\right). \tag{1.5a}$$

Here, $E = E(\cdot|\mathcal{A}_{t_0})$ denotes the conditional expectation given the information $\mathcal{A}_{t_0}$ about the control process up to the considered starting time point t_0. Moreover, $f = f\left(\omega, z(\cdot), u(\cdot)\right)$ denote the stochastic total costs arising along the trajectory $z = z(t, \omega)$ and at the terminal point $z_f = z(t_f, \omega)$, cf. [3, 34]. Hence,

$$f\Big(\omega, z(\cdot), u(\cdot)\Big) := \int_{t_0}^{t_f} L\Big(t, \omega, z(t), u(t)\Big)dt + G\Big(t_f, \omega, z(t_f)\Big), \tag{1.5b}$$

$z(\cdot) \in Z, u(\cdot) \in U$. Here,

$$\begin{aligned} L &: [t_0, t_f] \times \Omega \times \mathbb{R}^m \times \mathbb{R}^n \to \mathbb{R}, \\ G &: [t_0, t_f] \times \Omega \times \mathbb{R}^m \qquad\quad\; \to \mathbb{R} \end{aligned}$$

are given measurable cost functions. We suppose that $L(t, \omega, \cdot, \cdot)$ and $G(t, \omega, \cdot)$ are *convex functions* for each $(t, \omega) \in [t_0, t_f] \times \Omega$, having continuous partial derivatives $\nabla_z L(\cdot, \omega, \cdot, \cdot)$, $\nabla_u L(\cdot, \omega, \cdot, \cdot)$, $\nabla_z G(\cdot, \omega, \cdot)$. Note that in this case

$$\Big(z(\cdot), u(\cdot)\Big) \to \int_{t_0}^{t_f} L\Big(t, \omega, z(t), u(t)\Big)dt + G\Big(t_f, \omega, z(t_f)\Big) \tag{1.5c}$$

is a convex function on $Z \times U$ for each $\omega \in \Omega$. Moreover, assume that the expectation $F\Big(u(\cdot)\Big)$ exists and is finite for each admissible control $u(\cdot) \in D$.

In the case of feedback control $u = u(t, y(t)), t_0 \le t \le t_f$, where $y = y(t)$ denotes the state $z = z(t)$ or a certain state observation, the objective function $F = F\Big(u(\cdot, \cdot)\Big)$ reads

$$F\Big(u(\cdot, \cdot)\Big) := Ef\Bigg(\omega, S\Big(\omega, u(\cdot, \cdot)\Big), u(\cdot, \cdot)\Bigg). \tag{1.5d}$$

Example 1.1 (Tracking Problems) If a trajectory $z_f = z_f(t, \omega)$, e.g., the trajectory of a moving target, known up to a certain stochastic uncertainty, must be followed or reached during the control process, then the cost function L along the trajectory can be defined by

$$L\Big(t, \omega, z(t), u\Big) := \Big\| \Gamma_z\Big(z(t) - z_f(t, \omega)\Big)\Big\|^2 + \varphi(u). \tag{1.6a}$$

Here, Γ_z is a weight matrix, and $\varphi = \varphi(u)$ denotes the control costs, as, e.g.,

$$\varphi(u) = \|\Gamma_u u\|^2 \tag{1.6b}$$

with a further weight matrix Γ_u.

If a random target $z_f = z_f(\omega)$ has to be reached at the terminal time point t_f only, then the terminal cost function G may be defined, e.g., by

$$G\Big(t_f, \omega, z(t_f)\Big) := \Big\| G_f\Big(z(t_f) - z_f(\omega)\Big)\Big\|^2 \tag{1.6c}$$

with a weight matrix G_f.

Example 1.2 (Active Structural Control, Control of Robots) In case of active structural control or for optimal regulator design of robots, cf. [22, 33], the total cost function f is given by defining the individual6a cost functions L and G as follows:

$$L(t, \omega, z, u) := z^T Q(t, \omega) z + u^T R(t, \omega) u \tag{1.7a}$$

$$G(t_f, \omega, z) := G(\omega, z). \tag{1.7b}$$

Here, $Q = Q(t, \omega)$ and $R = R(t, \omega)$, resp., are certain positive (semi)definite $m \times m, n \times n$ matrices which may depend also on (t, ω), where in case of active structural control the external load term depends on the control $u(t)$. For endpoint control, the cost function G is given by

$$G(\omega, z) = (z - z_f)^T G_f(\omega)(z - z_f) \tag{1.7c}$$

with a certain desired, possibly random terminal point $z_f = z_f(\omega)$ and a positive (semi)definite, possibly random weight matrix $G_f = G_f(\omega)$.

1.1.2.1 Optimal Control Under Stochastic Uncertainty

For finding *optimal controls being robust with respect to stochastic parameter variations* $u^\star(\cdot), u^\star(\cdot, \cdot)$, resp., in this chapter we are presenting now several methods for approximation of the following minimum expected total cost problem:

$$\min F\Big(u(\cdot)\Big) \qquad \text{s.t. (subject to) } u(\cdot) \in D, \tag{1.8a}$$

for feedback problems the time-function $u(t)$ is replaced by $u = u(t, y(t))$:

$$\min F\Big(u(\cdot, \cdot)\Big) \qquad \text{s.t. } u(\cdot, \cdot) \in D. \tag{1.8b}$$

Information set $\mathcal{A}_t$ at time t:

The information set $\mathcal{A}_t \subset \mathcal{A}, t \geq t_0$, is defined by the σ-algebra of events $A \in \mathcal{A}$ until time t.

In many cases, as, e.g., for $PD-$ and $PID-$ (feedback) controllers, the information σ-algebra $\mathcal{A}_t$ is given by $\mathcal{A}_t = \mathcal{A}(y(t, \cdot))$, where $y = y(t, \omega)$ denotes the $\bar{m}-$vector function of *state-measurements or -observations* at time t. Then, an $\mathcal{A}_t-$ measurable control $u = u(t, \omega)$ has the representation, cf. [4],

$$u(t, \omega) = \eta(t, y(t, \omega)) \tag{1.8c}$$

with a measurable function $\eta(t, \cdot) : \mathbb{R}^{\bar{m}} \to \mathbb{R}^m$.

Since parameter-insensitive optimal controls can be obtained by stochastic optimization methods incorporating random parameter variations into the optimization procedure, see [24], the aim is to determine *stochastic optimal controls*:

Definition 1.1 An optimal solution of the expected total cost minimization problem (1.8a,b), providing a robust optimal open-loop, optimal feedback control, resp., is called an **optimal control under stochastic uncertainty** or—for short—a **stochastic optimal OL/FB control**.

Note For controlled processes working on a time range $t_b \le t \le t_f$ with an intermediate starting time point t_b, the objective function $F = F(u(\cdot))$ is defined also by (1.5a), but with the conditional expectation operator $E = E(\cdot|\mathcal{A}_{t_b})$, where $\mathcal{A}_{t_b}$ denotes the information set about the controlled process available up to time t_b.

Problem (1.8a) is of course equivalent $\left(E = E(\cdot|\mathcal{A}_{t_0})\right)$ to the *optimal control problem under stochastic uncertainty*:

$$\min E\left(\int_{t_0}^{t_f} L\Big(t,\omega,z(t),u(t)\Big)dt + G\Big(t_f,\omega,z(t_f)\Big)\Big|\mathcal{A}_{t_0}\right) \tag{1.9a}$$

s.t.

$$\dot{z}(t) = g\Big(t,\omega,z(t),u(t)\Big),\ t_0 \le t \le t_f,\ \text{a.s.} \tag{1.9b}$$

$$z(t_0,\omega) = z_0(\omega),\ \text{a.s.} \tag{1.9c}$$

$$u(\cdot) \in D, \tag{1.9d}$$

cf. [19, 20]. For further details see also [28].

Remark 1.2 Similar representations can be obtained also for stochastic optimal feedback control problems.

Remark 1.3 (State Constraints) In addition to the plant differential equation (dynamic equation) and the control constraint we may still have some stochastic state constraints

$$h_I\Big(t,\omega,z(t,\omega)\Big) \le (=)\, 0 \text{ a.s.} \tag{1.10a}$$

as well as state constraints involving (conditional) expectations

$$Eh_{II}\Big(t,\omega,z(t,\omega)\Big) = E\left(h_{II}\Big(t,\omega,z(t,\omega)\Big)\Big|\mathcal{A}_{t_0}\right) \le (=)\, 0. \tag{1.10b}$$

Here, $h_I = h_I(t, \omega, z), h_{II} = h_{II}(t, \omega, z)$ are given vector functions of (t, ω, z). By means of (penalty) cost functions, the random condition can be incorporated into the objective function of the control problem. As explained in Sect. 1.3, the expectations arising in the mean value constraints and in the objective function can be computed approximately by means of Taylor expansion with respect to the vector $\vartheta = \vartheta(\omega) := (z_0(\omega), \theta(\omega))$ of random initial values and model parameters at the conditional mean $\overline{\vartheta} = \overline{\vartheta}^{(t_0)} := E\Big(\vartheta(\omega)|\mathcal{A}_{t_0}\Big)$. This yields then ordinary deterministic constraints for the extended deterministic trajectory (nominal state and sensitivity)

$$t \to \Big(z(t, \overline{\vartheta}), D_{\vartheta} z(t, \overline{\vartheta})\Big), t \geq t_0.$$

1.2 Control Laws

Control or guidance usually refers [3, 16, 18] to direct influence on a dynamic system to achieve desired performance. In optimal control of dynamic systems mostly the following types of *control laws* or *control policies* are considered:

(I) *Open-Loop Control (OL)*
Here, the control function $u = u(t)$ is a **deterministic** function depending only on the (a priori) information $\mathcal{I}_{t_0}$ about the system, the model parameters, resp., available at the starting time point t_0. Hence, for the optimal selection of optimal (OL) controls

$$u(t) = u\Big(t; t_0, \mathcal{I}_{t_0}\Big), \quad t \geq t_0, \tag{1.11a}$$

we get optimal control problems of type (1.9a–d).

(II) *Closed-Loop control (CL) or Feedback Control*
In this case the control function $u = u(t)$ is a **stochastic** function

$$u = u(t, \omega) = u(t, \mathcal{I}_t), \quad t \geq t_0 \tag{1.11b}$$

depending on time t and the total information $\mathcal{I}_t$ about the system available up to time t. Especially $\mathcal{I}_t$ may contain information about the state $z(t) = z(t, \omega)$ up to time t. Optimal (CL) or feedback controls are obtained by solving problems of type (1.8b).

Remark 1.4 (Information Set $\mathcal{A}_t$ at Time t) Often the information $\mathcal{I}_t$ available up to time t is described by the *information set* or $\sigma-$algebra $\mathcal{A}_t \subset \mathcal{A}$ of events A occurred up to time t. In the important case $\mathcal{A}_t = \mathcal{A}(y(t, \cdot))$, where $y = y(t, \omega)$ denotes the $\bar{m}-$vector function of *state-measurements or -observations* at time t, then an $\mathcal{A}_t-$ measurable control $u = u(t, \omega)$, see problem (1.8b), has the representation, cf. [4],

$$u(t, \omega) = \eta_t(y(t, \omega)) \tag{1.11c}$$

with a measurable function $\eta_t : \mathbb{R}^{\bar{m}} \to \mathbb{R}^m$. Important examples of this type are the $PD-$ and $PID-$controllers.

(III) *Open-Loop Feedback (OLF) Control/Stochastic Open-Loop Feedback (SOLF) Control*
Due to their large complexity, in general, optimal feedback control laws can be determined approximately only. A very efficient approximation procedure for optimal feedback controls, being functions of the information $\mathcal{I}_t$, is the approximation by means of optimal open-loop controls. In this combination of (OL) and (CL) control, at each intermediate time point $t_b := t$, $t_0 \le t \le t_f$, given the information $\mathcal{I}_t$ up to time t, first the open-loop control function for the remaining time interval $t \le s \le t_f$, see Fig. 1.1, is computed, hence,

$$u_{[t,t_f]}(s) = u\Big(s; t, \mathcal{I}_t\Big), \quad s \ge t. \tag{1.11d}$$

Then, an approximate feedback control policy, originally proposed by Dreyfus (1964), cf. [9], can be defined as follows:

Definition 1.2 The hybrid control law, defined by

$$u_{OLF}(t, \mathcal{I}_t) := u\Big(t; t, \mathcal{I}_t\Big), \quad t \ge t_0, \tag{1.11e}$$

is called *open-loop feedback (OLF) control law*.

Thus, the OL control $u_{[t,t_f]}(s)$, $s \ge t$, for the remaining time interval $[t, t_f]$ is used only at time $s = t$, see also [2, 9–11, 15, 17, 35]. Optimal (OLF) controls are obtained therefore by solving again control problems of the type (1.9a–d) at each intermediate starting time point $t_b := t$, $t \in [t_0, t_f]$.

A major issue in optimal control is the **robustness**, cf. [12], i.e. the insensitivity of an optimal control with respect to parameter variations. In case of random parameter variations robust optimal controls can be obtained by means of stochastic optimization methods, cf. [24], incorporating the probability distribution, i.e. the random characteristics, of the random parameter variation into the optimization process, cf. Definition 1.1.

Thus, constructing stochastic optimal open-loop feedback controls, hence, optimal open-loop feedback control laws being insensitive as far as possible with respect to random parameter variations, means that besides the optimality of the control policy also its insensitivity with respect to stochastic parameter variations should be guaranteed. Hence, in the following sections we also develop a **stochastic version** of the optimal open-loop feedback control method, cf. [23, 25–27]. A short overview on this novel stochastic optimal open-loop feedback control concept is given below:

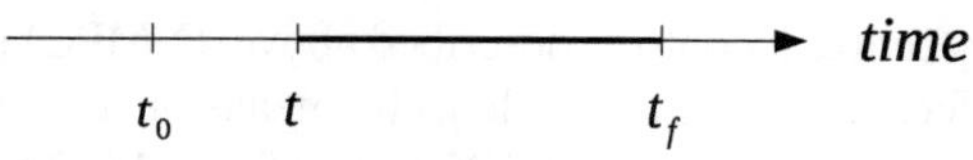

Fig. 1.1 Remaining time interval for intermediate time points t

At each intermediate time point $t_b = t \in [t_0, t_f]$, based on the given process observation $\mathcal{I}_t$, e.g., the observed state $z_t = z(t)$ at $t_b = t$, a stochastic optimal open-loop control $u^* = u^*(s) = u^*(s; t, \mathcal{I}_t), t \le s \le t_f$, is determined first on the remaining time interval $[t, t_f]$, see Fig. 1.1, by stochastic optimization methods, cf. [24].

Having a stochastic optimal open-loop control $u^* = u^*(s; t, \mathcal{I}_t), t \le s \le t_f$, on each remaining time interval $[t, t_f]$ with an arbitrary starting time point t, $t_0 \le t \le t_f$, a *stochastic optimal open-loop feedback (SOLF) control law* is then defined—corresponding to Definition 1.2—as follows:

Definition 1.3 The hybrid control law, defined by

$$u^*_{OLF}(t, \mathcal{I}_t) := u^*\Big(t; t, \mathcal{I}_t\Big), \quad t \ge t_0, \tag{1.11f}$$

is called the *stochastic optimal open-loop feedback (SOLF) control law*.

Thus, at time $t_b = t$ just the *first* control value $u^*(t) = u^*(t; t, \mathcal{I}_t)$ of $u^* = u^*(\cdot; t, \mathcal{I}_t)$ is used only.

For finding stochastic optimal open-loop controls, on the remaining time intervals $t_b \le t \le t_f$ with $t_0 \le t_b \le t_f$, the stochastic Hamilton function of the control problem is introduced. Then, the class of H-minimal controls, cf. [16], can be determined in case of stochastic uncertainty by solving a finite-dimensional stochastic optimization problem for minimizing the conditional expectation of the stochastic Hamiltonian subject to the remaining deterministic control constraints at each time point t. Having a $H-$minimal control, the related two-point boundary value problem with random parameters will be formulated for the computation of a stochastic optimal state and costate trajectory. In the important case of a linear-quadratic structure of the underlying control problem, the state and costate trajectory can be **determined analytically** to a large extent. Inserting then these trajectories into the H-minimal control, stochastic optimal open-loop controls are found on an arbitrary remaining time interval. According to Definition 1.3, these controls yield then immediately a stochastic optimal open-loop feedback control law. Moreover, the obtained controls can be realized in **real-time**, which is already shown for applications in optimal control of industrial robots, cf. [30].

(IV) *Nonlinear Model Predictive Control (NMPC)/Stochastic Nonlinear Model Predictive Control (SNMPC)*

Optimal open-loop feedback (OLF) control is the basic tool in *Nonlinear Model Predictive Control (NMPC)*. Corresponding to the approximation technique for feedback controls described above, (NMPC) is a method to solve complicated feedback control problems by means of stepwise computations of open-loop controls. Hence, in (NMPC), see [1, 13, 14, 29] optimal open-loop controls

$$u = u_{[t,t+T_p]}(s), \quad t \le s \le t + T_p, \tag{1.11g}$$

cf. (1.11c), are determined first on the time interval $[t, t + T_p]$ with a certain so-called *prediction time horizon* $T_p > 0$. In sampled-data MPC, cf. [13], optimal open-loop controls $u = u_{[t_i,t_i+T_p]}$ are determined at certain sampling instants $t_i, i = 0, 1, \ldots$, using the information $\mathcal{A}_{t_i}$ about the control process and its neighborhood up to time $t_i, i = 0, 1, \ldots$, see also [22]. The optimal open-loop control at stage i is applied then,

$$u = u_{[t_i,t_i+T_p]}(t), \quad t_i \le t \le t_{i+1}, \tag{1.11h}$$

until the next sampling instant t_{i+1}. This method is closely related to the *Adaptive Optimal Stochastic Trajectory Planning and Control (AOSTPC) procedure* described in [21, 22].

Corresponding to the extension of (OLF) control to (SOLF) control, (NMPC) can be extended to *Stochastic Nonlinear Model Predictive Control (SNMPC)*. For control policies of this type, robust (NMPC) with respect to stochastic variations of model parameters and initial values are determined in the following way:

- Use the a posteriori distribution $P(d\omega|\mathcal{A}_t)$ of the basic random element $\omega \in \Omega$, given the process information $\mathcal{A}_t$ up to time t, and
- apply stochastic optimization methods to incorporate random parameter variations into the optimal (NMPC) control design.

1.3 Computation of Expectations by Means of Taylor Expansions

Corresponding to the assumptions in Sect. 1.1.1, based on a parametric representation of the stochastic uncertainty of the control problem, using a finite dimensional random parameter vector $\theta = \theta(\omega)$, we suppose that the functions g, z_0, L, G are described by

$$g = \tilde{g}(t, \theta, z, u) \tag{1.12a}$$

$$z_0 = \tilde{z}_0(\theta) \tag{1.12b}$$

$$L = \tilde{L}(t, \theta, z, u) \tag{1.12c}$$

$$G = \tilde{G}(t, \theta, z). \tag{1.12d}$$

Here,

$$\theta = \theta(\omega), \omega \in (\Omega, \mathcal{A}, P), \tag{1.12e}$$

denotes the time-independent r-vector of random model parameters and random initial values, and g, z_0, L, G are sufficiently smooth functions of the variables indicated in (1.12a–d).

Again, for simplification, the conditional expectation $E(\dots|\mathcal{A}_{t_0})$ given the information $\mathcal{A}_{t_0}$ up to the considered starting time t_0 is denoted by E. Thus, let us denote

$$\overline{\theta} = \overline{\theta}^{t_0} := E\theta(\omega) = E\Big(\theta(\omega)|\mathcal{A}_{t_0}\Big) \tag{1.13a}$$

the conditional expectation of the random vector $\theta(\omega)$ given the information $\mathcal{A}_{t_0}$ at time point t_0. Taking into account the properties of the solution

$$z = z(t, \theta) = S\Big(z_0(\theta), \theta, u(\cdot)\Big)(t), \quad t \ge t_0, \tag{1.13b}$$

of the dynamic equation (1.1a,b), the expectations arising in the objective function of (1.9a) can be computed approximately by means of Taylor expansion with respect to θ at $\overline{\theta}$.

1.3.1 Complete Taylor Expansion

Considering first the costs L along the trajectory we obtain, cf. [24],

$$\begin{aligned} L\Big(t, \theta, z(t, \theta), u(t)\Big) &= L\Big(t, \overline{\theta}, z(t, \overline{\theta}), u(t)\Big) \\ &+\Big(\nabla_\theta L\Big(t, \overline{\theta}, z(t, \overline{\theta}), u(t)\Big) + D_\theta z(t, \overline{\theta})^T \nabla_z L\Big(t, \overline{\theta}, z(t, \overline{\theta}), u(t)\Big)\Big)^T (\theta - \overline{\theta}) \\ &+\frac{1}{2}(\theta - \overline{\theta})^T Q_L\Big(t, \overline{\theta}, z(t, \overline{\theta}), D_\theta z(t, \overline{\theta}), u(t)\Big)(\theta - \overline{\theta}) + \dots . \end{aligned} \tag{1.14a}$$

Retaining only 1st order derivatives of $z = z(t, \theta)$ with respect to θ, the approximate Hessian Q_L of $\theta \to L\Big(t, \theta, z(t, \theta), u\Big)$ at $\theta = \overline{\theta}$ is given by

$$Q_L\Big(t, \overline{\theta}, z(t, \overline{\theta}), D_\theta z(t, \overline{\theta}), u(t)\Big) := \nabla_\theta^2 L\Big(t, \overline{\theta}, z(t, \overline{\theta}), u(t)\Big)$$

$$+D_\theta z(t,\overline{\theta})^T \nabla^2_{\theta z} L\Big(t,\overline{\theta},z(t,\overline{\theta}),u(t)\Big) + \nabla^2_{\theta z} L\Big(t,\overline{\theta},z(t,\overline{\theta}),u(t)\Big)^T D_\theta z(t,\overline{\theta})$$

$$+D_\theta z(t,\overline{\theta})^T \nabla^2_{z} L\Big(t,\overline{\theta},z(t,\overline{\theta}),u(t)\Big) D_\theta z(t,\overline{\theta}). \tag{1.14b}$$

Here, $\nabla_\theta L, \nabla_z L$ denotes the gradient of L with respect to θ, z, resp., $D_\theta z$ is the Jacobian of $z = z(t,\theta)$ with respect to θ, and $\nabla^2_\theta L, \nabla^2_z L$, resp., denotes the Hessian of L with respect to θ, z. Moreover, $\nabla^2_{\theta z} L$ is the $r \times m$ matrix of partial derivatives of L with respect to θ_i and z_k, in this order.

Taking expectations in (1.14a), from (1.14b) we obtain the expansion

$$\begin{aligned}
EL\Big(t,\theta(\omega),z\big(t,\theta(\omega)\big),u(t)\Big) &= L\Big(t,\overline{\theta},z(t,\overline{\theta}),u(t)\Big) \\
&+\frac{1}{2}E\Big(\theta(\omega)-\overline{\theta}\Big)^T Q_L\Big(t,\overline{\theta},z(t,\overline{\theta}),D_\theta z(t,\overline{\theta}),u(t)\Big)\Big(\theta(\omega)-\overline{\theta}\Big)+\ldots \\
= L\Big(t,\overline{\theta},z(t,\overline{\theta}),u(t)\Big) &+ \frac{1}{2} tr Q_L\Big(t,\overline{\theta},z(t,\overline{\theta}),D_\theta z(t,\overline{\theta}),u(t)\Big)\mathrm{cov}\Big(\theta(\cdot)\Big)+\ldots .
\end{aligned} \tag{1.15}$$

For the terminal costs G, corresponding to the above expansion we find

$$\begin{aligned}
G\Big(t_f,\theta,z(t_f,\theta)\Big) &= G\Big(t_f,\overline{\theta},z(t_f,\overline{\theta})\Big) \\
&+\Big(\nabla_\theta G\Big(t_f,\overline{\theta},z(t_f,\overline{\theta})\Big) + D_\theta z(t_f,\overline{\theta})^T \nabla_z G\Big(t_f,\overline{\theta},z(t_f,\overline{\theta})\Big)\Big)^T(\theta-\overline{\theta}) \\
&+\frac{1}{2}(\theta-\overline{\theta})^T Q_G\Big(t_f,\overline{\theta},z(t_f,\overline{\theta}),D_\theta z(t_f,\overline{\theta})\Big)(\theta-\overline{\theta})+\ldots,
\end{aligned} \tag{1.16a}$$

where Q_G is defined in the same way as Q_L, see (1.14a). Taking expectations with respect to $\theta(\omega)$, we get

$$\begin{aligned}
EG\Big(t_f,\theta(\omega),z\big(t_f,\theta(\omega)\big)\Big) &= G\Big(t_f,\overline{\theta},z(t_f,\overline{\theta})\Big) \\
&+\frac{1}{2} tr Q_G\Big(t_f,\overline{\theta},z(t_f,\overline{\theta}),D_\theta z(t_f,\overline{\theta})\Big)\mathrm{cov}\Big(\theta(\cdot)\Big)+\ldots .
\end{aligned} \tag{1.16b}$$

Note Corresponding to (1.13a), for the mean and covariance matrix of the random parameter vector $\theta = \theta(\omega)$ we have

$$\overline{\theta} = \overline{\theta}^{(t_0)} := E\Big(\theta(\omega)|\mathcal{A}_{t_0}\Big)$$

$$\operatorname{cov}\Big(\theta(\cdot)\Big) = \operatorname{cov}^{(t_0)}\Big(\theta(\cdot)\Big) := E\Big(\Big(\theta(\omega) - \overline{\theta}^{(t_0)}\Big)\Big(\theta(\omega) - \overline{\theta}^{(t_0)}\Big)^T \Big| \mathcal{A}_{t_0}\Big).$$

1.3.2 Inner or Partial Taylor Expansion

Instead of a complete expansion of L, G with respect to θ, appropriate approximations of the expected costs EL,EG, resp., may be obtained by the inner 1st order approximation of the trajectory, hence,

$$L\Big(t, \theta, z(t, \theta), u(t)\Big) \approx L\Big(t, \theta, z(t, \overline{\theta}) + D_\theta z(t, \overline{\theta})(\theta - \overline{\theta}), u(t)\Big). \tag{1.17a}$$

Taking expectations in (1.17a), for the expected cost function we get the approximation

$$\begin{aligned} &EL\Big(t, \theta, z(t, \theta), u(t)\Big) \\ &\quad\approx EL\Big(t, \theta(\omega), z(t, \overline{\theta}) + D_\theta z(t, \overline{\theta})(\theta(\omega) - \overline{\theta}), u(t)\Big). \end{aligned} \tag{1.17b}$$

In many important cases, as, e.g., for cost functions L being quadratic with respect to the state variable z, the above expectation can be computed analytically. Moreover, if the cost function L is convex with respect to z, then the expected cost function EL is convex with respect to both, the state vector $z(t, \overline{\theta})$ and the Jacobian matrix of sensitivities $D_\theta z(t, \overline{\theta})$ evaluated at the mean parameter vector $\overline{\theta}$.

Having the approximate representations (1.15), (1.16b), (1.17b), resp., of the expectations occurring in the objective function (1.9a), we still have to compute the trajectory $t \to z(t, \overline{\theta})$, $t \geq t_0$, related to the mean parameter vector $\theta = \overline{\theta}$ and the sensitivities $t \to \frac{\partial z}{\partial \theta_i}(t, \overline{\theta})$, $i = 1, \ldots, r$, $t \geq t_0$, of the state $z = z(t, \theta)$ with respect to the parameters θ_i, $i = 1, \ldots, r$, at $\theta = \overline{\theta}$. According to [28], for $z = z(t, \overline{\theta})$ we have the system of differential equations

$$\dot{z}(t, \overline{\theta}) = g\Big(t, \overline{\theta}, z(t, \overline{\theta}), u(t)\Big), \quad t \geq t_0, \tag{1.18a}$$

$$z(t_0, \overline{\theta}) = z_0(\overline{\theta}). \tag{1.18b}$$

Moreover, assuming that differentiation with respect to θ_i, $i = 1, \ldots, r$, and integration with respect to time t can be interchanged, cf. [28], we obtain the following system of linear perturbation differential equation for the Jacobian $D_\theta z(t, \overline{\theta}) = \left(\frac{\partial z}{\partial \theta_1}(t, \overline{\theta}), \frac{\partial z}{\partial \theta_2}(t, \overline{\theta}), \ldots, \frac{\partial z}{\partial \theta_r}(t, \overline{\theta})\right)$, $t \geq t_0$:

$$\frac{d}{dt}\Big(D_\theta z(t,\overline{\theta})\Big) = D_z g\Big(t,\overline{\theta}, z(t,\overline{\theta}), u(t)\Big) D_\theta z(t,\overline{\theta}) + D_\theta g\Big(t,\overline{\theta}, z(t,\overline{\theta}), u(t)\Big), \quad t \geq t_0, \tag{1.19a}$$

$$D_\theta z(t_0,\overline{\theta}) = D_\theta z_0(\overline{\theta}). \tag{1.19b}$$

Note Equation (1.19a,b) is closely related to the perturbation equation (1.4e) for representing the derivative $D_u z$ of z with respect to the control u. Moreover, the matrix differential equation (1.19a) can be decomposed into the following r differential equations for the columns $\frac{\partial z}{\partial \theta_j}(t,\overline{\theta}),\ j = 1,\ldots,r$:

$$\frac{d}{dt}\Big(\frac{\partial z}{\partial \theta_j}(t,\overline{\theta})\Big) = \frac{\partial g}{\partial \theta_j}\Big(t,\overline{\theta}, z(t,\overline{\theta}), u(t)\Big) \frac{\partial z}{\partial \theta_j}(t,\overline{\theta}) + \frac{\partial g}{\partial \theta_j}\Big(t,\overline{\theta}, z(t,\overline{\theta}), u(t)\Big), \geq t_0, \quad j = 1,\ldots,r. \tag{1.19c}$$

Denoting by

$$\tilde{L} = \tilde{L}\Big(t,\theta, z(t,\overline{\theta}), D_\theta z(t,\overline{\theta}), u(t)\Big), \tag{1.20a}$$

$$\tilde{G} = \tilde{G}\Big(t_f,\theta, z(t_f,\overline{\theta}), D_\theta z(t_f,\overline{\theta})\Big), \tag{1.20b}$$

the approximation of the cost functions L, G by complete, partial Taylor expansion, for the optimal control problem under stochastic uncertainty (2.8a–d) we obtain now the following approximation:

Theorem 1.1 *Suppose that differentiation with respect to the parameters $\theta_i, i = 1,\ldots,r$, and integration with respect to time t can be interchanged in (1.4c). Retaining only 1st order derivatives of $z = z(t,\theta)$ with respect to θ, the optimal control problem under stochastic uncertainty (2.8a–d) can be approximated by the ordinary deterministic control problem:*

$$\min \int_{t_0}^{t_f} E\tilde{L}\Big(t,\theta(\omega), z(t,\overline{\theta}), D_\theta z(t,\overline{\theta}), u(t)\Big)dt + E\tilde{G}\Big(t_f,\theta(\omega), z(t_f,\overline{\theta}, D_\theta z(t,\overline{\theta})\Big) \tag{1.21a}$$

s.t.

$$\dot{z}(t,\overline{\theta}) = g\Big(t,\overline{\theta}, z(t,\overline{\theta}), u(t)\Big), \quad t \ge t_0, \tag{1.21b}$$

$$z(t_0,\overline{\theta}) = z_0(\overline{\theta}) \tag{1.21c}$$

$$\frac{d}{dt}\Big(D_\theta z(t,\overline{\theta})\Big) = D_z g\Big(t,\overline{\theta}, z(t,\overline{\theta}), u(t)\Big) D_\theta z(t,\overline{\theta})$$

$$+ D_\theta g\Big(t,\overline{\theta}, z(t,\overline{\theta}), u(t)\Big), \quad t \ge t_0, \tag{1.21d}$$

$$D_\theta z(t_0,\overline{\theta}) = D_\theta z_0(\overline{\theta}) \tag{1.21e}$$

$$u(\cdot) \in D. \tag{1.21f}$$

Remark 1.5 Obviously, the trajectory of above deterministic substitute control problem (1.21a–f) of the original optimal control problem under stochastic uncertainty (2.8a–d) can be represented by the $m(r+1)$-vector function:

$$t \to \xi(t) := \begin{pmatrix} z(t,\overline{\theta}) \\ \frac{\partial z}{\partial \theta_1}(t,\overline{\theta}) \\ \vdots \\ \frac{\partial z}{\partial \theta_r}(t,\overline{\theta}) \end{pmatrix}, \quad t_0 \le t \le t_f. \tag{1.22}$$

Remark 1.6 Constraints of the expectation type, i.e.

$$Eh_{II}\Big(t, \theta(\omega), z\big(t, \theta(\omega)\big)\Big) \le (=)\, 0,$$

can be evaluated as in (1.15) and (1.16b). This yields then deterministic constraints for the unknown functions $t \to z(t,\overline{\theta})$ and $t \to D_\theta z(t,\overline{\theta}),\ t \ge t_0$.

1.4 Taylor Approximation of Control Problems Under Stochastic Uncertainty: General Procedure

Let us denote $u = u(\cdot)$ the open-loop or closed-loop control input function defined on a certain time interval $[t_0, t_f]$. Moreover, let $z = z(\cdot)$ be the output, hence, the trajectory, of the underlying dynamic system and $a = a(\omega)$ the vector of random parameters, such as the dynamic parameters, the external load, and the initial state at time t_0. The control problem can then be represented formally by

$$\min_{z(\cdot)\in Z, u(\cdot)\in U} Ef(a(\omega), z(\cdot), u(\cdot)) \tag{1.23a}$$

s.t.

$$S(a(\omega), z(\cdot), u(\cdot)) = 0, \tag{1.23b}$$

where $f = f(a, z, u)$ denotes the costs along the trajectory and at the terminal point t_f, U is the set of admissible controls, and $S = S(a, z, u)$ is the dynamic input–output operator represented by a differential equation. Supposing that (1.23b) can be solved uniquely for $z(\cdot)$, we have

$$z(\cdot) = z(\cdot, a, u(\cdot)), \tag{1.23c}$$

where $\cdot$ refers to the time t. Putting (1.23c) into the objective function f, problem (1.23a) reads

$$\min_{u(\cdot)\in U} Ef(a(\omega), z(\cdot, a(\omega), u(\cdot)), u(\cdot)). \tag{1.24a}$$

Having uncertain, hence, random initial values $z_0 = z_0(\omega)$, the parameter vector reads

$$a(\omega) = \begin{pmatrix} p(\omega) \\ z_0(\omega) \end{pmatrix}, \tag{1.24b}$$

where $p = p(\omega)$ is the vector of dynamic parameters, load coefficients, and other exterior coefficients. In this case the open-loop control

$$u = u(\cdot, a) = u(\cdot, z_0) \tag{1.24c}$$

depends also on the random vector $z_0 = z_0(\omega)$ of initial values with respect to the current starting time $t_0 = t_b$. Then, the stochastic optimization problem (1.24a) reads

$$\min_{u(\cdot)\in U} Ef(a(\omega), z(\cdot, a(\omega), u(\cdot, a(\omega))), u(\cdot, a(\omega))). \tag{1.24d}$$

Since the expectations are defined here by multiple integrals, the objective function of (1.24a,d), resp., can be treated in general by means of approximation methods only. Supposing that the derivatives under consideration exist, in the following the expectation in (1.24a,d), resp., are determined by means of Taylor expansion of the integrand in (1.24a,d), resp., with respect to the parameter vector a at the mean vector $\overline{a} := Ea(\omega)$. For simplification we consider first problem (1.24a), where we assume that the control function $u(\cdot)$ does not depend explicitly on the parameter vector $a(\omega)$. Thus, by differentiation of

$$\tilde{f}(a) := f(a, z(\cdot, a, u(\cdot)), u(\cdot)) \tag{1.25}$$

with respect to a at $a = \overline{a}$, we get

$$\tilde{f}(\overline{a}) := f(\overline{a}, z(\cdot, \overline{a}, u(\cdot)), u(\cdot)) \tag{1.26a}$$

$$D_a\widetilde{f}(\overline{a})^T = \nabla\widetilde{f}(\overline{a})^T = \frac{\partial f}{\partial z}(z(\cdot,\overline{a},u(\cdot)),u(\cdot))\frac{\partial z}{\partial a}z(\cdot,\overline{a},u(\cdot)) \tag{1.26b}$$

$$D_a^2\widetilde{f}(\overline{a}) = \frac{\partial z}{\partial a}z(\cdot,\overline{a},u(\cdot))^T\frac{\partial^2 f}{\partial z^2}(z(\cdot,\overline{a},u(\cdot)),u(\cdot))\frac{\partial z}{\partial a}z(\cdot,\overline{a},u(\cdot))$$
$$+\sum_{k=1}^{m}\frac{\partial f}{\partial z_k}(z(\cdot,\overline{a},u(\cdot)),u(\cdot))\frac{\partial^2 z_k}{\partial a^2}z(\cdot,\overline{a},u(\cdot)). \tag{1.26c}$$

Moreover, the remainder term for the second order Taylor polynomial of f can be approximated by

$$|R_2(\overline{a},\Delta a)| \le \frac{\|\Delta a\|^3}{3!} sup_{0\le\theta\le 1}\sqrt{\sum_{i,j,k=1}^{\nu}\frac{\partial^3\widetilde{f}}{\partial a_i,\partial a_j,\partial a_k}(\overline{a}+\theta\Delta a)}. \tag{1.26d}$$

1.5 Control Problems with Linear and Sublinear Cost Functions

In the following we consider objective functions, cf. (1.5a,b),

$$F(u(\cdot)) = E\Big(\int_{t_0}^{t_f} Ldt + G\Big)$$

of the type

$$L = L(t,\theta,z,u) := L_I(t,\theta,z) + L_{II}(t,\theta,u), \tag{1.27a}$$

where L_I is a linear and/or sublinear [7] cost functions with respect to the state vector z:

$$L_I = L_I(t,\theta,z) := c(t,\theta)^T z + \max_{i\in I} c_i(t,\theta)^T z. \tag{1.27b}$$

Here, $c = c(t,\theta)$, $c_i = c_i(t,\theta)$, $i \in I$, $t_0 \le t \le t_f$, are vectors of cost coefficients depending on a random parameter vector $\theta = \theta(\omega)$ and a finite index set I. In the same way, the terminal cost function is defined by

$$G = G(t_f,\theta,z(t_f)) := c(t_f,\theta)^T z(t_f) + \max_{i\in I} c_i(t_f,\theta)^T z(t_f). \tag{1.27c}$$

Cost functions L, G of this type arise, e.g., in *active structural control*, see, e.g., [32, 33], where the state vector $z = \begin{pmatrix} q \\ \dot{q} \end{pmatrix}$ is composed of the vector $q = q(t)$ of structural displacements and the vector of velocities $\dot{q} = \dot{q}(t)$. Evaluating the performance of the controlled structure by the *compliance*, i.e. the work done by the external load $w = w(t, \theta)$ along the trajectory $z = z(t)$, we have

$$L_I(t, \theta, z) = C(t, \theta, q) := w(t, \theta)^T q, z = \begin{pmatrix} q \\ \dot{q} \end{pmatrix}. \tag{1.28}$$

For the expectation of the linear, sublinear, resp., cost function (1.27b) we get the representation, the lower bound, resp.,

$$\begin{aligned} EL_I(t, \theta(\omega), z(t, \theta(\omega), u(t, \cdot))) &= Ec(t, \theta(\omega))^T z(t, \theta(\omega), u(t, \cdot)) \\ &\quad + E \max_{i \in I} c_i(t, \theta(\omega))^T z(t, \theta(\omega), u(t, \cdot)) \\ (=) \geq \tilde{L}_I(t, \theta(\omega), z(t, \theta(\omega), u(t, \cdot))) &:= Ec(t, \theta(\omega))^T z(t, \theta(\omega), u(t, \cdot)) \\ &\quad + \max_{i \in I} Ec_i(t, \theta(\omega))^T z(t, \theta(\omega), u(t, \cdot)). \end{aligned} \tag{1.29}$$

The expectation of the scalar products $c(t, \theta(\omega))^T z(t, \theta(\omega), u(t, \cdot))$ and $c_i(t, \theta(\omega))^T z(t, \theta(\omega), u(t, \cdot))$ arising in (1.29) can be obtained approximately by Taylor expansion of $c(t, \theta)$ and $z = z(t, \theta, u(\cdot))$ with respect to θ at a nominal point $\overline{\theta}$. Using here only expansions up to first order, we get the approximation

$$\begin{aligned} &Ec(t, \theta(\omega))^T z(t, \theta(\omega), u(t, \cdot)) \\ &\approx E\Big(c(t, \overline{\theta}) + \frac{\partial c}{\partial \theta}(t, \overline{\theta})(\theta(\omega) - \overline{\theta})\Big)^T \Big(z(t, \overline{\theta}, u(t, \cdot)) + \frac{\partial z}{\partial \theta}(t, \overline{\theta}, u(t, \cdot))(\theta(\omega) - \overline{\theta})\Big) \\ &= c(t, \overline{\theta})^T z(t, \overline{\theta}, u(t, \cdot)) + E(\theta(\omega) - \overline{\theta})^T \frac{\partial c}{\partial \theta}(t, \overline{\theta})^T \frac{\partial z}{\partial \theta}(t, \overline{\theta}, u(t, \cdot))(\theta(\omega) - \overline{\theta}), \end{aligned} \tag{1.30}$$

assuming that $\overline{\theta} := E\theta(\omega)$. According to the results shown in [28], from (1.29), (1.30) we get the following result:

Theorem 1.2 *By means of first order Taylor expansion the expectation EL_I of the linear/sublinear cost function L_I can be approximated by*

$$\begin{aligned} &\tilde{L}_I(t, \theta(\omega), z(t, \theta(\omega), u(t, \cdot))) \\ &\quad = c(t, \overline{\theta})^T z(t, \overline{\theta}, u(t, \cdot)) + tr \frac{\partial c}{\partial \theta}(t, \overline{\theta})^T \frac{\partial z}{\partial \theta}(t, \overline{\theta}, u(t, \cdot)) cov(\theta(\cdot)) \end{aligned}$$

$$+\max_{i\in I}\Big(c_i(t,\overline{\theta})^T z(t,\overline{\theta},u(t,\cdot)) + tr\frac{\partial c_i}{\partial \theta}(t,\overline{\theta})^T \frac{\partial z}{\partial \theta}(t,\overline{\theta},u(t,\cdot))cov(\theta(\cdot))\Big). \tag{1.31}$$

Denoting by $R_{c,1} = R_{c,1}(t,\overline{\theta},\Delta\theta)$, $R_{z,1} = R_{z,1}(t,\overline{\theta},\Delta\theta), u(\cdot)$, $\Delta\theta = \theta - \overline{\theta}$, the remainders of the first order Taylor formulas, the error e_{L_1} of the first order Taylor approximation reads

$$\begin{aligned} e_{L_1} &= R_{c,1}(t,\overline{\theta},\Delta\theta) + R_{z,1}(t,\overline{\theta},\Delta\theta), u(\cdot) \\ &\le \frac{\|\Delta\theta\|^2}{2!}\left(\sup_{0\le\vartheta\le 1}\sqrt{\sum_{i,k=1}^{\nu}\|\frac{\partial^2 c}{\partial a_i \partial a_k}(t,\overline{a}+\vartheta\Delta\theta)\|}\right. \\ &\quad \left. + \sup_{0\le\vartheta\le 1}\sqrt{\sum_{i,k=1}^{\nu}\|\frac{\partial^2 z}{\partial a_i \partial a_k}(t,\overline{a}+\vartheta\Delta\theta)\|}\right). \end{aligned} \tag{1.32}$$

Concerning the reminder terms, the following special cases may often occur:

Lemma 1.3 *(i) For cost functions* $c(\theta) = B\theta$ *with a fixed matrix* B, *e.g.,* $B = I$, *we have* $R_{c,1} = 0$. *(ii) If* $c(\theta) = \big(\theta^T B^{(1)}, \dots, \theta^T B^{(n)}\big)$, *then* $R_{c,1}(t,\overline{\theta},\Delta\theta) \le \frac{\|\Delta\theta\|^2}{2!}\sum_{1\le k\le n}\|{B^{(k)}}^2\|$.

1.6 Stochastic Optimal Open-Loop Feedback Control of Tracking Systems

According to Sect. 1.2, a stochastic optimal open-loop feedback control (SOLF) is defined, see Definition 1.2, by

$$\varphi^*(t,\mathcal{I}_t) := u^*\Big(t; t, \mathcal{I}_t\Big), \quad t \ge t_0.$$

Here, $u^* = u^*(s; t, \mathcal{I}_t)$, $t \le s \le t_f$, is a stochastic optimal open-loop control on each remaining time interval $[t, t_f]$ with an arbitrary starting time point t, $t_0 \le t \le t_f$ and the set of information $\mathcal{I}_t$ up to the current time t. Thus, at time $t_b = t$ just the *first* control value $u^*(t) = u^*(t; t, \mathcal{I}_t)$ of $u^* = u^*(\cdot; t, \mathcal{I}_t)$ is used only.

Consider now tracking problems with a prescribed reference trajectory $q^R = q^R(t)$, $t_0 \le t \le t_f$. Moreover, let the feedforward control $u^R = u^R(t)$, $t_0 \le t \le t_f$, be defined by

$$F\left(\overline{p}_D^R, q^R(t), \dot{q}^R, \ddot{q}^R(t)\right) = u^R(t), t_0 \le t \le t_f, \tag{1.33a}$$

where $\overline{p}_D^R$ is an estimate or expectation of the stochastic vector of dynamic parameters.

For simplification we consider a PD-regulator. An open-loop control on a remaining interval $[t_b, t_f]$ has then the form:

$$\Delta u := \varphi(t; t_b, \Delta z_b), t_b \le t \le t_f, \tag{1.33b}$$

where the information set $\mathcal{I}_{t_b}$ is given here by the state deviation Δz_b between the actual state and the reference state at the initial time point t_b of the remaining time interval $[t_b, t_f]$, thus,

$$\Delta z_b = \begin{pmatrix} \Delta q_b \\ \Delta \dot{q}_b \end{pmatrix} := \Delta z(t_b) = z(t_b) - z^R(t_b) = \begin{pmatrix} q(t_b) \\ \dot{q}(t_b) \end{pmatrix} - \begin{pmatrix} q^R(t_b) \\ \dot{q}^R(t_b) \end{pmatrix}. \tag{1.33c}$$

If we start at time t_b with zero errors, i.e. with $\Delta z_b = 0$ and $p_D = \overline{p}_D^{(R)}$, then without a control correction $\Delta u(t)$ at any time $t, t_b \le t \le t_f$, the system follows the prescribed reference trajectory $z^R = z^R(t), t_b \le t \le t_f$. Consequently, in the following we consider open-loop controls with the property:

$$\varphi(t; t_b, 0) = 0, \quad t_b \le t \le t_f. \tag{1.34}$$

In the present case we have then, cf. [28], the following optimal regulator problem under stochastic uncertainty:

$$\min \mathrm{E} \left(\int_{t_b}^{t_f} \Delta q(t)^T C_q \Delta q(t) + \Delta \dot{q}(t) C_{\dot{q}} \Delta \dot{q}(t) \right.$$

$$\left. + \varphi\big(t; t_b, \Delta q(t), \Delta \dot{q}(t)\big)^T C_u \varphi\big(t; t_b, \Delta q(t), \Delta \dot{q}(t)\big) \Big) dt \, \Big| \mathcal{A}_{t_b} \right) \tag{1.35a}$$

s.t.

$$F\left(\overline{p}_D^{(R)} + \Delta p_D(\omega), q^{(R)}(t) + \Delta q(t), \dot{q}^{(R)}(t) + \Delta \dot{q}(t), \ddot{q}^{(R)}(t) + \Delta \ddot{q}(t)\right)$$

$$= u^{(R)}(t) + \varphi\left(t; t_b, \Delta q(t_b), \Delta \dot{q}(t_b)\right) \text{ a.s.}, t \ge t_b, \tag{1.35b}$$

$$\Delta q(t_b) = \Delta q_b(\omega), \Delta \dot{q}(t_b) = \Delta \dot{q}_b(\omega) \text{ a.s.} \tag{1.35c}$$

For the tracking error we obtain the representation:

$$\Delta z(t) = \Delta z\left(t, \Delta p_D, \Delta z_b\right) = \begin{pmatrix} \Delta q\left(t; \Delta p_D, \Delta z_b\right) \\ \Delta \dot{q}\left(t; \Delta p_D, \Delta z_b\right) \end{pmatrix}, t \ge t_b, \tag{1.35d}$$

and according to the system dynamics (1.35b,c) it holds

$$\Delta z\,(t, 0, 0) = 0. \tag{1.35e}$$

Having a stochastic optimal open-loop control $\varphi^* = \varphi^*\,(t; t_b, \Delta z(t_b))\,, t_b \le t \le t_f$, of (1.35a–c), the stochastic optimal open-loop feedback control reads

$$\varphi_{OLF}(t, \Delta z(t)) := \varphi^*\,(t; t, \Delta z(t))\,, t_0 \le t \le t_f. \tag{1.36}$$

1.6.1 *Approximation of the Expected Costs: Expansions of 1st Order*

According to (1.35a), we have to determine the following conditional expectation:

$$\mathrm{E}\left(f^{(b)}(t)|\mathcal{A}_{t_b}\right) :=$$

$$\mathrm{E}\Big(\Delta z(t)^T Q \Delta z(t) + \varphi\big(t; t_b, \Delta z(t_b)\big)^T C_u \varphi\big(t; t_b, \Delta z(t_b)\big)\Big|\mathcal{A}_{t_b}\Big), \tag{1.37a}$$

with

$$Q_0 := \begin{pmatrix} C_q & 0 \\ 0 & C_{\dot{q}} \end{pmatrix}. \tag{1.37b}$$

For the computation of the expectation in (1.37a) also the open-loop control function $\varphi = \varphi\big(t; t_b, \Delta z(t_b)\big)$ is approximated by means of Taylor expansion at $\Delta z_b = 0$. Because of $\varphi(t; t_b, 0) = 0$, see (1.34), we get

$$\varphi(t; t_b, \Delta z_b) = D_{\Delta z(t_b)}\varphi(t; t_b, 0)\Delta z_b + \ldots \tag{1.38}$$

with the unknown Jacobian $D_{\Delta z(t_b)}\varphi(t; t_b, 0)$ to be determined. Using first order Taylor approximation of the open-loop control $\varphi = \varphi(t; t_b, \Delta z_b)$, we find, cf. (1.37a)

$$\mathrm{E}\left(f^{(b)}(t)|\mathcal{A}_{t_b}\right) \approx \mathrm{E}\left(\tilde{f}^{(b)}|\mathcal{A}_{t_b}\right), \tag{1.39a}$$

where

$$\mathrm{E}\left(\tilde{f}^{(b)}|\mathcal{A}_b\right) := \mathrm{E}\Bigg(\Delta z(t)^T Q_0 \Delta z(t)$$

$$+\Delta z_b{}^T D_{\Delta z(t_b)}\varphi(t; t_b, 0)^T C_u D_{\Delta z(t_b)}\varphi(t; t_b, 0)\Delta z_b\Bigg|\mathcal{A}_{t_b}\Bigg). \tag{1.39b}$$

Define now the stochastic parameter vector

$$\Delta a(\omega) = \begin{pmatrix} \Delta p_D(\omega) \\ \Delta z_b(\omega) \end{pmatrix}. \tag{1.40a}$$

Because of (1.35e), by $1st$ order Taylor expansion of the tracking error we get the approximation:

$$\Delta z(t, \Delta a) \approx D_{\Delta a} \Delta z\,(t, 0)\, \Delta a, \quad t \geq t_b. \tag{1.40b}$$

Inserting (1.40b) into (1.39a,b), for $\mathrm{E}\left(f^{(b)}(t)|\mathcal{A}_b\right)$ we have the approximation:

$$\mathrm{E}\left(f^{(b)}(t)|\mathcal{A}_{t_b}\right) \approx \mathrm{E}\Big(\Delta a^T D_{\Delta a} \Delta z\,(t, 0)^T\, Q_0 D_{\Delta a} \Delta z\,(t, 0)\, \Delta a$$

$$+\Delta z_b{}^T D_{\Delta z_b} \varphi(t; t_b, 0)^T C_u D_{\Delta z_b} \varphi(t; t_b, 0) \Delta z_b \Big| \mathcal{A}_{t_b}\Big), t_b \leq t \leq t_f. \tag{1.41}$$

From (1.41) and [28] we obtain now the following result:

Theorem 1.3 *The integrand of the objective function of the optimal regulator problem under stochastic uncertainty and open-loop PD-control can be approximated by*

$$\begin{aligned} E\left(\tilde{f}^{(b)}\Big|\mathcal{A}_{t_b}\right) &= tr\ D_{\Delta a} \Delta z\,(t, 0)^T\, Q_0 D_{\Delta a} \Delta z\,(t, 0)\, E^{(b)} \Delta a(\cdot) \Delta a(\cdot)^T \\ &\quad +\ tr\ D_{\Delta z_b} \varphi(t; t_b, 0)^T C_u D_{\Delta z_b} \varphi(t; t_b, 0) E^{(b)} \Delta z_b(\cdot) \Delta z_b(\cdot)^T, \\ &\quad t \geq t_b, \end{aligned} \tag{1.42}$$

where $E^{(b)} \Delta a(\cdot) \Delta a(\cdot)^T$ is the symmetric matrix of the mixed second order moments of the random vector $\Delta a(\cdot)$.

While the Jacobian $D_{\Delta z_b} \varphi(t; t_b, 0)$ contains the parameters for the approximate optimal open-loop control $\varphi^* = \varphi^*\big(t; t_b, \Delta z(t_b)\big), t \geq t_b$, to be determined by the optimal regulator problem under stochastic uncertainty, according to (1.42) we still have to derive equations for the Jacobian $D_{\Delta a} \Delta z\,(t, \Delta a)$.

Taking the derivative of the initial value problem (1.35a–c) with respect to the parameter vector Δa at $\Delta a = 0$, for the derivative $D_{\Delta a} \Delta q\,(t, \Delta a)$ of the tracking error $\Delta q\,(t, \Delta a)$ at $\Delta a = 0$ we get the initial value problem

$$\begin{aligned} \big(Y^R(t), 0, 0\big) + K^R(t) D_{\Delta a} \Delta q\,(t, 0) + D^R(t) D_{\Delta a} \Delta \dot{q}\,(t, 0) + M^R(t) D_{\Delta a} \Delta \ddot{q}\,(t, 0) \\ = \big(0, D_{\Delta z_b} \varphi(t; t_b, 0)\big), t \geq t_b, \end{aligned} \tag{1.43a}$$

$$D_{\Delta a} \Delta z\,(t_b, 0) = \begin{pmatrix} 0 & I & 0 \\ 0 & 0 & I \end{pmatrix}. \tag{1.43b}$$

In (1.43a,b) the Jacobians M^R, K^R, D^R, Y^R of the vector function F are defined, cf. [28],

$$K^R(t) := D_q F(\overline{p}_D^R, q^R(t), \dot{q}^R(t), \ddot{q}^R(t)), \tag{1.44a}$$

$$D^R(t) := D_{\dot{q}} F(\overline{p}_D^R, q^R(t), \dot{q}^R(t), \ddot{q}^R(t)), \tag{1.44b}$$

$$Y^R(t) := D_{p_D} F(\overline{p}_D^R, q^R(t), \dot{q}^R(t), \ddot{q}^R(t)), \tag{1.44c}$$

$$M^R(t) := D_{\ddot{q}} F(\overline{p}_D^R, q^R(t), \dot{q}^R(t), \ddot{q}^R(t)). \tag{1.44d}$$

Using (1.35d), the second order linear initial value problem (1.43a,b) for the tracking error can be converted into a first order linear initial value problem:

Theorem 1.4 *The Jacobian of the tracking error and its time derivative $\Delta z(t) = \Delta z\,(t, \Delta p_D, \Delta z_b)$ with respect to $a = ({\Delta p_D}^T, {\Delta z_b}^T)^T$ at $a = 0$ fulfills the first order initial value problem:*

$$\frac{d}{dt} D_{\Delta a}\Delta z\,(t,0,0) = \begin{pmatrix} 0 & I \\ -M^R(t)^{-1}K^R(t) & -M^R(t)^{-1}D^R(t) \end{pmatrix} D_{\Delta a}\Delta z\,(t,0,0)$$

$$+ \begin{pmatrix} 0 & 0 \;\; 0 \\ -M^R(t)^{-1}Y^R(t) & -M^R(t)^{-1}D_{\Delta z_b}\varphi(t;t_b,0) \end{pmatrix}, t \ge t_b, \tag{1.45a}$$

$$D_{\Delta a}\Delta z\,(t_b,0,0) = \begin{pmatrix} 0 & I & 0 \\ 0 & 0 & I \end{pmatrix}. \tag{1.45b}$$

Based on the above shown results Theorems 1.3 and 1.4 we get this result:

Theorem 1.5 *Applying open-loop control under stochastic uncertainty yields the following approximate optimal regulator problem:*

$$\min \int_{t_0}^{t_f} \Big(tr\ D_{\Delta a}\Delta z\,(t,0)^T\ Q_0 D_{\Delta a}\Delta z\,(t,0)\ E^{(b)}\Delta a(\cdot)\Delta a(\cdot)^T$$

$$+\ tr\ D_{\Delta z_b}\varphi(t;t_b,0)^T C_u D_{\Delta z_b}\varphi(t;t_b,0) E^{(b)}\Delta z_b(\cdot)\Delta z_b(\cdot)^T \Big) dt \tag{1.46a}$$

s.t.

$$\frac{d}{dt} D_{\Delta a}\Delta z\,(t,0,0) = \begin{pmatrix} 0 & I \\ -M^R(t)^{-1}K^R(t) & -M^R(t)^{-1}D^R(t) \end{pmatrix} D_{\Delta a}\Delta z\,(t,0,0)$$

$$+ \begin{pmatrix} 0 & 0 \;\; 0 \\ -M^R(t)^{-1}Y^R(t) & -M^R(t)^{-1}D_{\Delta z_b}\varphi(t;t_b,0) \end{pmatrix}, t \ge t_b, \tag{1.46b}$$

$$D_{\Delta a}\Delta z\,(t_b,0,0) = \begin{pmatrix} 0 & I & 0 \\ 0 & 0 & I \end{pmatrix}. \tag{1.46c}$$

Remark 1.7 Comparison of Optimal open-loop control—Optimal feedback control under stochastic uncertainty

(i) Properties of the optimal control problems: According to (1.46a–c), for the computation of the optimal gain matrix $D_{\Delta z_b}\varphi(t; t_b, 0)$ of the optimal open-loop control we have a deterministic control problem with a quadratic objective function and linear constraints.

Comparing the present results for the optimal open-loop control under stochastic uncertainty with the corresponding results in Chap. 2, Theorem 2.2, we see that in case of optimal feedback control under stochastic uncertainty the objective function is also quadratic in the unknown gain matrices K_d, (K_i), K_p, resp., to be determined, but the perturbation differential equation for the determination of the sensitivities depends in a nonlinear way on the gain matrices K_d, (K_i), K_p, respectively.

(ii) Stability properties of the optimal controls: In the feedback control case the control system can be stabilized in easy way, see Chap. 2, Lemma 2.1, by an appropriate selection of the diagonal parameter matrices K_d, (K_i), K_p. On the other hand, in the open-loop case the stability of the resulting control system depends on the properties of the mass matrix $M^R(t)$, the damping matrix $D^R(t)$, and the stiffness matrix $K^R(t)$.

1.6.2 Approximate Computation of the Fundamental Matrix

As can be seen from (1.46a–c), the system matrix $A = A(t), t_0 \leq t \leq t_f$ of the linear dynamic equation (1.46b) depends on the matrices $M^R(t)$, $D^R(t)$, $K^R(t)$, hence, $A = A(t)$ is in general a function of time t. Consequently, the fundamental matrix $S = S(t)$ related to $A = A(t)$ cannot be determined explicitly in general. However, due to present concept of *open-loop feedback control*, see (1.2), the open-loop control function $\Delta u := \varphi(t; t_b, \Delta z_b), t_b \leq t \leq t_f$, has to be determined on relatively small intervals $[t_0, t_f] = [t_b, t_b + \delta], b = 0, 1, \ldots$. Hence, the homogeneous linear differential equation

$$\frac{d}{dt} D_{\Delta a}\Delta z\,(t, 0, 0) = A(t) D_{\Delta a}\Delta z\,(t, 0, 0)\,,\, t_0 \leq t \tag{1.47a}$$

related to (1.46b) can be approximated on $[t_0, t_f] = [t_b, t_b + \delta]$ by

$$\frac{d}{dt} D_{\Delta a}\Delta z\,(t, 0, 0) = A(t_b) D_{\Delta a}\Delta z\,(t, 0, 0)\,,\, t_b \leq t \leq t_b + \delta. \tag{1.47b}$$

Thus, the fundamental matrix $S = S(t)$ can be approximated on $[t_0, t_f] = [t_b, t_b+\delta]$ by

$$S(t) \approx e^{A(t_b)(t-t_b)},\, t_b \leq t \leq t_b + \delta. \tag{1.47c}$$

References

1. Allgöwer, F.E.A. (ed.): Nonlinear Model Predictive Control. Birkhäuser Verlag, Basel (2000)
2. Aoki, M.: Optimization of Stochastic Systems – Topics in Discrete-Time Systems. Academic Press, New York (1967)
3. Åström, K.J.: Introduction to Stochastic Control Theory. Elsevier, Amsterdam (1970)
4. Bauer, H.: Wahrscheinlichkeitstheorie und Grundzüge der Masstheorie. Walter de Gruyter & Co., Berlin (1968)
5. Bunke, H.: Gewöhnliche Differentialgleichungen mit zufälligen Parametern. Akademie-Verlag, Berlin (1972)
6. Carathéodory, C.: Vorlesungen über reelle Funktionen. Teubner, Leipzig (1918)
7. Demyanov, V., Rubinov, A.: Approximate Methods in Optimization Problems. American Elsevier, Publishing Company, Inc., New York (1970)
8. Dieudonné, J.: Foundations of Modern Analysis. Academic Press, New York (1969)
9. Dreyfus, S.: Some types of optimal control of stochastic systems. J. SIAM Control **2**(1), 120–134 (1964)
10. Dreyfus, S.: Dynamic Programming and the Calculus of Variations. Academic Press, New York (1965)
11. Dreyfus, S.E., Law, A.M.: The Art of Dynamic Programming. Academic Press, New York (1977)
12. Dullerud, G., Paganini, F.: A Course in Robust Control Theory. Springer, New York (2000)
13. Findeisen, R., et al.: Sampled–Data Nonlinear Model Predictive Control for Constrained Continuous Time Systems, pp. 207–235. Springer, Berlin (2007)
14. Garcia, C.E., et al.: Model predictive control: theory and practice - a survey. Automatica **25**(3), 335–348 (1989). https://doi.org/10.1016/0005-1098(89)90002-2
15. Gessing, R., Jacobs, O.L.R.: On the equivalence between optimal stochastic control and open-loop feedback control. Int. J. Control. **40**(1), 193–200 (1984). https://doi.org/10.1080/00207178408933267
16. Kalman, R., et al.: Topics in Mathematical System Theory. McGraw-Hill Book Company, New York (1969)
17. Ku, R., Athans, M.: On the adaptive control of linear systems using the open-loop feedback optimal approach. IEEE Trans. Autom. Control **18**, 489–493 (1973)
18. Luenberger, D.: Optimization by Vector Space Methods. J. Wiley, New York (1969)
19. Marti, K.: Convex approximations of stochastic optimization problems. Math. Meth. Oper. Res. **20**, 66–76 (1975)
20. Marti, K.: Approximationen Stochastischer Optimierungsprobleme. Hain, Konigstein/Ts (1979)
21. Marti, K.: Stochastic optimization methods in robust adaptive control of robots. In: Groetschel, M.E.A. (ed.) Online Optimization of Large Scale Systems, pp. 545–577. Springer, Berlin (2001)
22. Marti, K.: Adaptive Optimal Stochastic Trajectory Planning and Control (AOSTPC) for Robots, pp. 155–206. Springer, Berlin (2004)
23. Marti, K.: Stochastic nonlinear model predictive control (SNMPC). In: 79th Annual Meeting of the International Association of Applied Mathematics and Mechanics (GAMM), Bremen 2008, PAMM, vol. 8, Issue 1, pp. 10775–10776. Wiley-VCH, Weinheim (2008)
24. Marti, K.: Stochastic Optimization Methods, 2nd edn. Springer, Berlin (2008). https://doi.org/10.1007/978-3-540-79458-5
25. Marti, K.: Continuous-Time Control Under Stochastic Uncertainty. J. Wiley, Hoboken (2010). https://doi.org/10.1002/9780470400531.eorms0839
26. Marti, K.: Optimal control of dynamical systems and structures under stochastic uncertainty: stochastic optimal feedback control. Adv. Eng. Softw. **46**, 43–62 (2012). https://doi.org/10.1016/j.advengsoft.2010.09.008

27. Marti, K.: Stochastic optimal structural control: stochastic optimal open-loop feedback control. Adv. Eng. Softw. **44**(1), 26–34 (2012). https://doi.org/10.1016/j.advengsoft.2011.05.040
28. Marti, K.: Stochastic Optimization Methods: Applications in Engineering and Operations Research, 3rd edn. Springer, Berlin (2015). https://doi.org/10.1007/978-3-662-46214-0
29. Richalet, J., et al.: Model predictive heuristic control: applications to industrial processes. Automatica **14**, 413–428 (1978). https://doi.org/10.1016/0005-1098(78)90001-8
30. Schacher, M.: Stochastisch Optimale Regelung von Robotern. No. 1200 in Fortschritt-Berichte VDI, Reihe 8, Mess-, Steuerungs- und Regelungstechnik. VDI Verlag GmbH, Düsseldorf (2011)
31. Smirnov, W.: Lehrgang der Höheren Mathematik, Teil IV. Deutscher Verlag der Wissenschaft, Berlin (1966)
32. Soong, T.: Active structural control in civil engineering. Eng. Struct. **10**, 74–84 (1988)
33. Soong, T.: Active Structural Control: Theory and Practice. John Wiley, New York (1990)
34. Stengel, R.: Stochastic Optimal Control: Theory and Application. J. Wiley, New York (1986)
35. Tse, E., Athans, M.: Adaptive stochastic control for a class of linear systems. IEEE Trans. Autom. Control **17**(1), 38–52 (1972)
36. Walter, W.: Gewöhnliche Differentialgleichungen. Springer, Berlin (2000)

Chapter 2
Stochastic Optimization of Regulators

2.1 Introduction

The optimal design of regulators is often based on the use of given, fixed nominal values of initial conditions, load, and other model parameters. However, due to *variations of the material properties, measurement errors (e.g. in case of parameter identification), modeling errors (complexity of real systems), uncertainty on the working environment, the task to be executed, etc.*, the true initial conditions, external load, and further model parameters, like sizing parameters, mass values, gravity centers, moments of inertia, friction, tolerances, adjustment setting error, etc., are not known exactly in practice. Hence, a predetermined (optimal) regulator should be *robust*, i.e. the controller should guarantee satisfying results also in case of variations of the initial conditions, load, and other model parameters.

Robust controls have been considered up to now mainly for uncertainty models based on given fixed sets of parameters, like multiple intervals, assumed to contain the unknown, true parameter. In this case one requires then often that the controlled system fulfills certain properties, as, e.g., certain stability properties for all parameter vectors in the given parameter domain. If the required property can be described by a scalar criterion, then the controller design is based on a minimax criterion, such as the H^{∞}-criterion, see, e.g., [1, 4, 9, 10].

Since in many cases parameter uncertainty can be modeled more adequately by means of stochastic parameter models, in the following we suppose that the parameters involved in the regulator design problem are realizations of a random vector having a known or at least partly known joint probability distribution. The determination of an optimal controller under uncertainty with respect to model parameters, working neighborhood, modeling assumptions, etc. is a *decision theoretical problem*. Criteria of the type "*holds for all parameter vectors in a given set*" and the *minmax*-criterion are very pessimistic and often too strong. Indeed,

K. Marti, *Optimization Under Stochastic Uncertainty*, International Series in Operations Research & Management Science 296,
https://doi.org/10.1007/978-3-030-55662-4_2

in many cases the available a priori and empirical information on the dynamic system and its working neighborhood allows a more adequate, flexible description of the uncertainty situation by means of stochastic approaches. Thus, it is often more appropriate to model unknown and varying initial values, external loads, and other model parameters as well as modeling errors, e.g. incomplete representation of the dynamic system, by means of realizations of a random vector, a random function with a given or at least partly known probability distribution. Consequently, the optimal design of robust regulators is based on an optimization problem under stochastic uncertainty [2, 13, 14, 17].

Stochastic Optimal Design of Regulator
For the consideration of stochastic parameter variations as well as other uncertainties within the optimal design process of a regulator one has to introduce—as for any other optimization problem under (stochastic) uncertainty—an appropriate **deterministic substitute problem**. In the present case of stochastic optimal design of a regulator, hence, a map from the state or observation space into the space of control corrections, one has a **control problem under stochastic uncertainty**. Then, for the solution of the occurring deterministic substitute problems the methods of *stochastic optimization*, cf. [17], are available. As is well known [3, 20], the optimization of a regulator presupposes an optimal reference trajectory $q^R(t)$ and a corresponding feedforward control $u^R(t)$. In case of stochastic uncertainties, the guiding functions $\big(q^R(t), u^R(t)\big)$ can be determined also by stochastic optimization methods [3, 15, 21].

The computation of stochastic optimal regulators is based now on deterministic substitute control problems of the following type:

Minimize the expected total costs composed of (i) the costs arising from the deviation $\Delta z(t)$ between the (stochastic optimal) reference trajectory and the effective trajectory of the dynamic system and (ii) the costs for the control corrections $\Delta u(t)$

Subject to the following constraints:

- *dynamic equation of the underlying stochastic system with the total control input $u(t) = u^R(t) + \Delta u(t)$ being the sum of the feedforward control $u^R(t)$ and the control correction $\Delta u(t) = \varphi(t, \Delta z(t))$,*
- *stochastic initial conditions $q(t_0) = q_0(\omega)$, $\dot{q}(t_0) = \dot{q}_0(\omega)$ for the state of the system at the starting time point t_0,*
- *conditions for the feedback law $\varphi = \varphi(t, \Delta z(t))$, such as $\varphi(t, 0) = 0$ (if the effective state is equal to the state prescribed by the reference trajectory, then no control correction is needed).*

Here, as often in practice, we use quadratic cost functions. The resulting deterministic substitute problem can be interpreted again as a control problem for the unknown feedback control law $\Delta u(t) = \varphi(t, \Delta z(t))$. A main problem is then the computation of the (conditional) expectation arising in the objective function. Since the expectations are defined by multiple integrals, the expectations can be determined in general only approximately. In the following, approximations based on Taylor expansions with respect to the stochastic parameter vector $a(\omega)$ at its conditional mean $\overline{a}$ are taken into account. Using quadratic cost functions and first order Taylor expansions, hence, linearization, one has the advantage that often a certain part of the calculations can be done analytically!

An Example: Active Control Under Stochastic Uncertainty
As a typical application of the topic treated in this work, in Sect. 2.6 we present an illustrative example from *active control under stochastic uncertainty*: In order to stabilize mechanical structures under dynamic applied loads, active control strategies are taken into account [5, 22, 23]. The structures usually are stationary, safe, and stable without external dynamic disturbances, such as strong earthquakes, wind turbulences, water waves, etc. Thus, in case of dynamic disturbances, additional control elements can be installed enabling active control actions. Active control strategies for mechanical structures are applied in order to counteract heavy applied dynamic loads, such as earthquakes, wind, water waves, etc., which would lead to large vibrations causing possible damages of the structure. Modeling the structural dynamics by means of a system of first order random differential equations for the state vector (displacement vector q and time derivative of q), robust optimal controls are determined in order to cope with the stochastic uncertainty involved in the dynamic parameters, the initial values, and the applied loadings. The problem is modeled in the framework of stochastic optimal control for minimizing the expected total costs arising from the tracking error (deviation from the reference trajectory) of the structure and the regulation costs.

2.2 Regulator Design Under Stochastic Uncertainty

Feedforward and feedback control are based on the dynamic equation of the underlying control system

$$F(p_D, q(t), \dot{q}(t), \ddot{q}(t)) = u(t), \ t \geq t_0 \tag{2.1a}$$

$$q(t_0) = q_0, \quad \dot{q}(t_0) = \dot{q}_0. \tag{2.1b}$$

Here, $q = q(t)$ denotes the vector of configuration variables, and $u = u(t)$ is the control or input vector. Moreover, the state trajectory of the system

$$z(t) := \begin{pmatrix} q(t) \\ \dot{q}(t) \end{pmatrix}, t_0 \leq t \leq t_f \tag{2.1c}$$

is the solution of the dynamic equation (2.1a,b) related to (i) the initial state $z_0 := \begin{pmatrix} q_0 \\ \dot{q}_0 \end{pmatrix}$, (ii) the control or input vector $u = u(t)$, and (iii) the vector of dynamic parameters p_D.

For the control of a dynamic system one has [11, 14, 15] the following procedure: The control function $u = u(t), t_0 \leq t \leq t_f$, is represented by the equation

$$u(t) := u^R(t) + \Delta u(t), t_0 \leq t \leq t_f. \tag{2.2a}$$

Here, $u^R = u^R(t)$, $t_0 \leq t \leq t_f$, denotes the **feedforward control**, and $\Delta u = \Delta u(t), t_0 \leq t \leq t_f$, is a **control correction**, represented by a **feedback control**. In a preceding planning phase, the feedforward control $u^R = u^R(t), t_0 \leq t \leq t_f$, is selected (*tracking problem*) such that the effective trajectory $z = z(t), t_0 \leq t \leq t_f$, follows in the mean a given **(optimal) reference trajectory** $q^R = q^R(t), t_0 \leq t \leq t_f$. Hence, $u^R(\cdot)$ is defined by

$$u^R(t) := F(\overline{p_D}, q^R(t), \dot{q}^R(t), \ddot{q}^R(t)), \ t_0 \leq t \leq t_f, \tag{2.2b}$$

where

$$\overline{p_D} := \mathrm{E}\big(p_D(\omega)|\mathcal{A}_{t_0}\big) \tag{2.2c}$$

denotes the conditional expectation with respect to the information available at time t_0. For the correction (compensation) of the tracking error

$$\Delta z(t) := z(t) - z^R(t) = \begin{pmatrix} q(t) \\ \dot{q}(t) \end{pmatrix} - \begin{pmatrix} q^R(t) \\ \dot{q}^R(t) \end{pmatrix} = \begin{pmatrix} \Delta q(t) \\ \Delta \dot{q}(t) \end{pmatrix}, t_0 \leq t \leq t_f, \tag{2.3}$$

hence, the deviation between the effective state $z(t) = \begin{pmatrix} q(t) \\ \dot{q}(t) \end{pmatrix}$ and the reference trajectory $z^R(t) = \begin{pmatrix} q^R(t) \\ \dot{q}^R(t) \end{pmatrix}$ one considers the **PID-regulator**

$$\Delta u(t) := \varphi(t, \Delta q(t), \Delta q_I(t), \Delta \dot{q}(t)), t_0 \leq t \leq t_f, \tag{2.4a}$$

where

$$q_I(t) := \int_{t_0}^{t} q(\tau)\, d\tau, \ t_0 \leq t \leq t_f, \tag{2.4b}$$

denotes the integrated position, and

$$\Delta q_I(t) := \int_{t_0}^{t} \Delta q(\tau)\, d\tau = \int_{t_0}^{t} \Big(q(\tau) - q^R(\tau)\Big)\, d\tau$$

$$= \int_{t_0}^{t} q(\tau)\, d\tau - \int_{t_0}^{t} q^R(\tau)\, d\tau = q_I(t) - q_I^R(t) \tag{2.4c}$$

is then the deviation of the integrated position. Moreover,

$$\varphi = \varphi(t, \Delta z_I) = \varphi(t, \Delta q, \Delta q_I(t), \Delta \dot{q}) \tag{2.4d}$$

is the **feedback function (feedback law)** to be determined. In case of a zero tracking error, then, without further information about the control system, no control correction is needed, thus,

$$\varphi(t, 0) = 0, \quad t_0 \le t \le t_f. \tag{2.4e}$$

Working here with PID-regulators, the state $z = z(t)$, and the deviation $\Delta z(t)$ of the state from the reference state, resp., is replaced by the *generalized state*, the generalized tracking error $\Delta z_I(t)$, resp.,

$$z_I(t) := \begin{pmatrix} q(t) \\ q_I(t) \\ \dot{q}(t) \end{pmatrix}, \quad \Delta z_I(t) = \begin{pmatrix} \Delta q(t) \\ \Delta q_I(t) \\ \Delta \dot{q}(t) \end{pmatrix}$$

$$:= z_I(t) - z_I^R(t) = \begin{pmatrix} q(t) - q^R(t) \\ q_I(t) - q_I^R(t) \\ \dot{q}(t) - \dot{q}^R(t) \end{pmatrix}, t_0 \le t \le t_f. \tag{2.4f}$$

In former papers, cf. [14, 21], we assumed that the state $z(t) = \begin{pmatrix} q(t) \\ \dot{q}(t) \end{pmatrix}$ can be observed or measured **exactly**. Since in practice one has always some measurement error/noise, see, e.g., [6], we suppose now, cf. [16], that only a certain **estimate**

$$\hat{z}_I(t) := \begin{pmatrix} \hat{q}(t) \\ \hat{q}_I(t) \\ \hat{\dot{q}}(t) \end{pmatrix}, \quad t_0 \le t \le t_f, \tag{2.5a}$$

of the effective generalized state $z_I(t)$ is available, can be observed. Having then the *estimated/observed generalized tracking error*

$$\hat{\Delta} z_I(t) := \hat{z}_I(t) - z_I^R(t) = \begin{pmatrix} \hat{q}(t) - q^R(t) \\ \hat{q}_I(t) - q_I^R(t) \\ \hat{\dot{q}}(t) - \dot{q}^R(t) \end{pmatrix}, \quad t_0 \le t \le t_f, \tag{2.5b}$$

the control correction $\Delta u(t)$ is defined now (PID control), cf. (2.4a), by

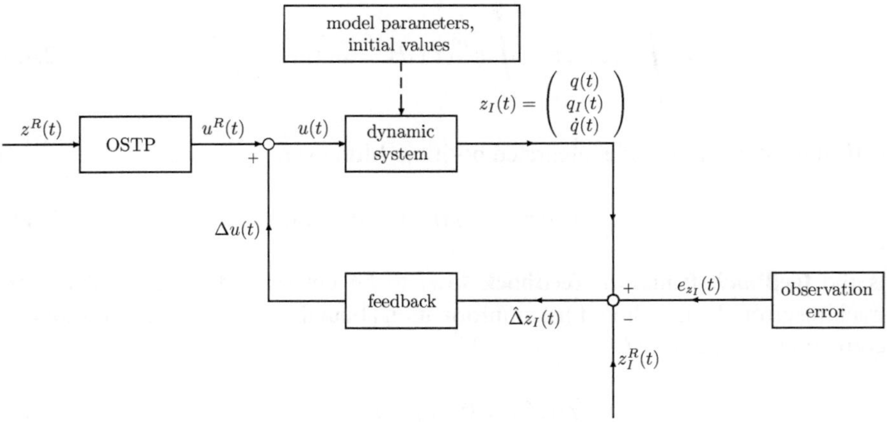

Fig. 2.1 Regulation under observation error

$$\begin{aligned}\Delta u(t) &:= \varphi(t, \hat{\Delta} z_I(t)) = \varphi(t, \hat{z}_I(t) - z_I^R(t)) \\ &= \varphi(t, \hat{q}(t) - q^R(t), \hat{q_I}(t) - q_I^R(t), \hat{\dot{q}}(t) - \dot{q}^R(t)), \quad t_0 \le t \le t_f. \end{aligned} \tag{2.5c}$$

For the further analysis of the problem, the estimate/observation $\hat{z}_I(t)$ of the effective generalized state $z_I(t)$ is represented (Fig. 2.1) as follows:

$$\hat{z}_I(t) = z_I(t) + e_{z_I}(t), \quad e_{z_I}(t) = \begin{pmatrix} e_q(t) \\ e_{q_I}(t) \\ e_{\dot{q}}(t) \end{pmatrix}, t_0 \le t \le t_f. \tag{2.6a}$$

Here, $e_q = e_q(t,\omega), e_{q_I} = e_{q_I}(t,\omega), e_{\dot{q}} = e_{\dot{q}}(t,\omega)$ denotes the possibly time-dependent random **estimation, observation or measurement error**, resp., arising at the determination of $q(t), q_I(t)$, and $\dot{q}(t)$. Considering additive and multiplicative errors [6, 16, 26], the random error vector $e_{z_I}(t)$ is defined by

$$e_{z_I}(t,\omega) = \begin{pmatrix} e_q(t,\omega) \\ e_{q_I}(t,\omega) \\ e_{\dot{q}}(t,\omega) \end{pmatrix} := e(t,\omega) + \mathcal{E}(t,\omega) z_I(t,\omega). \tag{2.6b}$$

Here, with a time-dependent matrix $\Gamma = \Gamma(t)$ and a corresponding random parameter vector $e_0 = e_0(\omega)$, the additive errors $e(t,\omega)$ are represented in the following *parametric form*:

$$e(t,\omega) = \Gamma(t) e_0(\omega) = \begin{pmatrix} \Gamma_q(t) e_{q0}(\omega) \\ \Gamma_{q_I}(t) e_{q_I 0}(\omega) \\ \Gamma_{\dot{q}}(t) e_{\dot{q}0}(\omega) \end{pmatrix} \tag{2.6c}$$

with submatrices $\Gamma_q(t), \Gamma_{q_I}(t), \Gamma_{\dot{q}}(t)$ of $\Gamma(t)$ and subvectors $e_{q0}(\omega), e_{q_I0}(\omega), e_{\dot{q}0}(\omega)$ of $e_0(\omega)$. Assuming that the estimate at each time point t is correct in the conditional mean, for the random vector $e_0(\omega)$ we have the condition:

$$E\big(e_0(\omega)|\mathcal{A}_t\big) = 0, \tag{2.6d}$$

where $\mathcal{A}_t$ denotes the σ-algebra of information (events) available up to the time point t.

Moreover, the block-diagonal matrix function of multiplicative errors

$$\mathcal{E} = \mathcal{E}(t,\omega) = \begin{pmatrix} \mathcal{E}_q(t,\omega) & 0 & 0 \\ 0 & \mathcal{E}_{q_I}(t,\omega) & 0 \\ 0 & 0 & \mathcal{E}_{\dot{q}}(t,\omega) \end{pmatrix} \tag{2.6e}$$

is composed of the diagonal blocks

$$\mathcal{E}_q(t,\omega) = \Lambda_q(t)H(e_{qm}(\omega)) \tag{2.6f}$$

$$\mathcal{E}_{q_I}(t,\omega) = \Lambda_{q_I}(t)H(e_{q_{Im}}(\omega)) \tag{2.6g}$$

$$\mathcal{E}_{\dot{q}}(t,\omega) = \Lambda_{\dot{q}}(t)H(e_{\dot{q}m}(\omega)) \tag{2.6h}$$

with the matrices $\Lambda_q(t), \Lambda_{q_I}(t), \Lambda_{\dot{q}}(t)$ and the matrices $H(e_{qm}), H(e_{q_{Im}}), H(e_{\dot{q}m})$ depending linearly on the random vectors $e_{qm}(\omega), q_{Im}(\omega), e_{\dot{q}m}(\omega)$, where, cf. (2.6d),

$$E\big(e_{qm}(\omega)|\mathcal{A}_t\big) = E\big(q_{Im}(\omega)|\mathcal{A}_t\big) = E\big(e_{\dot{q}m}(\omega)|\mathcal{A}_t\big) = 0. \tag{2.6i}$$

Additive Measurement Errors Only

In this part we suppose that we only have additive errors, hence, $\mathcal{E}(t,\omega) = 0$, cf. (2.6b), and therefore

$$e_{z_I}(t,\omega) = e(t,\omega). \tag{2.7}$$

Putting the representation (2.6a,c), (2.7) of $\hat{z}_I(t)$ into formula (2.5c), for the control correction then we get

$$\begin{aligned}\Delta u(t) = \varphi(t, \hat{z}_I(t) - z_I^R(t)) = \varphi(t, \Delta z_I(t) + e_{z_I}(t,\omega)) = \varphi(t, \Delta z_I(t) \\ + \Gamma(t)e_0(\omega)),\end{aligned} \tag{2.8a}$$

where the effective tracking error $\Delta z_I(t)$ is defined, cf. (2.4f), by

$$\Delta z_I(t) := z_I(t) - z_I^R(t). \tag{2.8b}$$

Using the control correction given by (2.5c), corresponding to the estimated tracking error (2.5b), the effective state $z = z(t) = \begin{pmatrix} q(t) \\ \dot{q}(t) \end{pmatrix}$, $t_0 \le t \le t_f$, and the related effective generalized state $z_I = z_I(t)$, $t_0 \le t \le t_f$, of the control system is the solution of the following system of second order differential equations:

$$
\begin{aligned}
F(p_D, q(t), \dot{q}(t), \ddot{q}(t)) = u^{(0)}(t) + \Delta u(t) = u^{(0)}(t) \\
+ \varphi(t, \Delta z_I(t) + \Gamma(t) e_0(\omega))), \ t_0 \le t \le t_f
\end{aligned} \tag{2.9a}
$$

$$
q(t_0) = q_0, \dot{q}(t_0) = \dot{q}_0. \tag{2.9b}
$$

According to the above consideration, the solution $q = q(t)$ of (2.9a,b) depends on the following quantities:

$$
q(t) = q(t, \underbrace{q_0, \dot{q}_0}_{\substack{\text{initial} \\ \text{values}}}, \underbrace{p_D}_{\substack{\text{dynamic} \\ \text{param.}}}, \underbrace{e_0(.)}_{\substack{\text{parameter} \\ \text{obs. error}}}), \quad t_0 \le t \le t_f. \tag{2.10}
$$

Additive and Multiplicative Measurement Errors

Putting the representation (2.6a–i) of $\hat{z}_I(t)$ into formula (2.5c) for the control correction, then we get

$$
\begin{aligned}
\Delta u(t) =& \varphi(t, \hat{z}_I(t) - z_I^R(t)) = \varphi(t, z_I(t) + e_{z_I}(t) - z_I^R(t)) \\
=& \varphi(t, (I + \mathcal{E}(t, \omega)) \Delta z_I(t) + \mathcal{E}(t, \omega) z_I^R(t) + \Gamma(t) e_0(\omega)),
\end{aligned} \tag{2.11}
$$

where $\Delta z_I(t) := z_I(t) - z_I^R(t)$ is the effective tracking error, cf. (2.8b).

Using the control correction given by (2.11), corresponding to the estimated tracking error (2.5b), the effective state $z = z(t) = \begin{pmatrix} q(t) \\ \dot{q}(t) \end{pmatrix}$, $t_0 \le t \le t_f$, and the related effective generalized state $z_I = z_I(t)$, $t_0 \le t \le t_f$, of the control system is the solution of a second order initial value problem as in the above additive case of measurement errors, see (2.9a,b).

For simplification of the presentation, in the following we consider *additive* measurement errors only.

2.3 Optimal Feedback Functions Under Stochastic Uncertainty

According to (2.5a–c) and (2.6c), (2.7), (2.9a,b), in case of observation errors, the arguments of the feedback function φ are given as follows:

$$
\begin{aligned}
\varphi =& \varphi(t, \hat{\Delta z}_I(t))) \\
:=& \varphi\big(t, \Gamma_q(t) e_{q0}(\omega) + \Delta q(t), \Gamma_{q_I}(t) e_{q_I 0}(\omega) + \Delta q_I(t), \Gamma_{\dot{q}}(t) e_{\dot{q}0}(\omega) + \Delta \dot{q}(t)\big).
\end{aligned} \tag{2.12}
$$

Due to (2.12), for the optimal selection of a PID-regulator $\varphi^* = \varphi^*(t, \Delta q, \Delta q_I, \Delta \dot{q})$ in case of stochastic uncertainty we have the following optimization problem [17]:

Definition 2.1 A stochastic optimal feedback control law $\varphi^* = \varphi^*(t, \Delta z_I(t))$ is a solution of the stochastic optimization problem:

$$\min \mathrm{E}\Bigg(\int_{t_0}^{t_f} \Big(c\big(t, \Delta q(t), \Delta q_I(t), \Delta \dot{q}(t)\big)$$

$$+\gamma\big(t, \widetilde{\varphi}(t, \Delta q(t), \Delta q_I(t), \Delta \dot{q}(t); e_0(\omega))\big)\Big) dt \Bigg| \mathcal{A}_{t_0} \Bigg) \tag{2.13a}$$

s.t.

$$F(p_D(\omega), q(t), \dot{q}(t), \ddot{q}(t)) = u^R(t)$$

$$+\varphi\big(t, \Gamma_q(t) e_{q0}(\omega) + \Delta q(t), \Gamma_{q_I}(t) e_{q_I 0}(\omega) + \Delta q_I(t), \Gamma_{\dot{q}}(t) e_{\dot{q}0}(\omega) + \Delta \dot{q}(t)\big),$$

$$t_0 \le t \le t_f, \ a.s. \tag{2.13b}$$

$$q(t_0, \omega) = q_0(\omega), \tag{2.13c}$$

$$\dot{q}(t_0, \omega) = \dot{q}_0(\omega), \tag{2.13d}$$

$$\varphi(t, \mathbf{0}) = 0. \tag{2.13e}$$

Here, the term $c\big(t, \Delta q(t), \Delta q_I(t), \Delta \dot{q}(t)\big)$ in the objective function (2.13a) describes the costs resulting from the tracking error, and the costs for the control corrections are represented by $\gamma\big(t, \Delta u(t)\big) = \gamma\big(t, \varphi(t, \cdot, \cdot, \cdot)\big)$. Possible loss functions are convex quadratic functions, different norms, and—more general—sublinear functionals [7].

Having the stochastic optimal feedback control $\varphi^* = \varphi^*(t, \Delta q, \Delta q_I, \Delta \dot{q})$, the effective trajectory $q = q(t), \ t \ge t_0$, of the stochastic optimally controlled dynamic system is then given by the solution of the initial value problem (2.13b–d). In the following we use the definitions:

$$\Delta p_D(\omega) := p_D(\omega) - \overline{p_D} = p_D(\omega) - \mathrm{E}\big(p_D | \mathcal{A}_{t_0}\big) \tag{2.14a}$$

$$\Delta q_0 := q(t_0) - q^R(t_0) = q_0 - \mathrm{E}\big(q_0 | \mathcal{A}_{t_0}\big), \tag{2.14b}$$

$$\Delta \dot{q}_0 := \dot{q}(t_0) - \dot{q}^R(t_0) = \dot{q}_0 - \mathrm{E}\big(\dot{q}_0 | \mathcal{A}_{t_0}\big), \tag{2.14c}$$

where $\Delta p_D(\omega)$ denotes the deviation of the effective dynamic parameter vector $p_D = p_D(\omega)$ from its nominal value or mean $\overline{p_D}$. Here, $\overline{p_D} := \mathrm{E}\big(p_D(\omega) | \mathcal{A}_{t_0}\big)$ denotes the (conditional) expectation of the random vector $p_D = p_D(\omega)$.

The constraints of the regulator optimization problem (2.13a–e) can be represented then also by

$$F(\overline{p_D} + \Delta p_D(\omega), q^R(t) + \Delta q(t), \dot{q}^R(t) + \Delta \dot{q}(t), \ddot{q}^R(t) + \Delta \ddot{q}(t)) = u^R(t)$$
$$+\varphi\big(t, \Gamma_q(t)e_{q0}(\omega) + \Delta q(t), \Gamma_{q_I}(t)e_{q_I0}(\omega) + \Delta q_I(t), \Gamma_{\dot{q}}(t)e_{\dot{q}0}(\omega) + \Delta \dot{q}(t)\big),$$
$$t_0 \le t \le t_f, \; a.s., \tag{2.15a}$$
$$\Delta q(t_0, \omega) = \Delta q_0(\omega), \tag{2.15b}$$
$$\Delta \dot{q}(t_0, \omega) = \Delta \dot{q}_0(\omega), \tag{2.15c}$$
$$\varphi(t, \mathbf{0}) = 0. \tag{2.15d}$$

Remark 2.1 For given control law $\varphi = \varphi(t, \Delta q(t), \Delta q_I(t), \Delta \dot{q}(t))$, the solution of the initial value problem (2.15a–c) yields the position error function

$$\Delta q := \Delta q(t, a(\omega)), t_0 \le t \le t_f, \tag{2.16a}$$

depending on the random parameter vector

$$a = a(\omega) := \big(\Delta p_D(\omega)^T, \Delta q_0(\omega)^T, \Delta \dot{q}_0(\omega)^T, e_0(\omega)^T\big)^T. \tag{2.16b}$$

2.3.1 Quadratic Cost Functions

In the following we consider *quadratic cost functions*. Hence, with symmetric, positive (semi-) definite matrices $C_q, C_{q_I}, C_{\dot{q}}$, the costs $c(t, \cdot, \cdot)$ arising from the tracking error are defined by

$$c\big(t, \Delta q(t), \Delta q_I(t), \Delta \dot{q}(t)\big)$$
$$:= \Delta q(t)^T C_q \Delta q(t) + \Delta q_I(t)^T C_{q_I} \Delta q_I(t) + \Delta \dot{q}(t)^T C_{\dot{q}} \Delta \dot{q}(t). \tag{2.17a}$$

Moreover, with a positive (semi-) definite matrix C_u the regulator costs $\gamma(t, \cdot, \cdot; \cdot, \cdot)$ are defined by

$$\gamma\big(t, \varphi(t, \Delta z_I(t) + \Gamma(t)e_0(\omega))\big) := \Delta u(t)^T C_u \Delta u(t)$$
$$= \varphi(t, \Delta z_I(t) + \Gamma(t)e_0(\omega))^T C_u \varphi(t, \Delta z_I(t) + \Gamma(t)e_0(\omega)). \tag{2.17b}$$

The total cost function is then defined by

$$f(t) := \Delta q(t)^T C_q \Delta q(t) + \Delta q_I(t)^T C_{q_I} \Delta q_I(t)$$
$$+\Delta \dot{q}(t)^T C_{\dot{q}} \Delta \dot{q}(t) + \Delta u(t)^T C_u \Delta u(t). \tag{2.18}$$

Thus, for the objective function of the regulator optimization problem (2.13a-e) we obtain

$$\mathrm{E}\Big(\int_{t_0}^{t_f} f(t)\,dt\Big|\mathcal{A}_{t_0}\Big) = \int_{t_0}^{t_f} \mathrm{E}\Big(f(t)\Big|\mathcal{A}_{t_0}\Big)\,dt$$

$$= \int_{t_0}^{t_f} \mathrm{E}\Big(\Delta q(t)^T C_q \Delta q(t) + \Delta q_I(t)^T C_{q_I} \Delta q_I(t) + \Delta\dot{q}(t)^T C_{\dot{q}} \Delta\dot{q}(t)$$

$$+ \varphi(t, \Delta z_I(t) + \Gamma(t)e_0(\omega))^T C_u \varphi(t, \Delta z_I(t) + \Gamma(t)e_0(\omega))\Big|\mathcal{A}_{t_0}\Big)dt. \tag{2.19}$$

2.3.1.1 Computation of the Expectation by Taylor Expansion

We use now Taylor expansion method to compute approximatively the arising conditional expectations arising above. Thus, by linearization of the feedback function φ at zero tracking error and zero observation error, hence, $\Delta z_I = 0$ and $e_0(\omega) = 0$, we get

$$\varphi(t, \Delta z_I(t) + \Gamma(t)e_0(\omega)) \approx \varphi_L(t, \Delta z_I(t) + \Gamma(t)e_0(\omega))$$

$$:= D_{\Delta q}\varphi(t, \mathbf{0})\big(\Gamma_q(t)e_{q0}(\omega) + \Delta q(t)\big) + D_{\Delta q_I}\varphi(t, \mathbf{0})\big(\Gamma_{q_I}(t)e_{q_I 0}(\omega) + \Delta q_I(t)\big)$$

$$+ D_{\Delta\dot{q}}\varphi(t, \mathbf{0})\big(\Gamma_{\dot{q}}(t)e_{\dot{q}0}(\omega) + \Delta\dot{q}(t)\big) = D_{\Delta z_I}\varphi(t, 0)\big(\Delta z_I(t) + \Gamma(t)e_0(\omega)\big), \tag{2.20a}$$

since $\varphi(t, 0) = 0$, cf. (2.4e). Here, we have the *gain matrices*, represented by the Jacobians, see, e.g., [8],

$$D_{\Delta q}\varphi(t, \mathbf{0}),\ D_{\Delta q_I}\varphi(t, \mathbf{0}),\ D_{\Delta\dot{q}}\varphi(t, \mathbf{0}), \tag{2.20b}$$

of φ with respect to $\Delta q, \Delta q_I, \Delta\dot{q}$, resp., at $(t, \mathbf{0})$.

Thus, the unknown feedback function is replaced, cf. Definition 2.1, in the optimal regulator problem under stochastic uncertainty (2.13a–e) by the total gain matrix:

Definition 2.2 Linearizing the feedback function $\varphi = \varphi(t, \Delta q, \Delta q_I, \Delta\dot{q})$ at $\Delta z_I = 0$, the unknowns of the stochastic optimal regulator problem (2.13a-g) are represented by the *total gain matrix*

$$D_{\Delta z_I}\varphi(t, \mathbf{0}) = \Big(D_{\Delta q}\varphi(t, \mathbf{0}), D_{\Delta q_I}\varphi(t, \mathbf{0}), D_{\Delta\dot{q}}\varphi(t, \mathbf{0})\Big), \tag{2.20c}$$

composed of the Jacobians of φ with respect to $\Delta q, \Delta q_I, \Delta\dot{q}$, resp., at $(t, \mathbf{0})$.

Taking into account the linearization (2.20a–c) of φ, for the expected total cost function we get now the approximation:

$$E\left(f\middle|\mathcal{A}_{t_0}\right) \approx E\left(\tilde{f}\middle|\mathcal{A}_{t_0}\right), \tag{2.21}$$

where

$$\begin{aligned}E\left(\tilde{f}|\mathcal{A}_{t_0}\right) &:= E\Bigg(\Delta q(t)^T C_q \Delta q(t) + \Delta q_I(t)^T C_{q_I} \Delta q_I(t) + \Delta\dot{q}(t)^T C_{\dot{q}} \Delta\dot{q}(t)\\ &+ \Big(D_{\Delta z_I}\varphi(t,0)\big(\Delta z_I(t) + \Gamma(t)e_0(\omega)\big)\Big)^T\\ &\times C_u\Big(D_{\Delta z_I}\varphi(t,0)\big(\Delta z_I(t) + \Gamma(t)e_0(\omega)\big)\Big)\Bigg|\mathcal{A}_{t_0}\Bigg).\end{aligned} \tag{2.22}$$

Defining the *positive (semi-)definite* matrices

$$Q_\varphi(t) := D_{\Delta z_I}\varphi(t,\mathbf{0})^T C_u D_{\Delta z_I}\varphi(t,\mathbf{0}), \tag{2.23a}$$

$$Q_0 := \begin{pmatrix} C_q & 0 & 0 \\ 0 & C_{q_I} & 0 \\ 0 & 0 & C_{\dot{q}} \end{pmatrix}, \tag{2.23b}$$

according to (2.22) we find

$$\begin{aligned}\tilde{f} &:= \Delta z_I(t)^T (Q_0 + Q_\varphi(t))\Delta z_I(t) + 2e_0^T \Gamma(t)^T Q_\varphi(t)\Delta z_I(t)\\ &+ e_0^T \Gamma(t)^T Q_\varphi(t)\Gamma(t)e_0.\end{aligned} \tag{2.24}$$

Since C_q, C_{q_I}, $C_{\dot{q}}$, and C_u are positive (semi-)definite matrices, we have $\tilde{f} \geq 0$.

In order to determine the expectation of $\tilde{f}$, in addition to the Taylor approximation of the feedback law $\varphi = \varphi\big(t, \Delta q(t), \Delta q_I(t), \Delta\dot{q}(t)\big)$, we need also the Taylor expansion of

$$\Delta z_I = \Delta z_I(t,\omega) = \Delta z_I(t, a(\omega)) \tag{2.25}$$

with respect to the parameter vector a, cf. (2.16b), at zero deviation and zero observation error. Because of Eqs. (2.2b,c) and conditions (2.15a–c) we have

$$\Delta q(t,\mathbf{0}) = 0,\ t \geq t_0, \tag{2.26a}$$

$$\Delta\dot{q}(t,\mathbf{0}) = 0,\ t \geq t_0. \tag{2.26b}$$

Hence, the tracking error vector can be approximated as follows:

$$\begin{aligned}\Delta q(t) &\approx D_{\Delta p_D}\Delta q(t,\mathbf{0})\Delta p_D(\omega) + D_{\Delta q_0}\Delta q(t,\mathbf{0})\Delta q_0(\omega) + D_{\Delta\dot{q}_0}\Delta q(t,\mathbf{0})\Delta\dot{q}_0(\omega)\\ &+ D_{e_0}\Delta q(t,\mathbf{0})e_0(t,\omega),\end{aligned} \tag{2.27a}$$

$$\Delta\dot{q}(t) \approx D_{\Delta p_D}\Delta\dot{q}(t,\mathbf{0})\Delta p_D(\omega) + D_{\Delta q_0}\Delta\dot{q}(t,\mathbf{0})\Delta q_0(\omega) + D_{\Delta\dot{q}_0}\Delta\dot{q}(t,\mathbf{0})\Delta\dot{q}_0(\omega) + D_{e_0}\Delta\dot{q}(t,\mathbf{0})e_0(t,\omega), \tag{2.27b}$$

$$\Delta q_I(t) \approx D_{\Delta p_D}\Delta q_I(t,\mathbf{0})\Delta p_D(\omega) + D_{\Delta q_0}\Delta q_I(t,\mathbf{0})\Delta q_0(\omega) + D_{\Delta\dot{q}_0}\Delta q_I(t,\mathbf{0})\Delta\dot{q}_0(\omega) + D_{e_0}\Delta q_I(t,\mathbf{0})e_0(\tau,\omega). \tag{2.27c}$$

Consequently, we also have

$$\Delta z_I(t,a) \approx D_{\Delta p_D}\Delta z_I(t,\mathbf{0})\Delta p_D(\omega) + D_{\Delta q_0}\Delta z_I(t,\mathbf{0})\Delta q_0(\omega) + D_{\Delta\dot{q}_0}\Delta z_I(t,\mathbf{0})\Delta\dot{q}_0(\omega) + D_{e_0}\Delta z_I(t,\mathbf{0})e_0(\omega) = D_a\Delta z_I(t,\mathbf{0})a(\omega). \tag{2.28}$$

Inserting this into Eq. (2.24), we get the approximation:

$$\tilde{f} \approx a^T D_a\Delta z_I(t,\mathbf{0})^T(Q_0 + Q_\varphi(t))D_a\Delta z_I(t,\mathbf{0})a + 2e_0^T\Gamma(t)^T Q_\varphi D_a\Delta z_I(t,\mathbf{0})a + e_0^T\Gamma(t)^T Q_\varphi\Gamma(t)e_0. \tag{2.29}$$

2.3.1.2 Approximation of the Expectation of the Total Cost Function

Without limitation we may suppose the following properties of the occurring random parameters:

Assumption The random vectors $\Delta q_0(\omega)$, $\Delta p_D(\omega)$ as well as $\Delta p_D(\omega)$, $\Delta\dot{q}_0(\omega)$ are stochastically independent. Moreover, the error vector $e_0 = e_0(t,\omega)$ is stochastically independent from the other random variables, i.e. $\Delta p_D(\omega)$, $\Delta q_0(\omega)$, $\Delta\dot{q}_0(\omega)$.

For the conditional expectation of $\tilde{f}$ we obtain now the approximation

$$E\left(\tilde{f}\middle|\mathcal{A}_{t_0}\right) \approx E\Big(a(\omega)^T D_a\Delta z_I(t,\mathbf{0}^T(Q_0 + Q_\varphi(t))D_a\Delta z_I(t,\mathbf{0})a(\omega) + 2e_0(\omega)^T\Gamma(t)^T Q_\varphi D_a\Delta z_I(t,\mathbf{0})a(\omega) + e_0(\omega)^T\Gamma(t)^T Q_\varphi\Gamma(t)e_0(\omega)\Big|\mathcal{A}_{t_0}\Big). \tag{2.30}$$

In order to evaluate the above conditional expectation, let P denote one of the three matrices arising in (2.30). Moreover, "tr" denotes the trace of a quadratic matrix. Abbreviating then the random variables $a = a(\omega)$, $e_0 = e_0(\omega)$ occurring in (2.30) by α, β, resp., for the conditional expectations of one of the occurring terms in (2.30) we have

$$\begin{aligned}
&\mathrm{E}\Big(\alpha(\omega)^T P\beta(\omega)\Big|\mathcal{A}_{t_0}\Big)\\
&= \sum_{k,l=1}^{\nu}[P]_{kl}\Big[\mathrm{E}\Big(\beta(\omega)\alpha(\omega)^T\Big|\mathcal{A}_{t_0}\Big)\Big]_{lk} = \sum_{k=1}^{\nu}\Big(\sum_{l=1}^{\nu}[P]_{kl}\Big[\mathrm{E}\Big(\beta(\omega)\alpha(\omega)^T\Big|\mathcal{A}_{t_0}\Big)\Big]_{lk}\Big)\\
&= \sum_{k=1}^{\nu}\Big(\text{ k-th row of } P\Big)\cdot\Big(\text{ k-th column of } \mathrm{E}\Big(\beta(\omega)\alpha(\omega)^T\Big|\mathcal{A}_{t_0}\Big)\Big)\\
&= \text{ tr } P\mathrm{E}\Big(\beta(\omega)\alpha(\omega)^T\Big|\mathcal{A}_{t_0}\Big). \qquad (2.31a)
\end{aligned}$$

Note that in case of random vectors α, β with zero mean,

$$\mathrm{cov}\Big(\beta,\alpha\Big|\mathcal{A}_{t_0}\Big) := \mathrm{E}\Big(\beta(\omega)\alpha(\omega)^T\Big|\mathcal{A}_{t_0}\Big) \qquad (2.31b)$$

is the matrix of all covariances the vectors α, β.

Using now (2.31a) in the approximation (2.30), for the objective function (2.18) we find the approximation:

Theorem 2.1 *Based on the Taylor expansions of the feedback law and the tracking error, for the expectation of the total cost function* $E\Big(f\Big|\mathcal{A}_{t_0}\Big)$ *we get the following approximation:*

$$\begin{aligned}
E\Big(f(t,\omega)\Big|\mathcal{A}_{t_0}\Big) \approx\ & tr\ D_a\Delta z_I(t,\mathbf{0})^T(Q_0+Q_\varphi(t))D_a\Delta z_I(t,\mathbf{0})cov\Big(a(\cdot)\Big|\mathcal{A}_{t_0}\Big)\\
&+2tr\Gamma(t)^T Q_\varphi(t)D_a\Delta z_I(t,\mathbf{0})E\Big(a(\cdot)e_0(\cdot)^T\Big|\mathcal{A}_{t_0}\Big)\\
&+tr\Gamma(t)^T Q_\varphi(t)\Gamma(t)\ cov\Big(e_0(\cdot)\Big|\mathcal{A}_{t_0}\Big). \qquad (2.32)
\end{aligned}$$

2.4 Calculation of the Tracking Error Rates (Sensitivities)

Because of the expansions (2.27a–c), (2.32), resp., we need the *tracking error rates*, i.e. the derivatives of the tracking error $\Delta z_I(t,a)$ with respect to the parameter vectors Δp_D, Δq_0, $\Delta\dot{q}_0$, e_0. The calculation of the derivatives is based on the initial value problem (2.13b–d). Due to Eqs. (2.15a–d), the constraints of the regulator optimization problem (2.13a–e) will be replaced now by the following equivalent ones:

$$\begin{aligned}
&F(\overline{p_D}+\Delta p_D(\omega), q^R(t)+\Delta q(t), \dot{q}^R(t)+\Delta\dot{q}(t), \ddot{q}^R(t)+\Delta\ddot{q}(t)) = u^R(t)\\
&+\varphi\big(t, \Gamma_q(t)e_{q0}(\omega)+\Delta q(t), \Gamma_{q_I}(t)e_{q_I0}(\omega)+\Delta q_I(t), \Gamma_{\dot{q}}(t)e_{\dot{q}0}(\omega)+\Delta\dot{q}(t)\big),
\end{aligned}$$

$$t_0 \le t \le t_f,\ a.s., \tag{2.33a}$$

$$\Delta q(t_0, \omega) = \Delta q_0(\omega), \tag{2.33b}$$

$$\Delta \dot{q}(t_0, \omega) = \Delta \dot{q}_0(\omega), \tag{2.33c}$$

$$\varphi(t_0, \mathbf{0}) = 0. \tag{2.33d}$$

By partial differentiation of the Eq. (2.33a–c) with respect to the above mentioned parameter vectors we obtain five systems of differential equations with related initial values for the sensitivities or partial derivatives of $\Delta z_I = \Delta z_I(t, a)$ with respect to the total parameter vector a needed in (2.32).

For the representation of these matrix functions we introduce the following definition:

Definition 2.3 The Jacobians of the vector function F in (2.33a) taken at $\left(\overline{p_D}, q^R(t), \dot{q}^R(t), \ddot{q}^R(t)\right)$ are denoted by

$$K^R(t) := D_q F(\overline{p_D}, q^R(t), \dot{q}^R(t), \ddot{q}^R(t)), \tag{2.34a}$$

$$D^R(t) := D_{\dot{q}} F(\overline{p_D}, q^R(t), \dot{q}^R(t), \ddot{q}^R(t)), \tag{2.34b}$$

$$Y^R(t) := D_{p_D} F(\overline{p_D}, q^R(t), \dot{q}^R(t), \ddot{q}^R(t)), \tag{2.34c}$$

$$M^R(t) := M(\overline{p_D}, q^R(t)) = D_{\ddot{q}} F(\overline{p_D}, q^R(t), \dot{q}^R(t), \ddot{q}^R(t)). \tag{2.34d}$$

2.4.1 Partial Derivative with Respect to Δp_D

As was shown in [16], by means of differentiation of (2.33a–c) with respect to the parameter subvector Δp_D and using that $\Delta q(t) = \Delta q(t, a) = \Delta q(t, \Delta p_D, \Delta q_0, \Delta \dot{q}_0, e_0)$, for the partial derivative

$$\eta(t) := D_{\Delta p_D} \Delta q(t, \mathbf{0}),\ t_0 \le t \le t_f, \tag{2.35}$$

we obtain, cf. (2.12), the following initial value problem:

$$\begin{aligned} Y^R(t) + K^R(t)\eta(t) + D^R(t)\dot{\eta}(t) + M^R(t)\ddot{\eta}(t) & \\ = D_{\Delta q}\varphi(t, \mathbf{0})\eta(t) + D_{\Delta q_I}\varphi(t, \mathbf{0}) \int_{t_0}^{t} \eta(\tau)\, d\tau &+ D_{\Delta \dot{q}}\varphi(t, \mathbf{0})\dot{\eta}(t), \end{aligned} \tag{2.36a}$$

$$\eta(t_0) = 0,\ \dot{\eta}(t_0) = 0. \tag{2.36b}$$

Since the matrix $M^R = M^R(t)$ is regular, see, e.g., [14], the unknown Jacobians of φ are represented as follows:

Definition 2.4 With given fixed matrices K_p, K_i, K_d, represent the gain matrices $G_p := D_{\Delta q}\varphi(t, \mathbf{0})$, $G_d := D_{\Delta \dot{q}}\varphi(t, \mathbf{0})$, $G_i := D_{\Delta q_I}\varphi(t, \mathbf{0})$ by

$$K_d =: M^R(t)^{-1}\big(D^R(t) - D_{\Delta \dot{q}}\varphi(t, \mathbf{0})\big), \tag{2.37a}$$

$$K_p =: M^R(t)^{-1}\big(K^R(t) - D_{\Delta q}\varphi(t, \mathbf{0})\big), \tag{2.37b}$$

$$K_i =: M^R(t)^{-1}\big(- D_{\Delta q_I}\varphi(t, \mathbf{0})\big). \tag{2.37c}$$

Remark 2.2 In many cases the matrices K_p, K_i, K_d in (2.37a–c) are supposed to be diagonal matrices.

With the definitions (2.37a–c), the initial value problem (2.36a,b) reads

$$\ddot{\eta}(t) + K_d\dot{\eta}(t) + K_p\eta(t) + K_i \int_{t_0}^{t} \eta(\tau)\, d\tau = -M^R(t)^{-1}Y^R(t),\ t \geq t_0, \tag{2.38a}$$

$$\eta(t_0) = 0,\ \dot{\eta}(t_0) = 0. \tag{2.38b}$$

2.4.2 Partial Derivative with Respect to Δq_0

Let us denote I the unit matrix. Defining

$$\xi(t) := D_{\Delta q_0}\Delta q(t, \mathbf{0}),\ \ t \geq t_0, \tag{2.39}$$

and using again definitions (2.37a–c), for $\xi = \xi(t)$ we get the initial value problem:

$$\ddot{\xi}(t) + K_d\dot{\xi}(t) + K_p\xi(t) + K_i \int_{t_0}^{t} \xi(\tau)\, d\tau = 0,\ t \geq t_0, \tag{2.40a}$$

$$\xi(t_0) = I,\ \dot{\xi}(t_0) = 0. \tag{2.40b}$$

2.4.3 Partial Derivative with Respect to $\Delta \dot{q}_0$

Also the partial derivative with respect to the deviations $\Delta \dot{q}_0$ in the initial velocities $\dot{q}_0$ follows in the same way: With a corresponding differentiation and evaluation at $a = (\Delta p_D, \Delta q_0, \Delta \dot{q}_0, e_0(\cdot)) = 0$, taking into account (2.37a–c), for the derivative

$$\zeta(t) := D_{\Delta \dot{q}0} \Delta q(t, 0),\ t \geq t_0, \tag{2.41}$$

we finally get the initial value problem

$$\ddot{\zeta}(t) + K_d \dot{\zeta}(t) + K_p \zeta(t) + K_i \int_{t_0}^{t} \zeta(\tau)\, d\tau = 0,\ t \geq t_0, \tag{2.42a}$$

$$\zeta(t_0) = 0, \dot{\zeta}(t_0) = I. \tag{2.42b}$$

2.4.4 *Partial Derivative with Respect to* e_0

The partial derivatives with respect to the sub vectors $e_{q0}, e_{q_I 0}, e_{\dot{q}0}$, resp., are given as follows:

2.4.4.1 Partial Derivative with Respect to e_q

For the partial derivative with respect to e_{q0}

$$\lambda(t) := D_{e_{q0}} \Delta q(t, \mathbf{0}),\ t \geq t_0, \tag{2.43}$$

with the same methods as applied above, and using again definitions (2.37a–c), for the partial derivative of Δq with respect to e_{q0} we find the following initial value problem:

$$\ddot{\lambda}(t) + K_d \dot{\lambda}(t) + K_p \lambda(t) + K_i \int_{t_0}^{t} \lambda(\tau)\, d\tau = M^R(t)^{-1} D_{\Delta q} \varphi(t, \mathbf{0}) \Gamma_q(t),\ t \geq t_0, \tag{2.44a}$$

$$\lambda(t_0) = 0, \dot{\lambda}(t_0) = 0. \tag{2.44b}$$

For the remaining derivatives

$$\mu(t) := D_{e_{q_I 0}} \Delta q(t, \mathbf{0}),\ \ \nu(t) := D_{e_{\dot{q}0}} \Delta q(t, \mathbf{0}), t \geq t_0, \tag{2.45}$$

we obtain again an integro-differential equation of type (2.44a,b) where the expression $D_{\Delta q} \varphi(t, \mathbf{0}) \Gamma_q(t)$ is replaced by $D_{\Delta q_I} \varphi(t, \mathbf{0}) \Gamma_{q_I}(t)$, $D_{\Delta \dot{q}} \varphi(t, \mathbf{0}) \Gamma_{\dot{q}}(t)$, respectively.

According to the form of the above given linear initial value problems for the sensitivities of the deviation Δq with respect to the parameters

$\Delta p_D, \Delta q_0, \Delta \dot{q}_0, e_{q_0}, e_{\dot{q}_0}$, for the stability of the above linear systems of 3rd order we have, cf. [1, 18, 21], the following result:

Lemma 2.1 *Suppose that K_d, K_p, K_i are diagonal matrices with positive diagonal elements $[K_p]_{jj}, [K_d]_{jj}, [K_i]_{jj}, j = 1, \ldots, j_K$. If the Hurwitz-criterion*

$$[K_p]_{jj}[K_d]_{jj} - [K_i]_{jj} > 0, j = 1, \ldots, j_K \tag{2.46}$$

holds, then the above perturbation integro-differential equations for the tracking error rates (sensitivities) are input-output stable, hence, the sensitivities are bounded, provided that the corresponding right hand sides are bounded functions.

2.5 The Approximate Regulator Optimization Problem

In the following we consider the position error rate function, hence, the matrix function $w = w(t)$ defined by the partial derivative of the position error function Δq with respect to the parameter vector a, cf. (2.16a,b) and (2.35), (2.39), ..., (2.45):

$$w(t) := D_a \Delta q(t, \mathbf{0}) = \big(D_{\Delta p_D} \Delta q(t, \mathbf{0}), D_{\Delta q_0} \Delta q(t, \mathbf{0}), D_{\Delta \dot{q}_0} \Delta q(t, \mathbf{0}), D_{e_{q0}} \\ \Delta q(t, \mathbf{0}), D_{e_{q I0}} \Delta q(t, \mathbf{0}), D_{e_{\dot{q}0}} \Delta q(t, \mathbf{0})\big) = \big(\eta(t), \xi(t), \zeta(t), \lambda(t), \mu(t), \nu(t)\big). \tag{2.47}$$

According to the integro-differential equations for the (sensitivities) derived in Sect. 2.4, for the matrix function $w = w(t)$ defined by (2.47) we have the following joint integro-differential equation:

$$\ddot{w}(t) + \mathbf{K_d} \dot{w}(t) + \mathbf{K_p} w(t) + \mathbf{K_i} \int_{t_0}^{t} w(\tau)\, d\tau = \psi(t), \quad t_0 \le t \le t_f, \tag{2.48a}$$

$$w(t_0) = w_0, \dot{w}(t_0) = \dot{w}_0. \tag{2.48b}$$

Here, $\mathbf{K_p}, \mathbf{K_d}, \mathbf{K_i}$ are the block-diagonal matrices having the diagonal blocks K_p, K_d, K_i, respectively. Moreover, $w_0, \dot{w}_0$ are the matrices containing the initial conditions and $\psi = \psi(t), t_0 \le t \le t_f$, is the matrix composed of the right hand sides of Eqs. (2.35), (2.39), ..., (2.45).

By means of (2.47), the tracking error rate function can then be represented, cf. (2.4f), (2.28), by

$$D_a \Delta z_I(t, \mathbf{0}) = \begin{pmatrix} w(t) \\ \int_{t_0}^{t} w(\tau)d\tau \\ \dot{w}(t) \end{pmatrix}. \tag{2.49a}$$

In order to replace system (2.48a,b) by an equivalent system of first order differential equations, according to (2.49a), we introduce the matrix function:

$$\mathbf{A} := \begin{pmatrix} 0 & 0 & I \\ I & 0 & 0 \\ -K_p & -K_i & -K_d \end{pmatrix}. \tag{2.49b}$$

Defining the matrix

$$W(t) := D_a \Delta z_I(t, \mathbf{0}) = \begin{pmatrix} w(t) \\ \int_{t_0}^{t} w(\tau)d\tau \\ \dot{w}(t) \end{pmatrix}, t_0 \le t \le t_f, \tag{2.49c}$$

according to (2.49a) and (2.48a,b) we find that the linear integro-differential system (2.48a,b) is equivalent to the following first order system of differential equations:

$$\dot{W}(t) = \mathbf{A}W(t) + \begin{pmatrix} 0 \\ 0 \\ \psi(t) \end{pmatrix}, t_0 \le t \le t_f, \quad W(t_0) := \begin{pmatrix} w_0 \\ 0 \\ \dot{w}_0 \end{pmatrix}. \tag{2.49d}$$

Consequently, the approximation of the regulator optimization problem under stochastic uncertainty yields the following deterministic optimal control problem:

Theorem 2.2 *Let $W(t) = D_a \Delta z_I(t, 0)$, $t_0 \le t \le t_f$, denote the error rate function according to (2.49c). Then, the approximate regulator optimization problem under stochastic uncertainty can be represented by the error rate minimization problem in terms of weighted deviation and correction costs:*

$$\min \int_{t_0}^{t_f} \Big(trW(t)^T (Q_0 + Q_\varphi(t)) W(t) cov\Big(a(\cdot)\Big|\mathcal{A}_{t_0}\Big)$$
$$+ 2tr\Gamma(t)^T Q_\varphi(t) W(t) E\Big(a(\cdot)e_0(\cdot)^T\Big|\mathcal{A}_{t_0}\Big)$$
$$+ tr\Gamma(t)^T Q_\varphi(t)\Gamma(t)\ cov\Big(e(\cdot)\Big|\mathcal{A}_{t_0}\Big) \Big) dt \tag{2.50a}$$

s.t.

$$\dot{W}(t) = \mathbf{A}W(t) + \begin{pmatrix} 0 \\ 0 \\ \psi(t) \end{pmatrix}, \quad t_0 \le t \le t_f \tag{2.50b}$$

$$W(t_0) := \begin{pmatrix} w_0 \\ 0 \\ \dot{w}_0 \end{pmatrix}. \tag{2.50c}$$

Remark 2.3 According to (2.22), (2.23a,b), (2.24), the objective function (2.50a) is *nonnegative*. Moreover, for given gain matrix function $D_{\Delta z_I}\varphi(t, 0)$, $t_0 \le t \le t_f$, in $Q_\varphi(t)$, the objective function (2.50a) is convex in the error rate matrix function $W = W(t)$.

2.6 Active Structural Control Under Stochastic Uncertainty

Because of the development of high strength materials and more efficient methods in structural analysis and design, larger and more complex mechanical structures, like tall buildings, offshore platforms, have been constructed. Constructions of this type provide only small damping in alleviating vibrations under heavy environmental loads such as strong earthquakes, wind turbulences, water waves, etc. The structures usually are stationary, safe, and stable without external dynamic disturbances, and external dynamic loads are the main sources inducing structural vibrations that should be controlled. Obviously, environmental loads, such as earthquakes, wind, waves, etc., are stochastic processes, having unknown time paths. In order to omit severe structural damages and therefore high compensation (recourse) costs, in recent years active control techniques were developed in structural engineering, see, e.g., [5, 12, 22, 23].

Basically, active control depends on the supply of external energy to counteract the dynamic response of a structure. An active control system consists therefore [19, 22, 23, 25] of (i) sensors installed at suitable locations of the structure to measure either external excitations or structural response quantities, (ii) devices to process the measured information and to compute necessary control forces based on a given control algorithm, and (iii) actuators to produce the required control forces. Possible technical devices, concepts, actuators to realize active structural controls are, e.g., electro hydraulic servomechanisms, passive/active/hybrid/semi active damping strategies, viscoelastic dampers, tuned mass dampers, aerodynamic appendages, gas pulse generators, gyroscopes, active structural members, and joints.

Active structural enhancement consists of the use of active control to modify structural behavior. This enhancement can be used to actively stiffen, or strengthen (against Euler buckling) a given structure. Therefore, actively controlled structures can adaptively modify their stiffness properties to be either stiff or flexible as demanded. For example, optimal control strategies maximize the critical buckling load using sensors and actuators. The aim is then to actively stabilize the structure to prevent it from collapsing.

Vibration control of large space structures [23]: Large space structures face difficult problems of vibration control. Because they require low weight, such structures will lack the stiffness and damping necessary for the passive control of vibration. Hence current research is directed towards the design of active vibration control to reduce the mean square response of the system to a desired level within a reasonable span of time. A great deal of research is currently in progress on designing active vibration control system for large flexible space structures.

While the actual time path of the random external load is not known at the planning stage, we may assume that the probability distribution or at least the occurring/needed moments of the applied load are known. The underlying dynamic control system with random parameters is described often, cf. [22, 24], by a second order linear system of differential equations

$$M\ddot{q}(t) + D\dot{q}(t) + Kq(t) = Cu(t) + Ef(t), t_0 \leq t \leq t_f, \tag{2.51}$$

where $q = q(t)$ is the vector of structural displacements, M, D, K denotes the mass, damping, stiffness matrix, resp., and C, E is the matrix denoting the locations of the control forces, the excitations/applied loads, resp., on the structure, see also the basic system of differential equations (2.1a–c) used in this paper as well as the perturbation equations (2.39), …, (2.45) derived in Sect. 2.4. The evaluation of the performance of the control system, the feedback law, resp., can be done by means of convex cost functions. In the following we use weighted quadratic cost function for evaluating the deviation from the reference trajectory as well as for evaluation of the control corrections, see (2.17a,b), (2.18). The approximation/solution method presented here can then be applied directly to the optimal control problems under stochastic uncertainty, as occurring in active structural control under stochastic uncertainty. An example is given in the following.

2.6.1 Example

We consider now the structure according to Fig. 2.2, see [5], where we want to control the supplementary active system while minimizing the expected total costs for the tracking errors and the costs for the control correction.

Corresponding to [24], the behavior of the vector of displacements $q = q(t, \omega)$ can be described by means of a system of differential equations of second order:

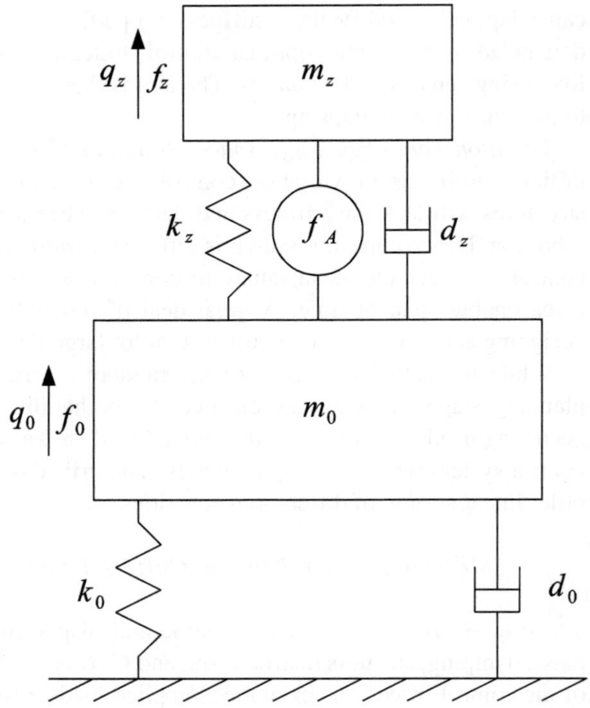

Fig. 2.2 Principle of active structural control

$$M\begin{pmatrix}\ddot{q}_0(t,\omega)\\ \ddot{q}_z(t,\omega)\end{pmatrix}+D\begin{pmatrix}\dot{q}_0(t,\omega,t\\ \dot{q}_z(t,\omega)\end{pmatrix}+K\begin{pmatrix}q_0(t,\omega)\\ q_z(t,\omega)\end{pmatrix}=f_L(t,\omega)+f_A(t) \tag{2.52a}$$

with the matrices and vectors

$$M=\begin{pmatrix}m_0 & 0\\ 0 & m_z\end{pmatrix} \qquad \text{mass matrix} \tag{2.52b}$$

$$D=\begin{pmatrix}d_0+d_z & -d_z\\ -d_z & d_z\end{pmatrix} \qquad \text{damping matrix} \tag{2.52c}$$

$$K=\begin{pmatrix}k_0+k_z & -k_z\\ -k_z & k_z\end{pmatrix} \qquad \text{stiffness matrix} \tag{2.52d}$$

$$f_A(t):=u(t)=\begin{pmatrix}u_1(t)\\ u_2(t)\end{pmatrix} \qquad \text{actuator force} \tag{2.52e}$$

$$f_L(t,\omega) = \begin{pmatrix} f_0(t,\omega) \\ f_z(t,\omega) \end{pmatrix} := f_{L0} + \Lambda(t)\big(\overline{p_D} + \Delta p_D(\omega)\big), \qquad \text{applied load} \tag{2.52f}$$

where f_{L0} is a constant load vector, $\Lambda = \Lambda(t)$ denotes a given time-dependent matrix, and $p_D = p_D(\omega) = \overline{p_D} + \Delta p_D(\omega)$ is random vector with a known probability distribution. Moreover, $u = u(t)$ is the control function, and the vector of displacements and its time derivatives reads

$$q(t,\omega) = \begin{pmatrix} q_0(t,\omega) \\ q_z(t,\omega) \end{pmatrix}, \quad \dot{q}(t,\omega) = \begin{pmatrix} \dot{q}_0(t,\omega) \\ \dot{q}_z(t,\omega) \end{pmatrix}. \tag{2.52g}$$

According to (2.52a), the function F of the dynamic equation (2.1a) reads

$$F(p_D, q(t), \dot{q}(t), \ddot{q}(t)) = M\ddot{q} + D\dot{q} + Kq - (f_{L0} + \Lambda(t)p_D). \tag{2.53}$$

Based on the definition (2.2b) of the feedforward control $u^R = u^R(t)$, $t_0 \le t \le t_f$, we get

$$u^R(t) = -\big(f_{L0} + \Lambda(t)\overline{p_D}\big) \ \textit{for} \ q^R(t) = 0, t_0 \le t \le t_f. \tag{2.54}$$

Thus, in the *optimal* case of zero displacements $q^R(t,\omega) = 0, \dot{q}^R(t) = 0, t_0 \le t \le t_f$, we have the feedforward control according to (2.54).

In order to see more details, in the following we use the following *simplifications*: First we consider a *PD control* only. Hence, we have, cf. (2.3), (2.4a-f), $z(t) = \begin{pmatrix} q(t) \\ \dot{q}(t) \end{pmatrix}$ and therefore, see (2.54),

$$\Delta u(t) := \varphi(t, \Delta z(t)) = \varphi(t, \Delta q(t), \Delta\dot{q}(t)), t_0 \le t \le t_f,$$

where

$$\Delta z(t) := z(t) - z^R(t) = \begin{pmatrix} \Delta q(t) \\ \Delta\dot{q}(t) \end{pmatrix} = \begin{pmatrix} q(t) \\ \dot{q}(t) \end{pmatrix} - \begin{pmatrix} q^R(t) \\ \dot{q}^R(t) \end{pmatrix} = z(t), \quad t_0 \le t \le t_f.$$

In addition we suppose that the deviations of $p_D, q_0, \dot{q}_0$ from their nominal values are hence, $\Delta p_D = 0, \Delta q_0 = 0, \Delta\dot{q}_0 = 0$. Thus, the random parameter vector $a = a(\omega)$ consists of the error vector e_0 only. Consequently, according to (2.29), the approximate total cost function reads

$$\begin{aligned} \tilde{f} \approx\ & e_0^T D_{e_0}\Delta z(t,\mathbf{0})^T(Q_0 + Q_\varphi(t))D_{e_0}\Delta z(t,\mathbf{0})e_0 + 2e_0^T\Gamma(t)^T Q_\varphi D_{e_0}\Delta z(t,\mathbf{0})e_0 \\ & + e_0^T\Gamma(t)^T Q_\varphi\Gamma(t)e_0. \end{aligned} \tag{2.55a}$$

Rearranging terms, we also have

$$\tilde{f} \approx e_0^T \Big(\Gamma(t) + D_{e_0}\Delta z(t, \mathbf{0})\Big)^T Q_\varphi(t) D_{e_0}\Big(\Gamma(t) + D_{e_0}\Delta z(t, \mathbf{0})\Big) e_0$$
$$+ e_0^T D_{e_0}\Delta z(t, \mathbf{0})^T Q_0 D_{e_0}\Delta z(t, \mathbf{0}) e_0. \tag{2.55b}$$

Finally, due to Theorem 2.1 and (2.32), in the present case for the objective function we have this approximation:

Corollary *Based on the Taylor expansions of the feedback law and the tracking error, in case* $\Delta p_D = 0$, $\Delta q_0 = 0$, $\Delta \dot{q}_0 = 0$ *the approximate objective function of the regulator optimization problem under stochastic uncertainty reads*

$$E\Big(f(t,\omega)\Big|\mathcal{A}_{t_0}\Big)$$
$$\approx \operatorname{tr}\Big(\Gamma(t) + D_{e_0}\Delta z(t, \mathbf{0})\Big)^T Q_\varphi(t)\Big(\Gamma(t) + D_{e_0}\Delta z(t, \mathbf{0})\Big) cov\Big(e_0(\cdot)\Big|\mathcal{A}_{t_0}\Big)$$
$$+ \operatorname{tr} D_{e_0}\Delta z(t, \mathbf{0})^T Q_0 D_{e_0}\Delta z(t, \mathbf{0})\; cov\Big(e_0\Big|\mathcal{A}_{t_0}\Big), \tag{2.55c}$$

where the unknowns, i.e. the gain matrix $G := D_{\Delta z}\varphi(t, \mathbf{0})$ *is contained in the matrix function*

$$Q_\varphi(t) := D_{\Delta z}\varphi(t, \mathbf{0})^T C_u D_{\Delta z}\varphi(t, \mathbf{0}). \tag{2.55d}$$

According to (2.56b) the derivative $\lambda(t) := D_{e_0}\Delta z(t, \mathbf{0})$ of Δz with respect to e_0 at $(t, \mathbf{0})$ needed in the objective function (2.55c) is given by following initial value problem:

$$\ddot{\lambda}(t) + K_d\dot{\lambda}(t) + K_p\lambda(t) = M^{-1} D_{\Delta z}\varphi(t, \mathbf{0})\Gamma(t),\ t \geq t_0, \tag{2.56a}$$
$$\lambda(t_0) = 0, \dot{\lambda}(t_0) = 0, \tag{2.56b}$$

where $\Gamma(t)$ denotes the block-diagonal matrix composed of the matrices $\Gamma_q(t)$, $\Gamma_{\dot{q}}(t)$.

Summarizing the above considerations, according to Theorem 2.1, the stochastic optimal regulator problem for the present example from active control under stochastic uncertainty can be described approximatively as follows:

Corollary *Suppose that only the additive measurement error is random. Then the optimal gain matrix* $G(t) := D_{\Delta z}\varphi(t, \mathbf{0})$ *of the approximate active control problem under stochastic uncertainty is the solution of the following optimal control problem:*

$$\min_G \int_{t_0}^{t_f} \Big(\operatorname{tr} \Big(\Gamma(t) + \lambda(t) \Big)^T G(t)^T C_u G(t) \Big(\Gamma(t) + \lambda(t) \Big) cov\Big(e_0(\cdot) \Big| \mathcal{A}_{t_0} \Big)$$

$$+ \operatorname{tr} \lambda(t)^T Q_0 \lambda(t)\, cov\Big(e_0 \Big| \mathcal{A}_{t_0} \Big) \Big) dt \tag{2.57a}$$

s.t. the matrix initial value problem:

$$\ddot{\lambda}(t) + K_d \dot{\lambda}(t) + K_p \lambda(t) = M^{-1} G(t) \Gamma(t), \ t \geq t_0, \tag{2.57b}$$

$$\lambda(t_0) = 0, \dot{\lambda}(t_0) = 0, \tag{2.57c}$$

Remark 2.4 Using stochastic optimization methods, hence, incorporating random variations of model parameters, initial values, external forces, measurement errors, etc., into the optimization process for regulators, by formulating and solving appropriate deterministic substitute problems, *robust optimal feedback control laws* are obtained. Based on a (stochastic optimal) reference trajectory $q^R(t)$ and a corresponding feedforward control $u^R(t)$, for the computation of stochastic optimal regulators, deterministic substitute control problems of the following type are provided: Optimize the performance of the feedback control function $\varphi = \varphi(t, \Delta z(t))$ by minimizing the expected total costs resulting from a) the deviation $\Delta z(t)$ between the (stochastic optimal) reference trajectory and the effective trajectory of the underlying dynamic system and b) the control corrections $\Delta u(t) = \varphi(t, \Delta z(t))$. Due to their favorable properties, weighted quadratic cost functions are chosen. For finding a stochastic optimal feedback control law $\varphi(t, \cdot)$ or the corresponding gain matrices one has then a (deterministic) optimal control problem. The expectations, hence, multiple integrals, arising in the objective function have to be determined by approximation methods. Here, *Taylor expansions* with respect to the parameter vector $a = a(\omega)$ at the conditional expectation $\bar{a}^{t_0}$ are used. For the computation of the needed derivatives (tracking error rates or sensitivities) with respect to the parameter vector a, linear integro-differential equations have been derived, which can be transformed into systems of ordinary differential equations. According to the structure of the perturbation equations one obtains a stochastic optimal parametric representation of the gain matrices guaranteeing the stability of the differential equations for the tracking error rates. Moreover, an explicit analytical form is obtained for the calculation of the expectation-valued objective function. Finally, the resulting approximate deterministic optimal control problem can be solved by standard numerical solution procedures of optimal control. The given stochastic optimization procedure for regulators is then applied to the important field of *active structural control under stochastic uncertainty*. An example is given.

References

1. Ackermann, J.: Robuste Regelung: Analyse und Entwurf von linearen Regelungssytemen mit unsicheren physikalischen Parametern. Springer, Berlin (1993)
2. Åström, K.J.: Introduction to Stochastic Control Theory. Elsevier, Amsterdam (1970)
3. Aurnhammer, A.: Optimale Stochastische Trajektorienplanung und Regelung von Industrierobotern. No. 1032 in Fortschrittberichte VDI, Reihe 8. VDI-Verlag GmbH, Düsseldorf (2004)
4. Basar, T., Bernhard, P.: H^{∞}-Optimal Control and Related Minimax Design. Birkhäuser, Boston (1995)
5. Block, C.: Aktive Minderung personeninduzierter Schwingungen an weit gespannten Strukturen im Bauwesen. No. 336 in Fortschrittberichte VDI, Reihe 11, Schwingungstechnik. VDI-Verlag GmbH, Düsseldorf (2008)
6. Buonaccorsi, J.: Measurement Error: Models, Methods and Application. Chapman and Hall/CRC Press, Boca Raton (2010)
7. Demyanov, V., Rubinov, A.: Approximate Methods in Optimization Problems. American Elsevier, Publishing Company, Inc., New York (1970)
8. Dieudonné, J.: Foundations of Modern Analysis. Academic Press, New York (1969)
9. Dullerud, G., Paganini, F.: A Course in Robust Control Theory. Springer, New York (2000)
10. Guo, L.: H_{∞} output feedback control for delay systems with nonlinear and parametric uncertainties. IEE Proc. Control Theory Appl. **149**(3), 226–236 (2002). https://doi.org/10.1049/ip-cta:20020336
11. Kalman, R., et al.: Topics in Mathematical System Theory. McGraw-Hill Book Company, New York (1969)
12. Korkmaz, S.: Review: a review of active structural control: challenges for engineering informatics. Comput. Struct. **89**(23–24), 2113–2132 (2011). https://doi.org/10.1016/j.compstruc.2011.07.010
13. Marti, K.: Approximationen Stochastischer Optimierungsprobleme. Konigstein/Ts, Hain (1979)
14. Marti, K.: Stochastic optimization methods in robust adaptive control of robots. In: Groetschel, M.E.A. (ed.) Online Optimization of Large Scale Systems, pp. 545–577. Springer, Berlin (2001)
15. Marti, K.: Adaptive Optimal Stochastic Trajectory Planning and Control (AOSTPC) for Robots, pp. 155–206. Springer, Berlin (2004)
16. Marti, K.: Optimal design of regulators subject to stochastic uncertainty. In: Topping, B., Iványi, P. (eds.) Proceedings of the Twelfth International Conference on Computational Structures Technology, Paper 135. Civil-Comp Press, Stirlingshire (2014). https://doi.org/10.4203/ccp.106.135
17. Marti, K.: Stochastic Optimization Methods: Applications in Engineering and Operations Research, 3rd edn. Springer, Berlin (2015). https://doi.org/10.1007/978-3-662-46214-0
18. Müller, P.: Stabilität und Matrizen: Matrizenverfahren in der Stabilitätstheorie linearer dynamischer Systeme. Springer, Berlin (1977)
19. Nagarajaiah, S., Narasimhan, S.: Optimal control of structures. In: Arora, J. (ed.) Optimization of Structural and Mechanical Systems, pp. 221–244. World Scientific, New Jersey (2007)
20. Pfeiffer, F., Johanni, R.: A concept for manipulator trajectory planning. IEEE J. Robot. Autom. **3**(2), 115–123 (1987)
21. Schacher, M.: Stochastisch optimale Regelung von Robotern. No. 1200 in Fortschritt-Berichte VDI, Reihe 8, Mess-, Steuerungs- und Regelungstechnik. VDI Verlag GmbH, Düsseldorf (2011)
22. Soong, T.: Active Structural Control: Theory and Practice. John Wiley, New York (1990)

23. Soong, T., Costantinou, M.: Passive and Active Structural Vibration Control in Civil Engineering. CISM Courses and Lectures, vol. 345. Springer, Wien (1994)
24. Soong, T., Spencer, B.: Active, semi-active and hybrid control of structures. Bull. N. Z. Soc. Earthq. Eng. **33**(3), 387–402 (2000)
25. Spencer, B., Nagarajaiah, S.: State of the art of structural control. J. Struct. Eng. **129**(7), 845–856 (2003). https://doi.org/10.1061/(ASCE)0733-9445(2003)129:7(845)
26. Wei, G., Wang, Z., Shen, B.: Error-constrained finite-horizon tracking control with incomplete measurements and bounded noises. Int. J. Robust Nonlinear Control **22**(2), 223–238 (2012). https://doi.org/10.1002/rnc.1728

Chapter 3
Optimal Open-Loop Control of Dynamic Systems Under Stochastic Uncertainty

As shown in former sections on optimal tracking problems, optimal regulator problems, active structural optimization problems under stochastic uncertainty, approximate optimal feedback controls under stochastic uncertainties can be obtained by means of open-loop feedback control methods used also in model predictive control under stochastic uncertainty. In order to simplify the notation, the intermediate starting points t_b of the open-loop control problems are here abbreviated here just by t_0.

In the following the computation of approximate optimal open-loop controls is considered for more general optimal control problems under stochastic uncertainty. As a key tool stochastic optimization methods are used for the incorporation random variations of model parameters, initial values, external forces, measurement errors, etc. into the optimization process for finding robust optimal controls. Hence, corresponding deterministic substitute control problems are formulated as follows:

Optimize the performance of the open-loop control $u = u(t, z_0)$, depending on the time t and the initial state z_0, by minimizing the expected total costs. The expectations, hence, multiple integrals, arising in the objective function have to be determined by approximation methods. Here, *Taylor expansions* with respect to the parameter vector $a = a(\omega)$ at the conditional expectation $\bar{a}^{t_0}$ are used. For the computation of the needed derivatives (tracking error rates or sensitivities) with respect to the parameter vector a, linear integro-differential equations are derived, which can be transformed into systems of ordinary differential equations.

K. Marti, *Optimization Under Stochastic Uncertainty*, International Series in Operations Research & Management Science 296,
https://doi.org/10.1007/978-3-030-55662-4_3

3.1 Optimal Control Problems Under Stochastic Uncertainty

Corresponding to the former chapters, dealing especially with optimal regulator, tracking and active structural control problems, here we consider following more general optimal control problem under stochastic uncertainty:

Definition 3.1 A stochastic optimal open-loop control $u^* = u^*(t; t_0, z_0), t_0 \leq t \leq t_f$, is a solution of the stochastic optimization problem:

$$\min E\Big(\int_{t_0}^{t_f} L\big(t, a(\omega), q(t), \dot{q}(t), u(t, z_0(\omega))\big)dt + G\big(a(\omega), q(t_f)\big) \Big| \mathcal{A}_{t_0} \Big) \tag{3.1a}$$

s.t.

$$F(t, p(\omega), q(t), \dot{q}(t), \ddot{q}(t), u(t, z_0(\omega))) = 0, t_0 \leq t \leq t_f, \ a.s. \tag{3.1b}$$

$$q(t_0, \omega) = q_0(\omega), \tag{3.1c}$$

$$\dot{q}(t_0, \omega) = \dot{q}_0(\omega) \tag{3.1d}$$

$$u(t) \in U_t, t_0 \leq t \leq t_f, \tag{3.1e}$$

where the open-loop control $u(t) = u(t; t_0, z_0)$ depends on the initial time $t_0(t_b)$ and on the initial state $z_0(z_b) := \begin{pmatrix} q_0 \\ \dot{q}_0 \end{pmatrix}$.

3.1.1 Computation of the Expectation of the Cost Functions *L*, *G*

A first main problem is the computation of the expectations EL, EG of the cost function L along the trajectory and the terminal cost function G. In the following, let denote $a = a(\omega)$ the stochastic data vector available at time t_0

$$a(\omega) := a(\omega) := \begin{pmatrix} p(\omega) \\ z_0(\omega) \end{pmatrix} = \begin{pmatrix} p(\omega) \\ q_0(\omega) \\ \dot{q}_0(\omega) \end{pmatrix}. \tag{3.2a}$$

Moreover, let denote

$$z = z(t, a) := \begin{pmatrix} q_0(t, a) \\ \dot{q}_0(t, a) \end{pmatrix}, \ a^N := \begin{pmatrix} p^N \\ z_0^N \end{pmatrix} = \begin{pmatrix} p^N \\ q_0^N \\ \dot{q}_0^N \end{pmatrix} \tag{3.2b}$$

the state vector, a nominal data vector, resp., e.g. the expectation $a^N = Ea(\omega)$ of $a(\omega)$.

For simplification of the presentation we suppose that

$$L = L(t, z, u) = L_I(t, z) + L_{II}(t, u(t)), G = G(t_f, z(t_f)), \text{ with} \qquad (3.3)$$
$$z(t) = z(t, a), u(t) = u(t, a) = u(t, z_0),$$

hence, L, G do not depend explicitly on the random parameter vector a.

Using second order Taylor expansions of L with respect to a at a^N, we get

$$L(t, z(t, a), u(t, z_0)) \approx L_I(t, z(t, a^N)) + L_{II}(t, u(t, z_0^N)) \qquad (3.4a)$$
$$+\frac{\partial L_I}{\partial z}(t, z(t, a^N))\frac{\partial z}{\partial a}(t, a^N)(a - a^N) + \frac{\partial L_{II}}{\partial u}(t, u(t, z_0^N))\Big)\frac{\partial u}{\partial z_0}(t, z_0^N)(z_0 - z_0^N)$$
$$+\frac{1}{2}(a - a^N)^T \frac{\partial z}{\partial a}(t, a^N)^T \frac{\partial^2 L_I}{\partial z^2}(t, z(t, a^N))\frac{\partial z}{\partial a}(t, a^N)(a - a^N)$$
$$+\frac{1}{2}(a - a^N)^T \frac{\partial L_I}{\partial z}(t, z(t, a^N)) \otimes \frac{\partial^2 z}{\partial a^2}(t, a^N)(a - a^N)$$
$$+\frac{1}{2}(z_0 - z_0^N)^T \frac{\partial u}{\partial z_0}(t, z_0^N)^T \frac{\partial^2 L_{II}}{\partial u^2}(t, u(t, z_0^N))\frac{\partial u}{\partial z_0}(t, z_0^N)(z_0 - z_0^N)$$
$$+\frac{1}{2}(z_0 - z_0^N)^T \frac{\partial L_{II}}{\partial u}(t, u(t, z_0^N)) \otimes \frac{\partial^2 u}{\partial {z_0}^2}(t, z_0^N)(z_0 - z_0^N).$$

Moreover, the second order Taylor expansions of G with respect to a at a^N reads

$$G(t_f, z(t_f, a)) \approx G(t_f, z(t_f, a^N)) + \frac{\partial G}{\partial z}(t_f, z(t_f, a^N))\frac{\partial z}{\partial a}(t_f, a^N)(a - a^N) \qquad (3.4b)$$
$$+\frac{1}{2}(a - a^N)^T \frac{\partial z}{\partial a}(t_f, a^N)^T \frac{\partial^2 G}{\partial z^2}(t_f, z(t, a^N))\frac{\partial z}{\partial a}(t_f, a^N)(a - a^N)$$
$$+\frac{1}{2}(a - a^N)^T \frac{\partial G}{\partial z}(t_f, z(t_f, a^N)) \otimes \frac{\partial^2 z}{\partial a^2}(t_f, a^N)(a - a^N).$$

Here, for $L = L_I, L_{II}, G$ we put

$$\frac{\partial L}{\partial z} \otimes \frac{\partial^2 z}{\partial a^2} := \sum_{k=1}^{n} \frac{\partial L}{\partial z_k}\frac{\partial^2 z_k}{\partial a^2}. \qquad (3.5)$$

From (3.4a,b), we obtain, cf. [1]:

Theorem 3.1 *Assume that* $a^N = \overline{a} := Ea(\omega)$*, and let denote* $cov(a(\cdot)), cov(z_0(\cdot))$*, resp., the covariance matrix of* $a = a(\omega), z_0 = z_0(\omega)$ *available at time* $t = t_0$*. Retaining only first order derivatives of* z, u *with respect to* a, z_0*, resp., the expected cost function* L *can be approximated by*

$$EL(t, z(t, a(\omega)), u(t)) \approx L_I(t, z(t, \overline{a}) + L_{II}(t, u(t, \overline{z}_0)) \tag{3.6a}$$
$$+\frac{1}{2}tr\frac{\partial z}{\partial a}(t, \overline{a})^T\frac{\partial^2 L_I}{\partial z^2}(t, z(t, \overline{a}))\frac{\partial z}{\partial a}(t, \overline{a})cov(a(\cdot))$$
$$+\frac{1}{2}tr\frac{\partial u}{\partial z_0}(t, \overline{z}_0)^T\frac{\partial^2 L_{II}}{\partial u^2}(t, u(t, \overline{z}_0))\frac{\partial u}{\partial z_0}(t, \overline{z}_0)cov(z_0(\cdot)).$$

Moreover, for EG *we have*

$$EG(t_f, z(t_f, a(\omega))) \approx G(t_f, z(t_f, \overline{a}) \tag{3.6b}$$
$$+\frac{1}{2}tr\frac{\partial z}{\partial a}(t_f, \overline{a})^T\frac{\partial^2 G}{\partial z^2}(t_f, z(t_f, \overline{a}))\frac{\partial z}{\partial a}(t_f, \overline{a})cov(a(\cdot)).$$

As can be seen from (3.6a,b), for the approximate computation of the expected cost function EL we need the trajectory $t \to z(t, \bar{a})$ and the Jacobian $\frac{\partial z}{\partial a}(t, \bar{a})$.

Having the initial value problem, cf. (3.1a–e),

$$F(t, p, q(t, a), \dot{q}(t, a), \ddot{q}(t, a), u(t, z_0)) = 0, t_0 \le t \le t_f, \tag{3.7a}$$
$$q(t_0, a) = q_0, \tag{3.7b}$$
$$\dot{q}(t_0, a) = \dot{q}_0, \tag{3.7c}$$

the basic trajectory $t \to z(t, \overline{a})$ is obtained by solving the initial value problem (3.7a–c) with the parameter vector $a := \overline{a}$. Based on this trajectory, an initial value problem for the Jacobian $\frac{\partial z}{\partial a}(t, \overline{a})$ of $z = z(t, a)$ with respect to the parameter vector a at $\overline{a}$ is obtained by partial differentiation of (3.7a–c) with respect to a at $\overline{a}$. Hence, define first the Jacobians $Y(t, \overline{a}), K(t, \overline{a}), D(t, \overline{a}), M(t, \overline{a}), B(t, \overline{a})$ of the vector function F with respect to $a, q, \dot{q}, \ddot{q}, u$ at $t, \overline{p}, q(t, \overline{a}), \dot{q}(t, \overline{a}), \ddot{q}(t, \overline{a}), u(t, \overline{z}_0)$:

$$Y_{(z,u)}(t, \overline{a}) := D_a F(t, \overline{p}, q(t, \overline{a}), \dot{q}(t, \overline{a}), \ddot{q}(t, \overline{a}), u(t, \overline{z}_0)), \tag{3.8a}$$
$$K_{(z,u)}(t, \overline{a}) := D_q F(t, \overline{p}, q(t, \overline{a}), \dot{q}(t, \overline{a}), \ddot{q}(t, \overline{a}), u(t, \overline{z}_0)), \tag{3.8b}$$
$$D_{(z,u)}(t, \overline{a}) := D_{\dot{q}} F(t, \overline{p}, q(t, \overline{a}), \dot{q}(t, \overline{a}), \ddot{q}(t, \overline{a}), u(t, \overline{z}_0)), \tag{3.8c}$$
$$M_{(z,u)}(t, \overline{a}) := D_{\ddot{q}} F(t, \overline{p}, q(t, \overline{a}), \dot{q}(t, \overline{a}), \ddot{q}(t, \overline{a}), u(t, \overline{z}_0)), \tag{3.8d}$$
$$B_{(z,u)}(t, \overline{a}) := D_u F(t, \overline{p}, q(t, \overline{a}), \dot{q}(t, \overline{a}), \ddot{q}(t, \overline{a}), u(t, \overline{z}_0)). \tag{3.8e}$$

Next to, for the Jacobian $\frac{\partial q}{\partial a}(t, \overline{a})$ we get the system of differential equations

$$Y_{(z,u)}(t,\overline{a}) + K_{(z,u)}(t,\overline{a})\frac{\partial q}{\partial a}(t,\overline{a}) + D_{(z,u)}(t,\overline{a})\frac{d}{dt}\frac{\partial q}{\partial a}(t,\overline{a}) \tag{3.9}$$

$$+M_{(z,u)}(t,\overline{a})\frac{d^2}{dt^2}\frac{\partial q}{\partial a}(t,\overline{a}) + B_{(z,u)}(t,\overline{a})\frac{\partial u}{\partial a}(t,\overline{z}_0) = 0,\ t_0 \leq t \leq t_f,$$

$$\frac{\partial q}{\partial a}(t_f,\overline{a}) = (0, I, 0),$$

$$\frac{d}{dt}\frac{\partial q}{\partial a}(t_f,\overline{a}) = (0, 0, I),$$

where $\frac{\partial u}{\partial a}(t,\overline{z}_0) := \left(0, \frac{\partial u}{\partial z_0}(t,\overline{z}_0)\right)$.

The second order linear initial value problem (3.9) can be converted into a first order linear initial value problem:

Theorem 3.2 *Given the parameter vector* $a = \overline{a} := Ea(\omega)$*, the matrix* $\frac{\partial z}{\partial a}(t,\overline{a}) := \begin{pmatrix} \frac{\partial q}{\partial a}(t,\overline{a}) \\ \frac{d}{dt}\frac{\partial q}{\partial a}(t,\overline{a}) \end{pmatrix}$ *composed of the Jacobian of* $q = q(t,\overline{a})$ *and its time derivative fulfills the first order initial value problem:*

$$\frac{d}{dt}\frac{\partial z}{\partial a}(t,\overline{a}) = \begin{pmatrix} 0 & I \\ -M_{(z,u)}(t,\overline{a})^{-1}K_{(z,u)}(t,\overline{a}) & -M_{(z,u)}(t,\overline{a})^{-1}D_{(z,u)}(t,\overline{a}) \end{pmatrix}\frac{\partial z}{\partial a}(t,\overline{a})$$

$$+\begin{pmatrix} O \\ -M_{(z,u)}(t,\overline{a})^{-1}Y_{(z,u)}(t,\overline{a}) \end{pmatrix} + \begin{pmatrix} O \\ -M_{(z,u)}(t,\overline{a})^{-1}B_{(z,u)}(t,\overline{a})\frac{\partial u}{\partial a}(t,\overline{z}_0) \end{pmatrix},$$

$$t_0 \leq t \leq t_f, \tag{3.10a}$$

$$\frac{\partial z}{\partial a}(t_f,\bar{a}) = \begin{pmatrix} 0 & I & 0 \\ 0 & 0 & I \end{pmatrix}. \tag{3.10b}$$

According to the above shown Theorem 3.1 and Theorem 3.2 we find the following result:

Theorem 3.3 *For the approximate computation of an optimal open-loop control* $u^* = u^*(t, z_0)$*, according to Definition 3.1, we have the optimization problem:*

$$\min \int_{t_0}^{t_f} \Big(L_I(t, z(t,\bar{a})) + L_{II}(t, u(t,\overline{z}_0)) \tag{3.11a}$$

$$+\frac{1}{2}tr\frac{\partial z}{\partial a}(t,\bar{a})^T\frac{\partial^2 L_I}{\partial z^2}(t, z(t,\bar{a}))\frac{\partial z}{\partial a}(t,\bar{a})cov(a(\cdot))$$

$$+\frac{1}{2}tr\frac{\partial u}{\partial z_0}(t,\overline{z}_0)^T\frac{\partial^2 L_{II}}{\partial u^2}(t, u(t,\overline{z}_0))\frac{\partial u}{\partial z_0}(t,\overline{z}_0)cov(z_0(\cdot))\Big)dt$$

$$+ G(t_f, z(t_f, \overline{a}) + \frac{1}{2} tr \frac{\partial z}{\partial a}(t_f, \overline{a})^T \frac{\partial^2 G}{\partial z^2}(t_f, z(t_f, \overline{a})) \frac{\partial z}{\partial a}(t_f, \overline{a}) cov(a(\cdot))$$

s.t.

$$F(t, \overline{p}, q(t, \overline{a}), \dot{q}(t, \overline{a}), \ddot{q}(t, \overline{a}), u(t, \overline{z}_0)) = 0, t_0 \leq t \leq t_f, \tag{3.11b}$$

$$q(t_0, \overline{a}) = \bar{q}_0, \tag{3.11c}$$

$$\dot{q}(t_0, \overline{a}) = \bar{\dot{q}}_0, \tag{3.11d}$$

$$\frac{d}{dt} \frac{\partial z}{\partial a}(t, \bar{a}) = \begin{pmatrix} 0 & I \\ -M_{(z,u)}(t, \overline{a})^{-1} K_{(z,u)}(t, \overline{a}) & -M_{(z,u)}(t, \overline{a})^{-1} D_{(z,u)}(t, \overline{a}) \end{pmatrix} \frac{\partial z}{\partial a}(t, \bar{a})$$

$$+ \begin{pmatrix} O \\ -M_{(z,u)}(t, \overline{a})^{-1} Y_{(z,u)}(t, \overline{a}) \end{pmatrix} + \begin{pmatrix} O \\ -M_{(z,u)}(t, \overline{a})^{-1} B_{(z,u)}(t, \overline{a}) \frac{\partial u}{\partial a}(t, \overline{z}_0) \end{pmatrix},$$

$$t_0 \leq t \leq t_f, \tag{3.11e}$$

$$\frac{\partial z}{\partial a}(t_0, \bar{a}) = \begin{pmatrix} 0 & I & 0 \\ 0 & 0 & I \end{pmatrix}. \tag{3.11f}$$

With the optimal solutions $u^* = u^*(t, \overline{z}_0)$ *and* $\frac{\partial u^*}{\partial a}(t, \overline{z}_0) := \left(0, \frac{\partial u^*}{\partial z_0}(t, \overline{z}_0)\right)$ *the approximate optimal open-loop control is given by*

$$u^*(t, z_0) = u^*(t, \overline{z}_0) + \frac{\partial u^*}{\partial z_0}(t, \overline{z}_0)(z_0 - \overline{z}_0). \tag{3.11g}$$

3.2 Solution of the Substitute Control Problem

The present substitute problem (3.11a–f) can be interpreted as a two-stage control problem:

In the *inner loop* or *second stage*, for given functions

$$z = z(t, \overline{a}), u = u(t, \overline{z}_0), t_0 \leq t \leq t_f, \tag{3.12a}$$

corresponding to the expected parameter vector $\overline{a}$, $\overline{z}_0$, resp., we have the optimal computation of the sensitivities

$$\frac{\partial z}{\partial a} = \frac{\partial z}{\partial a}(t, \bar{a}), \frac{\partial u}{\partial z_0} = \frac{\partial u}{\partial z_0}(t, \overline{z}_0), t_0 \leq t \leq t_f. \tag{3.12b}$$

Then, in the *outer loop* or *first stage* the optimal trajectory and control (3.12a) have to be determined.

A complete separation of the two loops occurs if the Hessians

$$\frac{\partial^2 L_I}{\partial z^2} = \frac{\partial^2 L_I}{\partial z^2}(t), \frac{\partial^2 L_{II}}{\partial u^2} = \frac{\partial^2 L_{II}}{\partial u^2}(t), t_0 \leq t \leq t_f, \tag{3.12c}$$

and the matrices

$$Y_{(z,u)}(t,\overline{a}) = Y(t,\overline{a}), K_{(z,u)}(t,\overline{a}) = K(t,\overline{a}), D_{(z,u)}(t,\overline{a}) = D(t,\overline{a}),$$
$$M_{(z,u)}(t,\overline{a}) = M(t,\overline{a}), B_{(z,u)}(t,\overline{a}) = B(t,\overline{a}) \tag{3.12d}$$

are independent of the guiding functions (3.12a).

This holds, e.g., if the vector function

$$F(t,p,q,\dot{q},\ddot{q},u) := Y(t)p + K(t)q + D(t)\dot{q} + M(t)\ddot{q} + B(t)u \tag{3.12e}$$

is linear in its arguments $(p,q,\dot{q},\ddot{q},u)$ and $L_I = L_I(t,z), L_{II} = L_{II}(t,u)$ are quadratic cost functions with constant Hessians.

An approximate decoupling can be obtained by solving first the following separate first- stage control problem with the mean parameter vector $\overline{a}$:

$$\min \int_{t_0}^{t_f} \big(L_I(t,z(t,\overline{a})) + L_{II}(t,u(t,\overline{z}_0))\big)dt + G(t_f,z(t_f,\overline{a})) \tag{3.13a}$$

s.t.

$$F(t,\overline{p},q(t,\overline{a}),\dot{q}(t,\overline{a}),\ddot{q}(t,\overline{a}),u(t,\overline{z}_0)) = 0, t_0 \leq t \leq t_f, \tag{3.13b}$$

$$q(t_0,\overline{a}) = \overline{q}_0, \tag{3.13c}$$

$$\dot{q}(t_0,\overline{a}) = \overline{\dot{q}}_0 \tag{3.13d}$$

$$u(t,\overline{z}_0) \in U_t, t_0 \leq t \leq t_f. \tag{3.13e}$$

Let then $z^* = z^*(t,\overline{a}), u = u^*(t,\overline{z}_0), t_0 \leq t \leq t_f$, denote an optimal solution of the control problem (3.13a–e). Inserting now the resulting functions $z*(\cdot), u*(\cdot)$ into the matrices Y, K, D, M, B, hence defining, c.f. (3.8a–e),

$$Y^*(t,\overline{a}) := Y_{(z^*,u^*)}(t,\overline{a}), \tag{3.14a}$$

$$K^*(t,\overline{a}) := K_{(z^*,u^*)}(t,\overline{a}), \tag{3.14b}$$

$$D^*(t,\overline{a}) := D_{(z^*,u^*)}(t,\overline{a}), \tag{3.14c}$$

$$M^*(t,\overline{a}) := M_{(z^*,u^*)}(t,\overline{a}), \tag{3.14d}$$

$$B^*(t,\overline{a}) := B_{(z^*,u^*)}(t,\overline{a}), \tag{3.14e}$$

$$t_0 \leq t \leq t_f,$$

we obtain the following separated inner loop or second stage control problem:

$$\min \int_{t_0}^{t_f} \Big(\frac{1}{2} tr \frac{\partial z}{\partial a}(t, \bar{a})^T \frac{\partial^2 L_I}{\partial z^2}(t, z(t, \bar{a})) \frac{\partial z}{\partial a}(t, \bar{a}) cov(a(\cdot)) \tag{3.15a}$$

$$+ \frac{1}{2} tr \frac{\partial u}{\partial z_0}(t, \overline{z}_0)^T \frac{\partial^2 L_{II}}{\partial u^2}(t, u(t, \overline{z}_0)) \frac{\partial u}{\partial z_0}(t, \overline{z}_0) cov(z_0(\cdot)) \Big) dt$$

$$+ \frac{1}{2} tr \frac{\partial z}{\partial a}(t_f, \overline{a})^T \frac{\partial^2 G}{\partial z^2}(t_f, z(t_f, \overline{a})) \frac{\partial z}{\partial a}(t_f, \overline{a}) cov(a(\cdot))$$

s.t.

$$\frac{d}{dt} \frac{\partial z}{\partial a}(t, \bar{a}) = \begin{pmatrix} 0 & I \\ -M^*(t, \overline{a})^{-1} K^*(t, \overline{a}) & -M^*(t, \overline{a})^{-1} D^*(t, \overline{a}) \end{pmatrix} \frac{\partial z}{\partial a}(t, \bar{a})$$

$$+ \begin{pmatrix} O \\ -M^*(t, \overline{a})^{-1} Y^*(t, \overline{a}) \end{pmatrix} + \begin{pmatrix} O \\ -M^*(t, \overline{a})^{-1} B^*(t, \overline{a}) \frac{\partial u}{\partial a}(t, \overline{z}_0) \end{pmatrix},$$

$$t_0 \leq t \leq t_f, \tag{3.15b}$$

$$\frac{\partial z}{\partial a}(t_0, \bar{a}) = \begin{pmatrix} 0 & I & 0 \\ 0 & 0 & I \end{pmatrix}. \tag{3.15c}$$

Adding the optimal values of the separated control problems (3.13a–e) and (3.15a–c), we get an upper bound of the optimal value of the original coupled control problem (3.11a–f).

3.3 More General Dynamic Control Systems

As stated in (1.1a,b), more general dynamic control systems can be described by a first order initial value problem

$$\dot{z}(t) = g\Big(t, p(\omega), z(t), u(t)\Big), t_0 \leq t \leq t_f, \tag{3.16a}$$

$$z(t_0) = z_0(\omega). \tag{3.16b}$$

Corresponding to (3.6a,b), for the Taylor expansion of the expected costs we need the Jacobian $\frac{\partial z}{\partial a}(t, \overline{a})$. Proceeding as above, for this we get, cf. Theorem 3.2, the following initial value problem

Theorem 3.4

$$\frac{d}{dt}\frac{\partial z}{\partial a}(t,\overline{a}) = \frac{\partial g}{\partial z}(t,\overline{p},z(t,\overline{a}),u(t,\overline{z_0}))\frac{\partial z}{\partial a}(t,\overline{a})$$

$$+\left(\frac{\partial g}{\partial p}(t,\overline{p},z(t,\overline{a}),u(t,\overline{z_0})),\ \frac{\partial g}{\partial u}(t,\overline{p},z(t,\overline{a}),u(t,\overline{z_0}))\frac{\partial u}{\partial z_0}(t,\overline{z_0})\right), \tag{3.17a}$$

$$\frac{\partial z}{\partial a}(t_0,\overline{a}) = \big(0, I\big). \tag{3.17b}$$

Reference

1. Marti, K.: Stochastic Optimization Methods: Applications in Engineering and Operations Research, 3rd edn. Springer, Berlin (2015). https://doi.org/10.1007/978-3-662-46214-0

Chapter 4
Construction of Feedback Control by Means of Homotopy Methods

Homotopy methods are known from topology [3] to describe *deformations* of certain mathematical objects O, e.g. geometrical bodies, sets, functions, etc., such that an initial object O_1 is continuously transformed into a terminal one O_2.

In the following we apply this procedure to the construction of feedback control laws by starting from open-loop controls which can be obtained in general much easier than feedback control laws. Thus, a given open-loop control $u_0 = \varphi(t, z_0)$, depending besides the time t on the initial state z_0 only, is continuously transformed ("deformed") into a feedback control $u_1 = \varphi(t, z)$.

We suppose now that the dynamic equation of the control system is modeled by a first order initial value problem:

$$\dot{z}(t) = f\left(t, \theta(\omega), z(t), u(t)\right), \quad t_0 \le t \le t_f, \tag{4.1a}$$

$$z(t_0) = z_0(\omega). \tag{4.1b}$$

Moreover, we consider the construction of a state-feedback control

$$u(t) = \varphi(t, z(t)), \quad t \ge t_0, \tag{4.2a}$$

with a feedback control law $\varphi = \varphi(t, z)$ optimizing a certain cost function to be defined later on. Using the homotopy theory, cf. [2, 4], the construction is based on a transfer

$$u_\epsilon(t) = \varphi_\epsilon(t, z) := \varphi_0(t, z) + \epsilon(\varphi(t, z) - \varphi_0(t, z)), \quad 0 \le \epsilon \le 1, \tag{4.2b}$$

from an open-loop control (put $\epsilon = 0$)

$$u_{OL} = \varphi_0(t, z) := u_0(t, z_0), \quad t_0 \le t \le t_f, \tag{4.2c}$$

K. Marti, *Optimization Under Stochastic Uncertainty*, International Series in Operations Research & Management Science 296,
https://doi.org/10.1007/978-3-030-55662-4_4

with a time-function $u_0 = u_0(t, z_0)$, to a feedback control (put $\epsilon = 1$)

$$u_{FB} = \varphi = \varphi(t, z), \quad t_0 \le t \le t_f. \tag{4.2d}$$

Supposing at the moment that the controls $u_0 = u_0(t, z_0)$ and $\varphi = \varphi(t, x)$ are given, and putting the hybrid control law (4.2b) into the dynamic equation (4.1a,b), for each ϵ, $0 \le \epsilon \le 1$, we get the initial value problem:

$$\dot{z}(t) = f\left(t, \theta(\omega), z(t), \varphi_\epsilon(t, z)\right), \quad t_0 \le t \le t_f, \tag{4.3a}$$

$$z(t_0) = z_0(\omega). \tag{4.3b}$$

Obviously, (4.3a,b) contains for $\epsilon = 0, 1$, resp., the open-loop, closed loop dynamic control system. Under standard analytical assumptions on the function $f = f(t, \theta, z, u)$, system (4.3a,b) yields a unique solution, analytical in the variables (t, ϵ), (t, θ, ϵ), resp., cf. [1]:

$$z = z(t, a, \epsilon), \qquad t_0 \le t \le t_f, \quad 0 \le \epsilon \le 1, \tag{4.4a}$$

where a is the total parameter vector

$$a := \begin{pmatrix} z_0 \\ \theta \end{pmatrix}. \tag{4.4b}$$

In case of an affine-linear function

$$f(t, \theta, z, u) = A(t, \theta)z + B(t, \theta)u + b(t, \theta), \tag{4.5a}$$

from (4.3a,b) we get the (partly linear) dynamical system

$$\dot{z}(t, a, \epsilon) = A(t, \theta)z(t, a, \epsilon) + B(t, \theta)\varphi_\epsilon(t, z(t, a, \epsilon)), + b(t, \theta),$$
$$t_0 \le t \le t_f, \quad 0 \le \epsilon \le 1, \tag{4.5b}$$

$$z(t_0, a, \epsilon) = z_0. \tag{4.5c}$$

In the following, for $\epsilon = 0$, $\epsilon = 1$, resp. we compare now the solutions

$$z_{OL} = z_{OL}(t, a) := z(t, a, 0), t_0 \le t \le t_f, \tag{4.6a}$$

$$z_{FB} = z_{FB}(t, a) := z(t, a, 1), t_0 \le t \le t_f, \tag{4.6b}$$

of the open-loop ($\epsilon = 0$), feedback ($\epsilon = 1$), resp., control system. Using first the mean value theorem [1], we have

$$z_{FB}(t, a) - z_{OL}(t, a) = z(t, a, 1) - z(t, a, 0) = \frac{\partial z}{\partial \epsilon}(t, a, \vartheta) \tag{4.7a}$$

with $0 < \vartheta = \vartheta_{(t,a)} < 1$. By means of Taylor expansion of $z = z(t, a, \epsilon)$ with respect to ϵ at $\epsilon = 0$, we get

$$z_{FB}(t, a) = z(t, a, 1) = z_{OL}(t, a) + \sum_{k=1}^{\infty} \frac{1}{k!} \frac{\partial^k z}{\partial \epsilon^k}(t, a, 0), \tag{4.7b}$$

where the partial derivatives $\frac{\partial^k z}{\partial \epsilon^k}$ of $z = z(t, a, \epsilon)$ at $\epsilon = 0$ can be obtained by successive differentiation of the initial value problem (4.3a,b) with respect to ϵ at $\epsilon = 0$. Moreover, the Lagrange remainder term related to a Taylor polynomial of order κ reads

$$R_\kappa(t, a, \vartheta) = \frac{1}{(\kappa+1)!} \frac{\partial^{\kappa+1} z}{\partial \epsilon^{\kappa+1}}(t, a, \vartheta), \quad 0 < \vartheta < 1. \tag{4.7c}$$

Thus, differentiating the initial value problem (4.3a,b) with respect to ϵ and using (4.2a,b) we obtain

$$\begin{aligned} \frac{d}{dt}\frac{\partial z}{\partial \epsilon}(t, a, \epsilon) =& \frac{\partial f}{\partial z}(t, a, z(t, a, \epsilon), \varphi_\epsilon(t, z(t, a, \epsilon))) \frac{\partial z}{\partial \epsilon}(t, a, \epsilon) \\ &+ \frac{\partial f}{\partial u}(t, a, z(t, a, \epsilon), \varphi_\epsilon(t, z(t, a, \epsilon)))\big(\varphi(t, z(t, a, \epsilon)) - u_0(t, z_0) \\ &+ \epsilon \frac{\partial \varphi}{\partial z}(t, z(t, a, \epsilon)) \frac{\partial z}{\partial \epsilon}(t, a, \epsilon)\big), \end{aligned} \tag{4.8a}$$

$$\frac{\partial z}{\partial \epsilon}(t_0, a, \epsilon) = 0. \tag{4.8b}$$

Putting $\epsilon = 0$ in (4.8a,b), we find the initial value problem for the first derivative $\frac{\partial z}{\partial \epsilon}(t, a, 0)$ in the Taylor expansion (4.7b):

$$\begin{aligned} \frac{d}{dt}\frac{\partial z}{\partial \epsilon}(t, a, 0) =& \frac{\partial f}{\partial z}(t, a, z(t, a, 0), u_0(t, z_0)) \frac{\partial z}{\partial \epsilon}(t, a, 0) \\ &+ \frac{\partial f}{\partial u}(t, a, z(t, a, 0), u_0(t, z_0))\big(\varphi(t, z(t, a, 0)) - u_0(t, z_0)\big), \end{aligned} \tag{4.9a}$$

$$\frac{\partial z}{\partial \epsilon}(t_0, a, 0) = 0, \tag{4.9b}$$

where the trajectory $z = z(t, a, 0)$ is defined by the initial value problem, cf. (4.1a,b)

$$\dot{z}(t, a, 0) = f\Big(t, \theta, z(t, a, 0), u_0(t, z_0)\Big), \quad t_0 \le t \le t_f, \tag{4.9c}$$

$$z(t_0, a, 0) = z_0, \tag{4.9d}$$

with $a = (z_0^T, \theta^T)^T$, see (4.4b).

In the linear case (4.5a) Eqs. (4.5b,c) and (4.9a–d) read

$$\begin{aligned}\frac{d}{dt}\frac{\partial z}{\partial \epsilon}(t,a,\epsilon) =& A(t,\theta)\frac{\partial z}{\partial \epsilon}(t,a,\epsilon) + B(t,\theta)\big(\varphi(t,z(t,a,\epsilon)) - u_0(t,z_0) \\ &+ \epsilon\frac{\partial \varphi}{\partial z}(t,z(t,a,\epsilon))\frac{\partial z}{\partial \epsilon}(t,a,\epsilon)\big), \qquad (4.10a)\\ \frac{\partial z}{\partial \epsilon}(t_0,a,\epsilon) =& 0, \qquad (4.10b)\end{aligned}$$

and

$$\begin{aligned}\frac{d}{dt}\frac{\partial z}{\partial \epsilon}(t,a,0) =& A(t,\theta)\frac{\partial z}{\partial \epsilon}(t,a,0) + B(t,\theta)\big(\varphi(t,z(t,a,0)) - u_0(t,z_0)\big), \qquad (4.11a)\\ \frac{\partial z}{\partial \epsilon}(t_0,a,0) =& 0, \qquad (4.11b)\\ \dot{z}(t,a,0) =& A(t,\theta)z(t,a,0) + B(t,\theta)u_0(t,z_0) + b(t,\theta), \qquad (4.11c)\\ z(t_0,a,0) =& z_0. \qquad (4.11d)\end{aligned}$$

Arbitrary higher derivatives of $z = z(t,a,\epsilon)$ with respect to ϵ can be obtained by further differentiation of system (4.9a,b) with respect to ϵ and putting then $\epsilon = 0$. For example, in the linear case, for the second order derivative $\frac{\partial^2 z}{\partial \epsilon^2}(t,a,0)$ we get the condition

$$\begin{aligned}\frac{d}{dt}\frac{\partial^2 z}{\partial \epsilon^2}(t,a,0) =& A(t,\theta)\frac{\partial^2 z}{\partial \epsilon^2}(t,a,0) + 2B(t,\theta)\frac{\partial \varphi}{\partial z}(t,z(t,a,0))\frac{\partial z}{\partial \epsilon}(t,a,0), \qquad (4.12a)\\ \frac{\partial^2 z}{\partial \epsilon^2}(t_0,a,0) =& 0. \qquad (4.12b)\end{aligned}$$

Proceeding this way and assuming a linear feedback law

$$\varphi(t,z) = \Gamma(t)z \qquad (4.12c)$$

with the gain matrix $\Gamma = \Gamma(t)$, for the determination of the higher derivatives $\frac{d}{dt}\frac{\partial^k z}{\partial \epsilon^k}(t,a,0),\ k \geq 2$, we get the following initial value problems:

$$\frac{d}{dt}\frac{\partial^k z}{\partial \epsilon^k}(t,a,0) = A(t,\theta)\frac{\partial^k z}{\partial \epsilon^k}(t,a,0) + kB(t,\theta)\Gamma(t)\frac{\partial^{k-1} z}{\partial^{k-1} \epsilon}(t,a,0), \qquad (4.12d)$$

$$\frac{\partial^k z}{\partial \epsilon^k}(t_0, a, 0) = 0. \tag{4.12e}$$

Consider now the optimal feedback control problem under stochastic uncertainty minimizing the expected costs

$$E\Bigg(\int_{t_0}^{t_f} L\big(t, \theta(\omega), z(t, a(\omega)), \varphi(t, z(t, a(\omega)))\big)dt + G\big(\theta(\omega), z(t_f, a(\omega))\big)\Bigg|\mathcal{A}_{t_0}\Bigg) \tag{4.13}$$

subject to the control system (4.1a,b) with the feedback control (4.2a). According to the above results, in a first approach we may approximate the trajectory $z = z_{FB}$(t,a) resulting from (4.1a,b), (4.2a) by, cf. (4.7a),

$$z_{FB}(t, a) = z(t, a, 1) \approx z(t, a, 0) + \frac{\partial z}{\partial \epsilon}(t, a, 0) = z_{OL}(t, a) + \frac{\partial z}{\partial \epsilon}(t, a, 0), \tag{4.14a}$$

where the subtrajectories $t \to z_{OL}(t, a)$, $\frac{\partial z}{\partial \epsilon}(t, a, 0)$ are defined by (4.9a–d). In the linear case (4.5a–c) and assuming the linear feedback law (4.12c) with the gain matrix $\Gamma = \Gamma(t)$, for

$$w(t) := \begin{pmatrix} z(t, a, 0) \\ \frac{\partial z}{\partial \epsilon}(t, a, 0) \end{pmatrix}, t \geq t_0, \tag{4.14b}$$

we get the initial value problem

$$\dot{w} = \begin{pmatrix} A(t, \theta) & 0 \\ B(t, \theta)\Gamma(t) & A(t, \theta) \end{pmatrix} w(t) + \begin{pmatrix} B(t, \theta) \\ -B(t, \theta) \end{pmatrix} u_0(t, z_0) + \begin{pmatrix} b(t, \theta) \\ 0 \end{pmatrix},$$

$$t_0 \leq t \leq t_f, \tag{4.15a}$$

$$w(t_0) = \begin{pmatrix} z_0 \\ 0 \end{pmatrix}. \tag{4.15b}$$

For minimizing the expected cost function (4.13) subject to the dynamic system (4.1a,b), we represent the cost function as follows:

$$\int_{t_0}^{t_f} L\big(t, \theta, z(t, a), \varphi(t, z(t, a))\big)dt + G\big(\theta, z(t_f, a)\big) \tag{4.16}$$

$$= \int_{t_0}^{t_f} L_{OL}\big(t, \theta, z_{OL}(t, a), u_0(t)\big)dt + G_{OL}\big(\theta, z_{OL}(t_f, a)\big)$$

$$+\int_{t_0}^{t_f} L_{FB}\big(t,\theta,\Delta(t,a),\Gamma(t)\big)dt + G_{FB}\big(\theta,\Delta(t_f,a)\big),$$

where $\Delta(t,a) := \frac{\partial z}{\partial \epsilon}(t,a,0)$, $t \geq t_0$. The optimization of the feedback control can then be split into two stages: Optimize first the open-loop control $u_0(\cdot)$ and then the gain matrix $\Gamma(\cdot)$:

Step I: Find $u_0^*(t)$, $t_0 \leq t \leq t_f$, solving

$$\min_{u_0(\cdot)} \mathrm{E}\left(\int_{t_0}^{t_f} L_{OL}\big(t,\theta,z_{OL}(t,a),u_0(t,z_0)\big)dt + G_{OL}\big(\theta,z_{OL}(t_f,a)\big)\right) \tag{4.17a}$$

s.t.

$$\dot{z}_{OL}(t,a) = A(t,\theta)z_{OL}(t,a) + B(t,\theta)u_0(t,z_0) + b(t,\theta), \tag{4.17b}$$

$$z_{OL}(t_0,a) = z_0. \tag{4.17c}$$

Step II: Having $u_0^*(t,z_0)$, $t_0 \leq t \leq t_f$, find the gain matrix $\Gamma^* = \Gamma^*(t)$ solving

$$\min_{\Gamma(\cdot)} \mathrm{E}\left(\int_{t_0}^{t_f} L_{FB}\big(t,\theta,\Delta(t,a),\Gamma(t)\big)dt + G_{FB}\big(\theta,\Delta(t_f,a)\big)\right) \tag{4.18a}$$

s.t.

$$\dot{\Delta}(t,a) = A(t,\theta)\Delta(t,a) + B(t,\theta)\big(\Gamma(t)z_{OL}^*(t,a) - u_0^*(t,z_0)\big), \tag{4.18b}$$

$$\Delta(t_0,a) = 0, \tag{4.18c}$$

where $z_{OL}^* = z_{OL}^*(t,a)$ denotes the optimal open-loop trajectory obtained in from Step I, and $\theta = \theta(\omega)$, $a = a(\omega)$ are the random vectors defined above.

Remark 4.1 The optimal open-loop type problems under stochastic uncertainty (4.17a–c) and (4.18a–c) can be solved also by the developed in Chap. 3.

References

1. Dieudonné, J.: Foundations of Modern Analysis. Academic Press, New York (1969)
2. Liao, S.: Homotopy Analysis Method in Nonlinear Differential Equations. Springer, Heidelberg (2012)
3. Schubert, H.: Topologie: eine Einführung. Teubner, Stuttgart (1971)
4. Whitehead, G.: Elements of Homotopy Theory. Springer, New York (1995)

Chapter 5
Constructions of Limit State Functions

5.1 Introduction

The observations or measurements from real devices or objects, as, e.g., in natural sciences (such as physics, chemistry, biology, earth sciences, meteorology) as well as in social sciences (such as economics, psychology, sociology) are analyzed and ordered in order to describe them (approximate) by mathematical concepts, in mathematical terms. An important application of these *mathematical models* is the *prediction* of the future behavior of the real devices or objects under consideration. Moreover, future measurements can be used to validate the model.

The mathematical model of a technical or economic system or structure Σ is based on the state m_y-vector y containing the minimum number of internal state variables $y_1, y_2, \ldots, y_{m_y}$, needed to describe the varying, time-varying properties, resp., of the device. Besides the m_y-vector y of state variables, several other variables and parameters occur in the description of the input–output behavior of the system/structure. Important further variables are the ν-vector a of model parameters a_j, $1 \leq j \leq \nu$, the r-vector x of nominal design variables x_k, $1 \leq j \leq r$, and, in some cases, the n-vector u of control variables u_j, $1 \leq j \leq n$, for handling/steering the underlying system/structure, as, e.g., by mounting additional devices to improve the performance of the modified system, see, e.g., the *active control* [17, 18, 23] of mechanical structures. Finally, there is often a vector z of output variables describing the (observable/measurable) response of the device.

Remark 5.1 Since the control vector u, if present, is an internal variable, system/structure Σ is represented parametrically $\Sigma = \Sigma_{a(\omega),x}$ by the vector $a = a(\omega)$ of model parameters and the vector x of nominal design variables.

Due to random variations of the material strengths (e.g., yield stress, modulus of elasticity), the external loadings, initial conditions, geometrical quantities (moments

K. Marti, *Optimization Under Stochastic Uncertainty*, International Series in Operations Research & Management Science 296,
https://doi.org/10.1007/978-3-030-55662-4_5

of inertia, cross- sections, sizing variables, etc.), manufacturing errors, etc., the vector of model parameters $a = a(\omega)$ is not a given fixed quantity, but has to be represented by a random vector on a certain probability space $(\Omega, \mathfrak{A}, P)$. On the other hand, the vectors x, u of nominal design and control variables are deterministic vectors.

Originally, in reliability analysis [9, 15] and reliability theory of structural systems [20], the *limit state function, failure function or also safety margin* $g = g(a, x)$ has been introduced to describe, to separate, resp., for given design vector x, the safe and unsafe domain (the regions of failure and survival) in the area A of random parameters (material strengths, external loads, sizing variables, geometrical quantities, etc.) $a = a(\omega), \omega \in (\Omega, \mathfrak{A}, P)$:

$$\begin{aligned} g(a, x) &\leq 0, \text{ a failure may occur at parameter vector } a \\ g(a, x) &= 0, \text{ the structure may be safe or unsafe at } a, \\ g(a, x) &\geq 0, \quad \text{the structure is safe with parameter } a. \end{aligned} \tag{5.1a}$$

Note that, besides the vectors a, x, the function $g = g(a, x, \cdot)$ may depend also on further variables, such as the time t.

An initial representation of the limit state function has been given by

$$g(a) = \gamma(R(a) - L(a)), \tag{5.1b}$$

where $R = R(a, \cdot)$ is the structural resistance, $L = L(a, \cdot)$ denotes the external loading, and $\gamma = \gamma(\delta)$ is an appropriate function, e.g. $\gamma(v) = v$, indicating a failure of the structure if the loadings exceeds the resistance of the structure/system.

In the literature on reliability analysis, one finds methods for the generation of safety margins for truss, frame and elasto-plastic structures [20] and procedures for the approximate representation of limit state functions, cf. [1, 4, 13]: Based on the evaluation of the status of the structural system of a sample of parameter vectors $a^l, l = 1, \ldots, l_0$, in A, by means of regression-type techniques an approximate polynomial surface is determined separating the unsafe and safe realizations of the random parameter vector $a = a(\omega)$. This way, the computation of the failure/survival probabilities can be simplified considerably.

In the following an optimization-oriented approach is suggested for the construction of an appropriate limit state or performance function.

5.2 Optimization-Based Construction of Limit State Functions

The methods for the *analysis and optimal design of structures and systems* are based, as already mentioned above, mostly on certain functions, called *performance functions* or *limit state functions* [20], indicating the existence of a safe, a failure state, respectively. Here, the state of a structure/system is described by certain observable, measurable variables characterizing the properties of the device or object under consideration. In mechanical systems the state is often described by the time, position variables, velocity of an object, forces. In electrical circuits state variables are the voltages in the nodes and the currents through the components as, e.g., inductors, capacitors, resistors. In thermodynamics possible state variables are: Temperature, pressure, volume, internal energy, enthalpy, entropy. Finally, in economic system the population sizes, resources (e.g., nutrients, organic material) are possible state variables.

Corresponding to (5.1a), the meaning of a scalar or vectorial limit state or performance function,

$$s^* = s^*(a, x), \tag{5.2a}$$

depending on the configuration variables a, x of the structure, i.e. the model parameter vector a and the nominal design vector x is: The safety/survival, failure, resp., of a system/structure is represented by a scalar or vectorial functional criterion of the following type:

$$s^*(a, x) \begin{cases} < 0, & \text{a safe state } y \text{ of } S \text{ exists,} \\ & \text{(for a certain control input vector } u \text{ - if } u \text{ is available),} \\ > 0 & \text{no safe state } y \text{ of } S \text{ exists,} \\ & \text{(for every control vector } u \text{ - if control inputs are available),} \\ = 0, & S \text{ a safe or unsafe state may occur,} \end{cases} \tag{5.2b}$$

see, e.g., in reliability analysis [9, 15, 20], reliability-based optimal design [15, 19], stochastic optimization [11]. In the vectorial case, the above relations are understood component-wise. Note that by a simple transformation

$$\tilde{s}^*(a, x) := -s^*(a, x), \tag{5.2c}$$

with the transformed function $\tilde{s}^* = \tilde{s}^*(a, x)$, state conditions having the same orientation as in (5.1a) can be obtained.

For a general definition of the limit state function $s^* = s^*\big(a(\omega), x\big)$, in static technical, economic systems/structures one has, after a possible discretization (e.g., by FEM), the following two basic relations [10]:

(I) **The state equation** (equilibrium, balance)

$$T\Big(y, u, \big(a(\omega), x\big)\Big) = 0, \, y \in Y_0, u \in U_0. \tag{5.3a}$$

Here, y is the state m_y-vector of the system/structure Σ, e.g. a stress-state vector or vector of displacements of a mechanical structure, lying in a range

$Y_0 \subset \mathbb{R}^{m_y}$; in case of dynamical systems, the state y may represent the trajectory of the system. Moreover, u denotes a control or input n-vector of the system/structure with range U_0. Furthermore, for given configuration vectors (a, x) of model parameters and design variables, $T = T\big(y, u, (a, x)\big)$ is a linear or nonlinear mapping from $\mathbb{R}^{m_y} \times \mathbb{R}^n$ into $\mathbb{R}^{m_T}$. Equation (5.3a) represents, e.g. the structural equilibrium in elastic or plastic structural design [7, 14] or the dynamical equation of a control system [24].

Remark 5.2 We assume that for each $\big(a(\omega), x\big)$ under consideration, Eq. (5.3a) has at least one solution y, (y, u), respectively. Depending on the application, (5.3a) may have a unique solution y, (y, u), resp., as in the case of elastic mechanical structures [14], or multiple solutions y, (y, u), resp., as in for plastic structures [7]. Note that condition (5.3a) may also include inequality state conditions $T\Big(y, u, \big(a(\omega), x\big)\Big) \le 0$.

In case of a linear system/structure with respect to the state y or a system/structure linearized with respect to y at $y = 0$, the state equation (5.3a) reads

$$K\big(u, a(\omega), x\big)y = L\big(u, a(\omega), x\big) \tag{5.3b}$$

with the equilibrium matrix $K\big(u, a(\omega), x\big)$ and the vector $L\big(u, a(\omega), x\big)$ of external loadings. In many cases K, L are independent of the control vector u, while in *active control* [17, 18] the load vector L depends also on u. Hence, in many applications the right hand side of (5.3b) reads

$$L(u, (a(\omega), x)) := F(a(\omega), x) + C(a(\omega), x)u \tag{5.3c}$$

with a vector function $F = F(a(\omega), x)$ and a matrix function $C = C(a(\omega), x)$.

Having then an equilibrium matrix $\widetilde{K} = \widetilde{K}\big(a(\omega), x\big)$ independent of the input u, then (5.3b) can be represented also by

$$\big(\widetilde{K}(a(\omega), x), -C(a(\omega), x)\big)\begin{pmatrix} y \\ u \end{pmatrix} = F(a(\omega), x). \tag{5.3d}$$

(II) **Admissibility condition (operating condition)**
In addition to the state equation (5.3a), (5.3b), resp., for the state vector y we have the admissibility or operating condition

$$y \in \mathrm{Y}\big(u, a(\omega), x\big) \text{ or } y \in \operatorname{int}\Big(\mathrm{Y}\big(u, a(\omega), x\big)\Big) \tag{5.4a}$$

with the admissible (feasible) state domain $Y = \mathrm{Y}\big(u, a(\omega), x\big)$ depending on the model parameter vector $a = a(\omega)$ and the nominal input vectors u, x. Using certain response variables, such as displacements, stress variables, forces, discrepancies, etc., a frequent representation of the feasible domain, e.g.

by means of lower and upper bounds of response variables reads

$$\mathrm{Y}\big(u, a(\omega), x\big) := \Big\{y \in \mathbb{R}^{m_y} : g\Big(y, u, \big(a(\omega), x\big)\Big) \leq 0\Big\}, \tag{5.4b}$$

where

$$g = g\Big(y, u, \big(a(\omega), x\big)\Big) \tag{5.4c}$$

denotes a scalar or m_g-vector **response function** of y, u, depending also on $\big(a(\omega), x\big)$, and having appropriate analytical properties (e.g., convexity, linearity).

In many applications or after linearization with respect to y at $y = 0$, the function g is affine–linear with respect to y, hence,

$$g\Big(y, u, \big(a(\omega), x\big)\Big) := H\big(u, a(\omega), x\big)y - h\big(u, a(\omega), x\big) \tag{5.4d}$$

with a matrix $(H, h) = \Big(H\big(u, a(\omega), x\big), h\big(u, a(\omega), x\big)\Big)$.

We assume that the zero state $y = 0$ is an interior point of the admissible state domain $Y\big(u, a(\omega), x\big)$, i.e.

$$0 \in \operatorname{int}\Big(Y\big(u, a(\omega), x\big)\Big) \text{ or } g\Big(0, u, a(\omega), x\Big) < 0 \tag{5.4e}$$

for all variables $\big(u, a(\omega), x\big)$ under consideration. Thus, in case (5.4d) we have

$$h\big(u, a(\omega), x\big) > 0. \tag{5.4f}$$

Representation by a Minkowski Functional If the feasible state domain $Y\big(u, a(\omega), x\big)$ is a closed convex set containing the zero state as an interior point, then the admissibility condition (5.4a) can be represented also by

$$\pi\big(y|Y(u, a(\omega), x)\big) \leq 1, \pi\big(y|Y(u, a(\omega), x)\big) < 1, \text{ resp.,} \tag{5.5a}$$

where $\pi = \pi\big(y|Y(u, a(\omega), x)\big)$ denotes the *Minkowski or distance functional* of the admissible state domain $Y = Y\big(u, a(\omega), x\big)$, defined, cf. [22], by

$$\pi(y|Y) = \inf\big\{\lambda > 0 : \frac{y}{\lambda} \in Y\big\}. \tag{5.5b}$$

Suppose that the state vector can be partitioned,

$$y = (y_i)_{1 \leq i \leq n_G}, \tag{5.5c}$$

into a certain number n_G of m_i-subvectors $y_i, i = 1, \ldots$, such that the admissible state domain $Y\big(u, a(\omega), x\big)$ is represented then by the product

$$Y\big(u, a(\omega), x\big) = \otimes_{i=1}^{n_G} Y_i\big(u, a(\omega), x\big) \tag{5.5d}$$

of n_G convex factor domains

$$Y_i\big(u, a(\omega), x\big) := \big\{y_i : g_i\big(y_i, u, a(\omega), x\big) \leq 0\big\} \tag{5.5e}$$

with the functions

$$g_i = g_i\big(y_i, u, a(\omega), x\big), i = 1, \ldots, n_G. \tag{5.5f}$$

In the above mentioned case, due to (5.5d), the distance functional of $Y\big(u, a(\omega), x\big)$ can be represented by

$$\pi\big(y|Y\big(u, a(\omega), x\big)\big) = \max_{1 \leq i \leq n_G} \pi_i\big(y_i|Y_i\big(u, a(\omega), x\big)\big), \tag{5.5g}$$

where

$$\pi_i\big(y_i|Y_i\big(u, a(\omega), x\big)\big) := \inf\Big\{\lambda > 0 : \frac{y_i}{\lambda} \in Y_i\big(u, a(\omega), x\big)\Big\} \tag{5.5h}$$

denotes the Minkowski functional of the factor domain $Y_i\big(u, a(\omega), x\big)$.

Example 5.1 Suppose that $m_i = 1, i = 1, \ldots, n_G(= m_y)$, and let $Y_i := [a_i, b_i]$ be intervals on the real line such that $a_1 \leq 0 \leq b_i, i = 1, \ldots, m_y$. Then, $\pi(y_i) = max\{\frac{y_i}{a_i}, \frac{y_i}{b_i}\}$, $y_i \in \mathbb{R}$, and $\pi(y) = \max_{1 \leq i \leq m_y} max\{\frac{y_i}{a_i}, \frac{y_i}{b_i}\}$.

5.3 The (Limit) State Function s^*

The role of the (limit) state function is to indicate whether a state vector y belongs to the safe or the failure domain. According to the conditions (I,II) stated in Sect. 5.1, for a given pair of configuration vectors $\big(a(\omega), x\big)$, the state function $s^* = s^*\big(a(\omega), x\big)$ can be defined by the minimum value function of one of the following optimization problems:

Problem A

$$\min_{s,y,u} s \tag{5.6a}$$

s.t.

$$T\Big(y, u, \big(a(\omega), x\big)\Big) = 0 \tag{5.6b}$$

$$g\Big(y, u, \big(a(\omega), x\big)\Big) \le s\mathbb{1} \tag{5.6c}$$

$$y \in Y_0 \tag{5.6d}$$

$$u \in U_0, \tag{5.6e}$$

where $\mathbb{1} := (1, \ldots, 1)^T$. Note that due to Remark 5.2, problem (5.6a–e) always has the feasible solutions $(s, y, u) = (\tilde{s}, \tilde{y}, \tilde{u})$ with any solution $(\tilde{y}, \tilde{u})$ of (5.3a), (5.6b) and $\tilde{s}$ with $g\Big(\tilde{y}, \tilde{u}\big(a(\omega), x\big)\Big) \le \tilde{s}\mathbb{1}$.

Obviously, if the state equation (5.3a) is linear in y, u, problem (5.6a–e) is convex, provided that the vector function $g = g\Big(y, u, \big(a(\omega), x\big)\Big)$ is convex with respect to y, u, and the ranges Y_0 and U_0 are convex.

Problem B: Linear State Equation, Linear Operating Condition, and Convex Polyhedral Domains of the State and Control Variables

Suppose that relations (5.3b,c) and (5.4d) hold, and the ranges Y_0, U_0 are closed convex polyhedrons:

$$Y_0 := \{y \in R^{m_y} : Ay \le b\},\ U_0 := \{u \in R^n : Bu \le c\} \tag{5.7}$$

with fixed matrices (A, b), (B, c). Moreover, assume that K, H, h are independent of u, and the right hand side of (5.3b) is given by (5.3c). Then (5.6a–e) can be represented by the linear program (LP):

$$\min_{s,y,u} s \tag{5.8a}$$

s.t.

$$K\big(a(\omega, x\big)y - C\big(a(\omega, x\big)u = F\big(a(\omega), x\big) \tag{5.8b}$$

$$H\big(a(\omega), x\big)y - h\big(a(\omega), x\big) \le s\mathbb{1} \tag{5.8c}$$

$$Ay \le b \tag{5.8d}$$

$$Bu \le c. \tag{5.8e}$$

The dual representation of the LP (5.8a–e) reads:

$$\max F\big(a(\omega), x\big)^T \theta - h\big(a(\omega), x\big)^T v - b^T w - c^T z \tag{5.9a}$$

s.t.

$$K\big(a(\omega), x\big)^T \theta - H\big(a(\omega), x\big)^T v - A^T w = 0 \tag{5.9b}$$

$$C^T \theta + B^T z = 0 \tag{5.9c}$$

$$1^T v \leq 1 \tag{5.9d}$$

$$\theta \text{ free}, v, w, z \geq 0. \tag{5.9e}$$

Remark 5.3 If the state vector y has full range $Y_0 = R^{m_y}$, then (5.8c) is canceled, and in (5.9a–e) the expressions $b^T w$, $A^T w$, w are canceled.

In case of a closed convex admissible state domain $Y\big(a(\omega), x\big)$, we also have the admissibility condition (5.5a) with the distance functional $\pi = \pi\Big(y, \big(a(\omega), x\big)\Big)$ of $Y\big(a(\omega), x\big)$. Then, the state function $s^* = s^*\big(a(\omega), x\big)$ can be defined by the minimum value function of the following optimization problem:

Problem C

$$\min_{s,y,u} s \tag{5.10a}$$

s.t.

$$T\Big(y, u, \big(a(\omega), x\big)\Big) = 0 \tag{5.10b}$$

$$\pi\Big(y, \big(a(\omega), x\big)\Big) - 1 \leq s \tag{5.10c}$$

$$y \in Y_0 \tag{5.10d}$$

$$u \in U_0. \tag{5.10e}$$

Special forms of (5.10a–e) are obtained for a linear state equation (5.3b) and/or for special admissible domains $Y\big(a(\omega), x\big)$, as e.g. closed convex polyhedrons, cf. (5.4b,d),

$$Y\big(a(\omega), x\big) = \big\{y : H\big(a(\omega), x\big)y \leq h\big(a(\omega), x\big)\big\}. \tag{5.11a}$$

If (H_i, h_i), $i = 1, \ldots, m_H$, denote the rows of (H, h), then, because of property (5.4f), the distance functional of (5.11a) is given by

$$\begin{aligned}\pi\Big(y, \big(a(\omega), x\big)\Big) &= \inf\Big\{\lambda > 0 : H_i\big(a(\omega), x\big)\frac{y}{\lambda} \leq h_i\big(a(\omega), x\big), i = 1, \ldots, n_H\Big\} \\ &= \inf\left\{\lambda > 0 : \frac{1}{h_i\big(a(\omega), x\big)} H_i\big(a(\omega), x\big)y \leq \lambda, i = 1, \ldots, n_H\right\} \\ &= \max_{0 \leq i \leq m_H} \widetilde{H}_i\big(a(\omega), x\big)y,\end{aligned} \tag{5.11b}$$

where

$$\tilde{H}_i\big(a(\omega), x\big) := \begin{cases} 0, & i = 0 \\ \frac{1}{h_i\left(a(\omega), x\right)} H_i\,\big(a(\omega), x\big), & 1 \leq i \leq m_H. \end{cases} \tag{5.11c}$$

Problem D: Scalar/Vectorial Operating Condition
If one has a scalar response function g, as in the above Problem C, then after elimination the variable s, problem (5.6a–e) reads:

$$\min_{y,u} g\Big(y, u, \big(a(\omega), x\big)\Big) \tag{5.12a}$$

s.t.

$$T\Big(y, u, \big(a(\omega), x\big)\Big) = 0 \tag{5.12b}$$

$$y \in Y_0 \tag{5.12c}$$

$$u \in U_0, \tag{5.12d}$$

where the state function $s^* = s^*(a, x)$ is then given by

$$s^*(a, x) := g\Big(y^*, u^*, \big(a(\omega), x\big)\Big) \tag{5.12e}$$

with a pair of optimal solutions (y^*, u^*).

In case of an m_g-vector response function $g = (g_i)_{1 \le i \le m_g}$, the vectorial conditions (5.6c) can be represented in the following *scalar* form:

$$\max_{1 \le i \le m_g} g_i\big(y, u, (a(\omega), x)\big) \le s. \tag{5.13}$$

Corresponding to (5.12a–d), in the vectorial case, problem (5.6a–e) can be represented also by

$$\min_{y,u} \max_{1 \le i \le m_g} g_i\big(y, u, (a(\omega), x)\big) \tag{5.14a}$$

s.t.

$$T\Big(y, u, \big(a(\omega), x\big)\Big) = 0 \tag{5.14b}$$

$$y \in Y_0 \tag{5.14c}$$

$$u \in U_0, \tag{5.14d}$$

Constrained Nonnegative Response Functions
Suppose that the response functions $g_i, i = 1, ..., m_g$, in the admissibility condition (II) are given by

$$g_i := \frac{\widetilde{g}_i}{d_i} - 1, i = 1, ..., m_g, \tag{5.15a}$$

with nonnegative primary response functions $\widetilde{g}_i \geq 0$ and given positive upper bounds $d_i > 0$, $i = 1, ..., m_g$ for $\widetilde{g}_i$. Then, we have

$$\max_{1 \leq i \leq m_g} g_i = \max_{1 \leq i \leq m_g} \left(\frac{\widetilde{g}_i}{d_i} \right) - 1 = \| \left(\frac{\widetilde{g}_i}{d_i} \right)_i \|_\infty - 1, \tag{5.15b}$$

and for the primary response variables $\widetilde{g}_i$ the admissible condition (II) reads:

$$\widetilde{g}_i \big(y, u, (a(\omega), x) \big) \leq d_i, 1 \leq i \leq m_g. \tag{5.15c}$$

With the p-norm $\|w\|_p := \left(\sum_{j=1}^{m} |w_j|^p \right)^{\frac{1}{p}}$ of an $m-$vector w, where $p \geq 1$, the maximum-norm can be approximated, cf. [6], by

$$m^{-\frac{1}{p}} \|w\|_p \leq \|w\|_\infty \leq \|w\|_p. \tag{5.15d}$$

Moreover, we have

$$\|w\|_p \to \|w\|_\infty, \; p \to \infty. \tag{5.15e}$$

Hence, using p-norms and the diagonal matrix D with the diagonal elements d_i, the maximum norm of the vector $D^{-1}\widetilde{g}$ occurring on the right hand side of (5.15b) can be estimated by

$$\|D^{-1}\widetilde{g}\|_\infty \leq \|D^{-1}\widetilde{g}\|_p \leq {m_g}^{\frac{1}{p}} \|D^{-1}\widetilde{g}\|_\infty, \; p \geq 1 \tag{5.15f}$$

with $\|D^{-1}\widetilde{g}\|_p := (\sum_{i=1}^{m_g} (\frac{\widetilde{g}_i}{d_i})^p)^{\frac{1}{p}}$. Consequently, taking the p-th power of the p-norm, the objective function of (5.14a–d) with the response functions (5.15a) can be represented approximately, see the following note, also by

$$\min_{y,u} \sum_{i=1}^{m_g} \left(\frac{1}{d_i} \widetilde{g}_i(y, u, (a(\omega), x)) \right)^p - 1. \tag{5.15g}$$

Admissibility conditions occurring frequently in practice are box constraints, as considered in the following:

Example 5.2 With given lower and upper bounds $a_i < b_i$, $1 \leq i \leq m_g$, consider the box conditions $a_i \leq \widetilde{g}_i \leq b_i$, $1 \leq i \leq m_g$ with the primal response functions $\widetilde{g}_i$, $1 \leq i \leq m_g$. Defining the quantities $c_i = \frac{a_i + b_i}{2}$ and $d_i = \frac{b_i - a_i}{2}$, $1 \leq i \leq m_g$, the box conditions can be represented by means of the following response variables:

$$g_i := \left| \frac{\widetilde{g}_i - c_i}{d_i} \right| - 1, 1 \leq i \leq m_g, \tag{5.15h}$$

which corresponds to the case (5.15a).

A further approximation is obtained by taking into account that $\max_{1\le i\le m_g} g_i \ge g_i$ for all $1 \le i \le m_g$. Thus, if $s_i^* = s_i^*\big(a(\omega), x\big)$ denotes the state functions of the systems/structures having the single performance functions g_i, $1 \le i \le m_g$, then

$$s^*\big(a(\omega), x\big) \ge \widetilde{s}^*\big(a(\omega), x\big) := \max_{1\le i\le m_g} s_i^*\big(a(\omega), x\big). \tag{5.15i}$$

5.3.1 *Characterization of Safe States*

Based on the state equation (I) and the admissibility condition (II), the safety of a structural system $\Sigma = \Sigma_{a(\omega),x}$ represented by the configuration variables $\big(a(\omega), x\big)$ can be defined as follows:

Definition 5.1 Let $a = a(\omega)$, x, resp., be given vectors of model parameters, design variables. A vector $\tilde{y} \in Y_0$ is called a **safe state** of system $\Sigma_{a(\omega),x}$ if—applicable—a control vector $\tilde{u} \in U_0$ exists such that $(\tilde{y}, \tilde{u})$ fulfills (i) the state equation (5.3a) or (5.3b), hence,

$$T\Big(\tilde{y}, \tilde{u}, \big(a(\omega), x\big)\Big) = 0,\ K\big(a(\omega), x\big)\tilde{y} = L\big(a(\omega), x, u\big), \text{resp.}, \tag{5.16a}$$

and (ii) the admissibility condition (5.4a,b) or (5.5a,b) in the standard form

$$g\Big(\tilde{y}, \big(a(\omega), x\big)\Big) \le 0,\ \pi\Big(\tilde{y}, \big(a(\omega), x\big)\Big) \le 1, \text{ resp.}, \tag{5.16b}$$

or in the strict form

$$g\Big(\tilde{y}, \big(a(\omega), x\big)\Big) < 0,\ \pi\Big(\tilde{y}, \big(a(\omega), x\big)\Big) < 1, \text{ resp.}, \tag{5.16c}$$

depending on the application.
System $\Sigma_{a(\omega),x}$ is called to be in a **safe, unsafe state**, resp., if a safe state $\tilde{y}$ exists, if no safe state vector $\tilde{y}$ exists, respectively.

Remark 5.4 If $\Sigma_{a(\omega),x}$ is in an unsafe state, then, due to violations of the basic safety conditions (I) and (II), failures and corresponding damages of $\Sigma_{a(\omega),x}$ may (will) occur. For example in mechanical structures (trusses, frames) high external loads may cause internal stresses in certain elements which damage then the structure, at least partly. Thus, in this case failures and therefore compensation and repair costs occur.

By means of the state function $s^* = s^*\big(a(\omega), x\big)$, the safety or failure of a system or structure $\Sigma_{a(\omega),x}$ having configuration $\big(a(\omega), x\big)$ can be described now, cf. Definition 5.1 as follows:

Theorem 5.1 *Let $s^* = s^*\big(a(\omega), x\big)$ be the state function of $\Sigma_{a(\omega),x}$ defined by the minimum function of one of the above optimization problems (A)–(D). Then, the following basic relations hold:*

- *If $s^*\big(a(\omega), x\big) < 0$, then a safe state vector $\tilde{y}$ exists. Hence, $\Sigma_{a(\omega),x}$ is in a safe state.*
- *If $s^*\big(a(\omega), x\big) > 0$, then no safe state vector exists. Thus, $\Sigma_{a(\omega),x}$ is in an unsafe state and failures and damages may occur.*
- *If a safe state vector $\tilde{\sigma}$ exists, then $s^*\big(a(\omega), x\big) \leq 0, s^*\big(a(\omega), x\big) < 0$, resp., corresponding to the standard, the strict admissibility condition (5.16b), (5.16c).*
- *If $s^*\big(a(\omega), x\big) \geq 0$, then no safe state vector exists with respect to the strict admissibility condition (5.16c).*
- *Suppose that the minimum (s^*, y^*, u^*) is attained in the selected optimization problem defining the state function s^*. If $s^*\big(a(\omega), x\big) = 0$, then a safe state vector exists in case of the standard admissibility condition (5.16b), and no safe state exists in case of the strict condition (5.16c).*

Proof

- If $s^*\big(a(\omega), x\big) < 0$, then according to the definition of the minimum value function s^*, there exists a feasible point $(\tilde{s}, \tilde{y}, \tilde{u}))$ in Problems (A), (B), and (C), resp., such that, cf. (5.6c), (5.8c), (5.10c), and (5.12a)

$$g\big(\tilde{y}, \big(a(\omega), x\big)\big) \leq \tilde{s}1, \qquad \tilde{s} < 0,$$

$$\pi\Big(\tilde{y}, \big(a(\omega), x\big)\Big) - 1 \leq \tilde{s}, \qquad \tilde{s} < 0.$$

  Consequently, $(\tilde{y}, \tilde{u}) \in Y_0 \times U_0$ fulfills the state equation (5.16a) and the strict admissibility condition (5.16c). Hence, according to Definition 5.1, the vector $\tilde{y}$ is a safe state vector fulfilling the strict admissibility condition (5.16c).
- Suppose that $s^*\big(a(\omega), x\big) > 0$, and assume that there exists a safe state vector $\tilde{y}$ with a related control vector $\tilde{u}$ for $\Sigma_{a(\omega),x}$. Thus, according to Definition 5.1, $(\tilde{y}, \tilde{u})$ fulfills the state equation (5.16a) and the admissibility condition (5.16b) or (5.16c). Consequently, $(\tilde{s}, \tilde{y}, \tilde{u})$ with $\tilde{s} = 0$ is feasible in Problems (A), (B), (C) or (D), and we obtain $s^*\big(a(\omega), x\big) \leq 0$ in contradiction to $s^*\big(a(\omega), x\big) > 0$. Thus, no safe state vector exists in this case.
- Suppose that $\tilde{y}$ is a safe state for $\Sigma_{a(\omega),x}$. According to Definition 5.1 there exists a vector $\tilde{u}$ such that $(\tilde{y}, \tilde{u})$ fulfills (5.16a) and condition (5.16b), (5.16c), respectively. Thus, $(\tilde{s}, \tilde{y}, \tilde{u})$ with $\tilde{s} = 0$, with a certain $\tilde{s} < 0$, resp., is feasible in Problems (A), (B) or (C), (D). Hence, we get $s^*\big(a(\omega), x\big) \leq 0$ in case of the standard and $s^*\big(a(\omega), x\big) < 0$ in case of the strict admissibility condition.
- This assertion follows directly from the third assertion.
- If $s^*\big(a(\omega), x\big) = 0$, then under the present assumption there exists $(y^*, u^*) \in Y_0 \times U_0$ such that $(s^*, y^*, u^*) = (0, y^*, u^*)$ is an optimal solution of Problems (A) and (B) or (C), (D). Hence, (y^*, u^*) fulfills the state equation (5.16a) and the admissibility condition (5.16b). Thus, y^* is a safe state vector with respect to the standard admissibility condition. On the other hand, because of (5.16c), no safe state vector $\tilde{\sigma}$ exists with respect to the strict admissibility condition.

□

5.4 Computation of the State Function for Concrete Cases

In this section the state function is computed for several applications and special cases.

5.4.1 *Mechanical Structures Under Stochastic Uncertainty*

5.4.1.1 Trusses

For the state y of a truss, i.e. the vector $y = (y_1, \ldots, y_B)^T$ of axial (normal) forces acting in the bars $i = 1, \ldots, B$, we have the linear state equation, cf. (5.3a–d),

$$Ky = F$$

with the $m \times B-$ equilibrium matrix K, where $m \leq B, rank K = m$, and the vector $F = F(a)$ of external forces depending on random parameter vector $a = a(\omega)$. Furthermore, let us denote $y^L = y^L(a), y^U = y^U(a)$, resp., the vector of bounds for the forces in compression, tension. Thus, for a truss under external load F we have the safety condition:

$$Ky = F(a), \tag{5.17a}$$

$$y^L(a) \leq y \leq y^U(a). \tag{5.17b}$$

Condition (5.17b) can be represented by

$$\frac{| y_i - c_i |}{d_i} \leq 1, i = 1, \ldots, B, \tag{5.18a}$$

with the vectors

$$c(a) = \frac{y^L + y^U}{2}, \ d(a) = \frac{y^U - y^L}{2}. \tag{5.18b}$$

Using the transformation $\widetilde{y} := y - c$, the safety condition reads

$$K\widetilde{y} = \widetilde{F}(a) := F(a) - Kc(a), \tag{5.18c}$$

$$\frac{| \widetilde{y}_i |}{d_i} \leq 1, i = 1, \ldots, B. \tag{5.18d}$$

Due to (5.18d) we introduce now the response function

$$g(y, (a, x)) := \max_{1 \leq i \leq B} \frac{| \widetilde{y}_i |}{d_i} - 1. \tag{5.18e}$$

For the treatment of the safety condition (5.18c,d) we consider now, cf. (5.2a–c), the limit state function, involving a projection problem with the maximum-norm,

$$s^*(a,x) := \min_{K\widetilde{y}=\widetilde{F}} \max_{1\le i\le B} \frac{|\widetilde{y}_i|}{d_i} - 1. \tag{5.19a}$$

Using the upper bound, see (5.18d),

$$\max_{1\le i\le B} \frac{|\widetilde{y}_i|}{d_i} \le \sqrt{\sum_{i=1}^{B} \left(\frac{\widetilde{y}_i}{d_i}\right)^2}, \tag{5.19b}$$

for the limit state function we get the upper bound

$$s^*(a,x) \le \widetilde{s}^*(a,x) := \min_{K\widetilde{y}=\widetilde{F}} \sqrt{\sum_{i=1}^{B} \left(\frac{\widetilde{y}_i}{d_i}\right)^2} - 1. \tag{5.19c}$$

The optimization problem, arising in (5.19c) can be represented by the minimization of a quadratic function subject to a linear constraint, hence, a projection problem with a quadratic norm. Denoting by D the diagonal matrix with the diagonal elements $d_i, i = 1,\dots,B$, for the upper bound $\widetilde{s} = \widetilde{s}^*(a,x)$ of the limit state function we get

$$\widetilde{s}^*(a,x) = \sqrt{\widetilde{F}^T(KD^2K^T)^{-1}\widetilde{F}} - 1. \tag{5.19d}$$

Obviously, negative values of $s^* = s^*(a,x)$ and therefore a safe truss result for larger values of the strengths $d_i, i = 1,\dots,B$ of the bars, and/or moderate values of the external load $\widetilde{F}$.

5.4.1.2 Elastic-Plastic Mechanical Structures

For mechanical structures made of elastic-plastic materials, the structural failure/survival can be described by means of the dual collapse theorems, cf. [12]. After a possible discretization of the structure by finite elements, the survival/failure condition can be formulated, see [21], by the equilibrium equation with the stress boundary conditions and the yield condition defining the elastic limits:

$$C\sigma = P \tag{5.20a}$$

$$N_i R_{id}^{-1}\sigma_i \le h_i, i = 1,2,\dots,n_G. \tag{5.20b}$$

Here, C denotes the $m \times n$ equilibrium matrix with $rankC = n, n = n_0 n_G, \sigma = (\sigma_1^T,\dots,\sigma_i^T,\dots,\sigma_{n_G}^T)^T$ is the stress-state $n-$vector involving the n_0-vectors σ_i of stress/stress-state components corresponding to the $i-$th reference (nodal, check or Gauss) points $x_i, i = 1, \ldots, n_G$, of the structure. P is the $m-$vector of external load components. The piecewise linearizations $\widetilde{K}_i^y$ of the closed, convex feasible domain (yield domains, plasticity domains) K_i^y corresponding to the reference points $x_i, i = 1, \ldots, n_G$, containing the origin as an interior point, are represented, cf. [3], by

$$\widetilde{K}_i^y = \{z \in \mathbb{R}^{n_0} : N_i z \leq h_i\} \tag{5.20c}$$

with the given matrices

$$(N_i, h_i) = \begin{pmatrix} N_{i1} h_{i1} \\ N_{i1} h_{i1} \\ \cdot \\ \cdot \\ \cdot \\ N_{im_y} h_{im_y} \end{pmatrix}, i = 1, 2, \ldots, n_G. \tag{5.20d}$$

Since also the piecewise linearized feasible domain $\widetilde{K}_i^y$ contains the origin as an interior point, we have

$$h_i > 0, i = 1, \ldots, n_G. \tag{5.20e}$$

Hence, one can put $h_i := (1, \ldots, 1)$. Moreover, $R_i = (R_{i1}, R_{i2}, \ldots, R_{in_0},), i = 1, \ldots, n_G$, denotes the n_0-vectors of positive material resistance (strength) parameters such as plastic capacities or yield stresses $R_{ij}, j = 1, \ldots, n_0$ at the i-th reference points $x_i, i = 1, \ldots, n_G$, and R_{id} is the diagonal matrix of the corresponding resistance parameters.

In the following we suppose that

$$\widetilde{K}_i^y \subset K_i^y \subset S(0; \eta_i). \tag{5.21}$$

Hence, $\widetilde{K}_i^y$ is an inner linearization of the admissible domain K_i^y lying in the sphere $S(0; \eta_i)$ with center in the origin 0 and radius η_i. Describing the admissibility conditions in the reference points $x_i, i = 1, \ldots, n_g$, by means of Minkowski functionals, cf. (5.5a,b), we have

$$\pi(z|\widetilde{K}_i^y) \geq \pi(z|K_i^y) \geq \frac{\|z\|}{\eta_i}. \tag{5.22a}$$

Thus, the approximate survival condition (5.20a,b) is sufficient for the exact survival condition containing the true yield domains $K_i^y, i = 1, \ldots, n_G$. Moreover, if (5.20b) holds, hence, if the admissibility condition holds, then (5.22a) yields

$$\|\sigma_i\| \leq \eta_i, i = 1, \ldots, n_G. \tag{5.22b}$$

By the above assumptions and derivations, for the admissibility condition (5.20b) we get the following, always valid lower bound:

Theorem 5.2 *Suppose that yield domains $K_i^y, i = 1, \ldots, n_G$ at the reference points of the structure are approximated from inside by the convex polygon $\widetilde{K}_i^y$ represented by (5.20c). Moreover, assume that the yield domains are bounded according to condition (5.21). Then the admissibility condition (5.20b) is equivalent to*

$$-L_i \leq N_i R_{id}^{-1} \sigma_i \leq h_i, i = 1, 2, \ldots, n_G, \tag{5.23a}$$

where $L_i := \eta_i \|R_{id}^{-1}\| \big(\|N_{ij}\|\big)$ and

$$\big(\|N_{ij}\|\big) := \begin{pmatrix} \|N_{i1}\| \\ \vdots \\ \|N_{im_y}\| \end{pmatrix}. \tag{5.23b}$$

Define now the block-diagonal matrices

$$N_d := \begin{bmatrix} N_1 & 0 & \ldots & 0 \\ 0 & N_2 & \ldots & 0 \\ \vdots & 0 & \ddots & \vdots \\ 0 & \ldots & 0 & N_{n_G} \end{bmatrix}, \; R_d := \begin{bmatrix} R_{1d} & 0 & \ldots & 0 \\ 0 & R_{2d} & \ldots & 0 \\ \vdots & 0 & \ddots & \vdots \\ 0 & \ldots & 0 & R_{n_G d} \end{bmatrix} \tag{5.24a}$$

and the vectors

$$h := \big(h_1^T, h_2^T \ldots, h_{n_G}^T\big)^T,$$
$$L := \big(L_1^T, L_2^T \ldots, L_{n_G}^T\big)^T. \tag{5.24b}$$

Using the definitions (5.24a,b), the admissibility condition (5.20b) can be represented by

Theorem 5.3 *Suppose that the assumptions of the above theorem hold. Then, the admissibility condition (5.20b) is equivalent to*

$$-\frac{h+L}{2} \leq N_d R_d^{-1} \sigma - \frac{h-L}{2} \leq \frac{h+L}{2}. \tag{5.25a}$$

Denoting by D_{h+L} the diagonal matrix with the components of the vector $h + L$ on the diagonal, condition (5.25a) can be represented also by

$$\|D_{h+L}^{-1}\big(N_d R_d^{-1} \sigma - \frac{h-L}{2}\big)\|_\infty \leq \frac{1}{2}, \tag{5.25b}$$

where $\| \cdot \|_\infty$ denotes the maximum-norm.

According to the representation (5.25b) of the admissibility condition (5.20b) the limit state function for elastic-plastic mechanical structures can be represented, (5.19a–d), (5.20a,b), as follows:

Corollary 5.1 *Suppose that the assumptions in Theorem 5.2, 5.3 hold. Then, the limit state functions of the above described elastic-plastic mechanical structures read*

$$s^*(a,x) = \min_{C\sigma=P} \|D_{h+L}^{-1}\big(N_d R_d^{-1}\sigma - \frac{h-L}{2}\big)\|_\infty - \frac{1}{2}. \tag{5.26}$$

Note 5.1 Due to the linear constraint in (5.26), the linear function in the max-norm and the max-norm, the minimum value in (5.26) can be determined by a linear program and its dual, see (5.3a–e), (5.4a–e).

Approximation of the Maximum-Norm by the Euclidean Norm Using the p-norm for $p = 2$, the max-norm can be approximated, see (5.18c,d), from above as follows:

$$\|D_{h+L}^{-1}\left(N_d R_d^{-1}\sigma - \frac{h-L}{2}\right)\|_\infty^2 \le \|D_{h+L}^{-1}\left(N_d R_d^{-1}\sigma - \frac{h-L}{2}\right)\|_2^2$$
$$\le \sigma^T Q\sigma - \sigma^T q + d, \tag{5.27a}$$

where

$$Q := R_d^{-1} N_d^T (D_{h+L}^{-1})^2 N_d R_d^{-1}, \tag{5.27b}$$

$$q := R_d^{-1} N_d^T (D_{h+L}^{-1})^2 \frac{h-L}{2}, \tag{5.27c}$$

$$d := \frac{(h-L)^T}{2} (D_{h+L}^{-1})^2 \frac{h-L}{2}. \tag{5.27d}$$

Note 5.2 In case of symmetric lower and upper bounds, hence, if $h = L$, cf. (5.23a), then $q = 0$, $d = 0$, and the right hand side of (5.27a) is a quadratic form in σ.

Taking the minimum in the inequality (5.27a), for the limit state function of the above described class of elastic-plastic mechanical structures we get the following upper bound:

Theorem 5.4 *Suppose that the assumptions in Theorem 5.2, 5.3 hold. For the exact limit state function (5.26) we have the upper bound*

$$\tilde{s}^*(a,x) = (CQ^{-1}q - P)^T (CQ^{-1}C^T)^{-1}(CQ^{-1}q - P) - q^T Q^{-1} q + d - \frac{1}{2}, \tag{5.28a}$$

where Q, q, d are defined by (5.27b–d). If $h = L$, then

$$\tilde{s}^*(a,x) = P^T (CQ^{-1}C^T)^{-1} P - \frac{1}{2}. \tag{5.28b}$$

5.4.2 Linear-Quadratic Problems with Scalar Response Function

With positive (semi) definite matrices $Q = Q(a,x)$, $R = R(a,x)$, vectors $q = q(a,x)$, $p = p(a,x)$, which may depend on the vectors a, x of model parameters and design variables, and a maximum response value $\overline{g}$, the response function $g = g(y, u, (a,x))$ is given by the convex, scalar function:

$$\begin{aligned} g(y, u, (a,x)) &:= y^T Q(a,x) y + q(a,x)^T y + u^T R(a,x) u + p(a,x)^T u - \overline{g} \\ &= \tilde{y}^T \tilde{Q}(a,x) \tilde{y} + \tilde{q}(a,x)^T \tilde{y} - \overline{g}, \end{aligned} \tag{5.29a}$$

where $\tilde{y} := \begin{pmatrix} y \\ u \end{pmatrix}$ is the modified state vector, and the matrix, vector $\tilde{Q}(a,x)$, $\tilde{q}(a,x)$, is defined then correspondingly. Considering then the limit state function generated by Problem D, hence (5.3a–d), where $Y_0 = \mathbb{R}^{m_y}$, $U_0 = \mathbb{R}^n$, and the operator T is given by (5.2a–c), the constraints of Problem D are reduced to the linear equation for $\tilde{y}$:

$$\tilde{K}\tilde{y} = F, \tag{5.29b}$$

where the matrix $\tilde{K}$ is defined by $\tilde{K} := \big(K, -C\big)$. Note that in most practical cases we can suppose that $\tilde{K}$ has full rank.

Consequently, under the above assumptions the minimization problem (D) can be represented by the quadratic program:

$$\min_{\tilde{y}} \; \tilde{y}^T \tilde{Q}(a,x)\tilde{y} + \tilde{q}(a,x)^T \tilde{y} - \overline{g} \;\text{ s.t. }\; \tilde{K}\tilde{y} = F. \tag{5.29c}$$

Supposing that $\tilde{Q}$ is a positive definite matrix, and $\tilde{K}$ has full rank, also the matrix $\tilde{K}\tilde{Q}^{-1}\tilde{K}^T$ is positive definite. Hence, in the present case the limit state function $s^* = s^*(a,x)$ reads

$$\begin{aligned} s^*(a,x) = &-\frac{1}{4}\tilde{q}^T \tilde{Q}^{-1}\tilde{q} + \left(F + \frac{1}{2}\tilde{K}\tilde{Q}^{-1}\tilde{q}\right)^T (\tilde{K}\tilde{Q}^{-1}\tilde{K}^T)^{-1} \left(F + \frac{1}{2}\tilde{K}\tilde{Q}^{-1}\tilde{q}\right) \\ &- \overline{g}, \end{aligned} \tag{5.29d}$$

where $\tilde{q}$, $\tilde{Q}$, $\tilde{K}$, F may depend on the vectors a and/or x of model parameters/design variables.

5.4.3 Approximation of the General Operating Condition

Assuming that $g = g(y, u, (a, x))$ is an m_g-vector function with components $g_j, j = 1, \dots, m_g$, the operating condition (5.4c) can also be represented by the scalar condition

$$\max_{1 \le j \le m_g} g_j(y, u, (a, x)) \le s. \tag{5.30a}$$

In many cases we may suppose that the functions g_j have a joint lower bound

$$g_j(y, u, (a, x)) \ge \underline{g}(a, x) \text{ for all } j = 1, \dots, m_g, \text{ and } y \in Y_0, u \in U_0 \tag{5.30b}$$

with a given function $\underline{g} = \underline{g}(a, x)$. Denoting by $\| \cdot \|_\infty$ the maximum-norm, then $\max_{1 \le j \le m_g} g_j(y, u, (a, x))$ can be represented by

$$\max_{1 \le j \le m_g} g_j(y, u, (a, x)) = \left\| g(y, u, (a, x)) - \underline{g}(a, x)\mathbf{1} \right\|_\infty + \underline{g}(a, x). \tag{5.30c}$$

Hence, for an exponent $p > 1$, because of (5.18d,e), Problem A, i.e. (2.6a–d), can be approximated by the minimum p-norm problem:

$$\min_{y,u} \left\| g(y, u, (a, x)) - \underline{g}(a, x)\mathbf{1} \right\|_p \tag{5.31a}$$

s.t.

$$T\Big(y, u, \big(a(\omega), x\big)\Big) = 0 \tag{5.31b}$$

$$y \in Y_0 \tag{5.31c}$$

$$u \in U_0. \tag{5.31d}$$

Consequently, if $\tilde{s}_p^* = \tilde{s}_p^*(a, x)$ the minimum value of (5.31a–d), then the limit state function can be approximated from above by

$$s^*(a, x) \le s_p^*(a, x) := \tilde{s}_p^*(a, x) + \underline{g}(a, x). \tag{5.32}$$

5.4.4 Two-Sided Constraints for the Response Functions

With the response functions $g_j(y, u, (a, x), j = 1, \dots, m_g$, the operating condition (5.4a–c) is often formulated in the following way:

$$\alpha \le g(y, u, (a, x)) \le \beta, \tag{5.33a}$$

where α, β are the m_g-vectors of lower, upper bounds, resp., for the m_g-vector g of response functions $g_j(y, u, (a, x), j = 1, \dots, m_g$. Defining the diagonal matrix $D_{\alpha,\beta}$ having the diagonal elements $\beta_j - \alpha_j$, $j = 1, \dots, m_g$, the two-sided condition (5.33a) can be represented also by

$$\|D_{\alpha,\beta}^{-1}\left(g(y, u, (a, x)) - \frac{\alpha+\beta}{2}\right)\|_{\infty} \leq \frac{1}{2}. \tag{5.33b}$$

Using again the p-norm approximation for the maximum-norm, corresponding to (5.18d) we have the p-norm minimization problem:

$$\min_{y,u} \|D_{\alpha,\beta}^{-1}\left(g(y, u, (a, x)) - \frac{\alpha+\beta}{2}\right)\|_p \tag{5.34a}$$

s.t.

$$T\Big(y, u, \big(a(\omega), x\big)\Big) = 0 \tag{5.34b}$$

$$y \in Y_0 \tag{5.34c}$$

$$u \in U_0. \tag{5.34d}$$

Finally, if s_p^* denotes the minimum value of (5.34a–d), then the limit state function corresponding to problems with the operating condition (5.33a) can be approximated from above by

$$s^*(a, x) \leq s_p^*(a, x). \tag{5.35}$$

5.5 Systems/Structures with Parameter-Dependent States

5.5.1 Dynamic Control Systems

In many practical cases, as, e.g., in the optimization and control of robots [11], active control of mechanical structures [17], analysis and control of vibrations of structures [18], etc., the state vector y depends on a certain parameter, such as the time t in dynamical problems, or dimensions, sizing variables, etc., in geometrical problems. Hence, e.g. in dynamical problems the state equation (2.6b) is given by a certain system of differential equations depending on random quantities:

$$\dot{y} = f(t, y(t), u(t), (a, x)), t \geq t_0, \quad y(t_0) = y_0(a, x), \tag{5.36a}$$

with a given function f and an initial value $y = y(t_0) = y_0(a, x)$. In case of a linear or linearized state differential equation we have

$$\dot{y} = A(t, (a, x))y(t) + B(t, (a, x))u(t) + b(t, (a, x)), \tag{5.36b}$$

with matrix and vector functions, resp., A, B, b. Often the state differential equation is also represented by a second order system of linear differential equation

$$M(t)\ddot{y} + D(t)\dot{y} + K(t)y(t) = L(t, u(u), (a, x)) \tag{5.36c}$$

with deterministic matrix functions $M = M(t)$, $D = D(t)$, $K = K(t)$ and a vector function $L = L(t, u(u), (a, x))$.

In dynamic control systems the response or performance function g in (2.6c) is very frequently represented in scalar way by

$$g(y(\cdot), u(\cdot), (a, x)) = \int_{t_0}^{t_f} \gamma\big(t, y(t), u(t), (a, x)\big)dt + \psi(t_f, y(t_f), (a, x)) \tag{5.37}$$

with a cost function along the trajectory $\gamma = \gamma(t, y, u, (a, x))$ and a terminal cost function $\psi = \psi(t_f, y(t_f), (a, x))$. In many cases convex quadratic functions are taken into account.

Consequently, in the present case, according to the optimization problem (2.6a–e), the limit state function $s^* = s^*(a, x)$ is given by the optimal value function of the *optimal control problem*:

$$\min_{y(\cdot),u(\cdot)} \int_{t_0}^{t_f} \gamma\big(t, y(t), u(t), (a, x)\big)dt + \psi(t_f, y(t_f), (a, x)) - g_{\max} \tag{5.38a}$$

s.t.

$$\dot{y}(t) = f(t, y(t), u(t), (a, x)), \tag{5.38b}$$

$$y(t_0) = y_0(a, x), \tag{5.38c}$$

$$y(t) \in Y_0(t), u(t) \in U_0(t), \ t_0 \le t \le t_f. \tag{5.38d}$$

5.5.2 *Variational Problems*

Corresponding to the representation of the limit state function $s^* = s^*(a, x)$ by an optimal control problem, also representations of s^* by using methods of *calculus of variation* [2] may occur. Corresponding problems may found in analytical mechanics and shape optimization. In a basic version the performance or response function $g = g(y, (a, x))$ is given by

$$g(g(y, (a, x)) := \int_{t_0}^{t_f} f\big(\tau, y(\tau), y'(\tau), (a, x)\big)d\tau, \tag{5.39a}$$

where τ is the (scalar) parameter in the state function $y = y(\tau)$, $y' = y'(\tau)$ denotes the derivative of the state function y with respect to τ, and $f = f(\tau, y, y')$ is a given cost function. Moreover, the state equation T reads then

$$T(y, (a, x)) = \begin{pmatrix} y(t_0) - y_0(a, x) \\ y(t_f) - y_f(a, x) \end{pmatrix}. \tag{5.39b}$$

In addition to the boundary conditions (5.39b) in some variational problems there is still a constraint of type:

$$\int_{t_0}^{t_f} h\big(\tau, y(\tau), y'(\tau), (a, x)\big) d\tau = \bar{h} \tag{5.39c}$$

with a given function h and given value $\bar{h}$.

Thus, according to the optimization problem (5.6a–e), here, the limit state function $s^* = s^*(a, x)$ is given by the optimal value function of the following *problem of calculus of variation*:

$$\min_{y(\cdot)} \int_{t_0}^{t_f} f\big(\tau, y(\tau), y'(\tau), (a, x)\big) d\tau - g_{\max} \tag{5.40a}$$

s.t.

$$\textstyle\int_{t_0}^{t_f} h\big(\tau, y(\tau), y'(\tau), (a, x)\big) d\tau = \bar{h} \tag{5.40b}$$

$$y(t_0) = y_0(a, x), \;\; y(t_f) = y_f(a, x). \tag{5.40c}$$

Remark 5.5 If the system differential equation (5.38b) can be solved uniquely to the control input $u(t)$, hence,

$$u(t) = h(t, y(t), \dot{y}(t), (a, x)) \tag{5.41}$$

with some function h, and putting (5.41) into the objective function (5.38a) of the control problem, then the control problem is transformed into a variational problem of the type (5.40a–c).

5.5.3 Example to Systems with Control and Variational Problems

5.5.3.1 Control Problems

Suppose that the (open-loop) control problem has, cf. (5.36b) the linear system differential equation

$$\dot{y} = Ay(t) + Bu(t), t \geq t_0, \tag{5.42a}$$

with fixed matrices A, B, and let the objective function be given (5.37) by

$$\int_{t_0}^{t_f} \gamma\big(t, y(t), u(t), (a, x)\big)dt = \frac{1}{2}\int_{t_0}^{t_f} u(t)^T Ru(t)dt \tag{5.42b}$$

with a positive definite matrix R. Moreover, assume the boundary conditions $y(t_0) = y_0(a, x)$, $y(t_f) = y_f(a, x)$. With the solution $P = P(t)$ of the Lyapunov equation, see, e.g., [8], the optimal open-loop control reads

$$u(t) = R^{-1}B^T e^{A^T(t_f - t)}P^{-1}(t_f)d(t_0, t_f) \tag{5.42c}$$

with

$$d(t_0, t_f) = d(t_0, t_f, (a, x)) := y_f(a, x) - e^{A(t_f - t_0)}y_0(a, x). \tag{5.42d}$$

Consequently, the related limit state function reads [8]

$$s^*(a, x) = \frac{1}{2}d(t_0, t_f, (a, x))^T P^{-1}(t_f)d(t_0, t_f, (a, x)) - g_{\max}. \tag{5.42e}$$

Similar results can be found for closed-loop control problems [8].

5.5.3.2 Variational Problems

Considering the problem of minimum distance of two points $(t_0, y_0(a, x))$, $(t_f, y_f(a, x))$ in the plane, the function f is given by

$$f(\tau, y(\tau), y'(\tau)) := \sqrt{1 + y'(\tau)^2}. \tag{5.43a}$$

By application of the Euler–Lagrange differential equation, for the optimal solution $y^* = y^s t(\tau)$ of the variational problem (5.40a,c) we have the line segment

$$y^* = y^s t(\tau) = \frac{y_f - y_0}{t_f - t_0}t + \frac{y_0 t_f - y_f t_0}{t_f - t_0}. \tag{5.43b}$$

Consequently, according to (5.40a,c), in the present case the limit state function reads

$$s^* = s^*(a, x) = \sqrt{(t_f - t_0)^2 + (y_f(a, x) - y_0(a, x))^2} - g_{\max}. \tag{5.43c}$$

5.5.3.3 Transformation of Control Problems into Variational Problems

As can be seen from the variational problem (5.40a–c), in many cases one has to minimize an explicit objective function (5.40a) subject to simple constraints (5.40c). Hence, using the methods of calculus of variation [2], the corresponding limit state function can be represented directly by the minimum value of the objective function (5.40a). As shown in the following, by a certain transformation of optimal control problems to variational problems, this procedure can also be applied to the representation of the limit state function of a certain class of optimal control problems.

5.5.4 *Discretization of Control Systems*

In the following we consider the approximate construction of limit state functions for optimal control systems by means of discretization techniques. Hence, first we have to treat the optimal control problem (5.38a–d). Starting with the time grid $t_0 \leq t_1 \leq \ldots \leq t_j \leq t_{j+1} \leq \ldots t_{N-1} \leq t_N := t_f$ with step lenghts $h_j := t_{j+1} - t_j, j = 0, 1, \ldots, N-1$, we define the state and control vectors

$$y_1 = y(t_1), \ldots, y_j = y(t_j), \ldots, y_{N-1} = y(t_{N-1}), y_N = y(t_N) = y(t_f), \tag{5.44a}$$

$$u_0 = u(t_0), \ldots, u_j = u(t_j), \ldots, u_{N-1} = u(t_{N-1}), \ldots, u_N = u(t_N). \tag{5.44b}$$

For the discretization of the initial value problem (5.38b,c) we consider the equivalent integral representation:

$$y(t) - y(t_0) = \int_{t_0}^{t} f(s, y(s), u(s), (a, x))ds, t_0 \leq t \leq t_f. \tag{5.45}$$

Numerous rules are available [16] for the discretization of the integral in (5.45). Applying the trapezoidal rule to the partial integrals over the subintervals $[t_j, t_{j+1}]$, for the state values defined in (5.44a) we get the approximations:

$$y_{j+1} - y_j = \frac{h_j}{2}\big(f(t_j, y_j, u_j, (a, x)) + f(t_{j+1}, y_{j+1}, u_{j+1}, (a, x))\big),$$
$$0 \leq j \leq N-1. \tag{5.46}$$

For given initial value y_0, inputs $u_j, j = 1, \ldots, N$, parameter vector a and design vector x, (5.46) is a system of equations for the approximations y_j of the states $y(t_j), j = 1, \ldots, N$.

In many cases the vector function f reads

$$f(t, y, u, (a, x)) = f_I(t, y) + b(t, u, (a, x)) \tag{5.47}$$

with given functions $f_I = f_I(t, y), b = b(t, u, (a, x))$. In this case system (5.46) can be represented by

$$y_1 - \frac{h_0}{2} f_I(t_1, y_1) = y_0 + \frac{h_0}{2}\big(f(t_0, y_0, u_0, (a, x)) + b(t_1, u_1, (a, x))\big), \tag{5.48a}$$

$$y_{j+1} - \frac{h_j}{2} f_I(t_{j+1}, y_{j+1}) - \left(y_j + \frac{h_j}{2} f_I(t_j, y_j)\right)$$

$$= \frac{h_j}{2}\big(b(t_j, u_j, (a, x)) + b(t_{j+1}, u_{j+1}, (a, x))\big), 1 \le j \le N-1. \tag{5.48b}$$

A sufficient condition for the solvability of the system (5.48a,b) for the state vectors $y_1, \ldots, y_N$ reads

Lemma 5.1 *Define the time, step size and control vectors* $t_D := (t_0, t_1, \ldots, t_{N-1}, t_N)$, $t_N = t_f$, $h_D := (h_0, h_1, \ldots, h_{N-1})$, $u_D := (u_0^T, u_1^T, \ldots, u_{N-1}^T, u_N^T)^T$, $y_D := (y_1^T, y_2^T, \ldots, y_N^T)^T$. *Moreover, suppose that*

$$\vartheta_j = \vartheta_j(y_j) := y_j - f_I(t_j, y_j), j = 1, 2, \ldots, N, \tag{5.48c}$$

are bijective mappings onto the state domain under consideration. Then, system (5.48a,b) can be solved uniquely for y_D*, hence,*

$$y_D = T_D^{-1}(b_D), \tag{5.48d}$$

where $b_D = b_D(t_D, h_D, y_0, u_D, (a, x))$ *is the* Nn*-vector containing the subvectors on the right hand side of (5.48a,b), and* $T_D = T_D(y_D; \tilde{t}_D, h_D)$, $\tilde{t}_D := (t_1, \ldots, t_N)$ *is the bijective operator defined by the left hand side of (5.48a,b).*

Proof The assertion follows from the bijectivity of the mappings $\vartheta_j, j = 1, 2, \ldots, N$, and the successive solution of (5.48a,b) for $y_1, \ldots, y_N$. □

Remark 5.6 Besides the trapezoidal rule, further well known methods for the numerical solution of initial value problems are: Euler polygonal-line, Heun, Runge–Kutta method, see [16].

Concerning the consistency, stability, and convergence of discretization methods for initial a vast amount of publications exist [5, 16].

If the function f_I is linear in y, i.e.

$$f_I(t, y) = A(t)y \tag{5.49}$$

with a matrix $A = A(t)$, then system (5.48a,b) reads

$$\left(I - \frac{h_0}{2}A(t_1)\right) y_1 = \left(I + \frac{h_0}{2}A(t_0)\right) y_0 + \frac{h_0}{2}\left(b(t_0, u_0, (a, x)) + b(t_1, u_1, (a, x))\right), \tag{5.50a}$$

$$\left(I - \frac{h_j}{2}A(t_{j+1})\right) y_{j+1} - \left(I + \frac{h_j}{2}A(t_j)\right) y_j$$

$$= \frac{h_j}{2}\big(b(t_j, u_j, (a, x)) + b(t_{j+1}, u_{j+1}, (a, x))\big), 1 \le j \le N-1. \tag{5.50b}$$

Concerning the properties of (5.50a,b) we get the following result:

Theorem 5.5 *Given the initial value y_0, system (5.50a) of equations for the vector y_D of state vectors on t_D is linear with a block-diagonal matrix $A_D = A_D\big(A(t_{j+1}), h_j, 0 \le j \le N-1\big)$. If the value $\frac{2}{h_j}$ is no eigenvalue of $A(t_{j+1})$ for each $j = 0, 1, \ldots, N-1$, then A_D is regular, and (5.50a,b) can be solved uniquely for*

$$y_D = A_D^{-1} b_D \tag{5.51}$$

with the vector $b_D = b_D(t_D, h_D, y_0, u_D, (a, x))$ of right hand sides in (5.50a,b).

(5.52)

Proof If the matrices $I - \frac{h_j}{2}A(t_{j+1}), 0 \le j \le N-1$, are regular, i.e. $\frac{2}{h_j}$ is no eigenvalue of $A(t_{j+1})$ for each $j = 0, 1, \ldots, N-1$, then the system $A_D y_D = b_D$ can be solved uniquely for y_D. Moreover, the N partial equations in (5.50a,b) can be solved sequentially for $y_1, y_2, \ldots, y_N$. □

Corresponding to the discretization of the dynamic system (5.36a), we may also discretize the integral $g_I(y(\cdot), u(\cdot), (a, x))$ in the objective function g, see (5.37), by the trapezoidal rule:

$$g_I(y(\cdot), u(\cdot), (a, x)) \approx g_{ID}(t_D, h_D, y_0, y_D, u_D, (a, x)), \tag{5.53a}$$

where

$$g_{ID}(t_D, h_D, y_0, y_D, u_D, (a, x)) := \frac{h_0}{2}\gamma(t_0, y_0, u_0, (a, x))$$

$$+ \sum_{j=0}^{N-2} \frac{h_j + h_{j+1}}{2}\gamma(t_{j+1}, y_{j+1}, u_{j+1}, (a, x)) + \frac{h_{N-1}}{2}\gamma(t_N, y_N, u_N, (a, x)). \tag{5.53b}$$

By means of the above discretizations of the control problem (5.38a–d), the corresponding limit state function $s^* = s^*(a, x)$ can be approximated by the optimal value $s_D^* = s_D^*(a, x)$ of the following parameter optimization problem:

Theorem 5.6 *Suppose that condition (5.47) holds. The approximate limit state function $s_D^* = s_D^*(a, x)$ is then given by the optimal value of the parameter optimization problem:*

$$\min_{y_D, u_D} g_{ID}(t_D, h_D, y_0, y_D, u_D, (a, x)) + \psi(t_N, y_N, (a, x)) - g_{\max} \tag{5.54a}$$

s.t.

$$T_D(y_D; \tilde{t}_D, h_D) = b_D(t_D, h_D, y_0, u_D, (a, x)) \tag{5.54b}$$

$$y_D \in Y_{0D}, u_D \in U_{0D}, \tag{5.54c}$$

where $Y_{0D} := Y_0(t_1) \times Y_0(t_2) \times \ldots \times Y_0(t_N)$, and U_{0D} is defined in the same way. Under the assumptions of Theorem 5.5 condition (5.54b) can be solved for y_D, hence,

$$y_D = A_D(\tilde{t}_D, h_D)^{-1} b_D(t_D, h_D, y_0, u_D, (a, x)). \tag{5.54d}$$

Remark 5.7 In many applications the objective function (5.54a) is quadratic in (y_D, u_D), and the right hand side b_D of (5.54b) is linear in u_D.

Putting (5.54d) into (5.54a), we get the following

Corollary 5.2 *Suppose that the assumptions in Theorem 5.2 and 5.3 hold. Furthermore, suppose that $Y_{0D} := \mathbb{R}^{Nm_y}$. Then (5.54a–c) is reduced to*

$$\min_{u_D \in U_{0D}} g_{ID}(t_D, h_D, y_0, y_D, u_D, (a, x)) + \psi(t_N, y_N, (a, x)) - g_{\max} \tag{5.55}$$

with $y_D = y_D(t_D, h_D, y_0, u_D, (a, x))$ given by (5.54d).

5.5.4.1 Control Problems with Quadratic Objective Functions

Suppose here that the cost functions γ, ψ are given by

$$\begin{aligned}\gamma(t, y, u, (a, x)) &:= y^T Q(t, (a, x))y + q(t, (a, x))^T y + \gamma_0(t, (a, x)) \\ &\quad + u^T W(t, (a, x))u \end{aligned} \tag{5.56a}$$

$$\psi(t, y, (a, x)) := y^T P(t, (a, x))y + p(t, (a, x))^T y + \psi_0(t, (a, x)), \tag{5.56b}$$

where $Q = Q(t, (a, x))$, $W = W(t, (a, x))$, $P = P(t, (a, x))$ are positive (semi)definite matrices, $q = q(t, (a, x))$, $p = p(t, (a, x))$ denote vectors, and $\gamma_0(t, (a, x))$, $\psi_0(t, (a, x))$ are scalars depending on the variables t, (a, x). According to (5.53a,b) we obtain then

$$
\begin{aligned}
g_{ID}(t_D, h_D, y_0, y_D, u_D, (a,x)) &:= \frac{h_0}{2}\big(y_0^T Q(t_0,(a,x))y_0 \\
&+ q(t_0,(a,x))^T y_0 + \gamma(t_0,(a,x)) + u_0^T W(t_0,(a,x))u_0\big) \\
&+ \sum_{j=0}^{N-2} \frac{h_j + h_{j+1}}{2}\big(y_{j+1}^T Q(t_{j+1},(a,x))y_{j+1} \\
&+ q(t_{j+1},(a,x))^T y_{j+1} + \gamma(t_{(j+1)},(a,x)) + u_{j+1}^T W(t_{j+1},(a,x))u_{j+1}\big) \\
&+ \frac{h_{N-1}}{2}\big(y_N^T Q(t_N,(a,x))y_N + q(t_N,(a,x))^T y_N \\
&+ \gamma(t_N,(a,x)) + u_N^T W(t_N,(a,x))u_N\big) \\
&= \frac{h_0}{2}(y_0^T Q(t_0,(a,x))y_0 + q(t_0,(a,x))^T y_0 + \gamma(t_0,(a,x))) \\
&+ y_D^T Q_D(\tilde{t}_D, h_D,(a,x))y_D \\
&+ q_D(\tilde{t}_D, h_D,(a,x))^T y_D + \gamma_D(\tilde{t}_D, h_D,(a,x)) + u_D^T W(t_D, h_D,(a,x))u_D
\end{aligned} \tag{5.57a}
$$

with block-diagonal matrices $Q_D(\tilde{t}_D, h_D, (a,x))$, $W(t_D, h_D, (a,x))$, a vector $q_D(\tilde{t}_D, h_D, (a,x))$ and a scalar $\gamma_D(\tilde{t}_D, h_D, (a,x))$. Moreover, the end point costs read:

$$
\psi(t_N, y_N, (a,x)) := y_N^T P(t_N,(a,x))y_N + p(t_N,(a,x))^T y_N + \psi_0(t_N,(a,x)). \tag{5.57b}
$$

Assuming that the vector $b = b(t, u, (a,x))$ in the representation (5.47) of f is given by

$$
b(t,u,(a,x)) := B(t,(a,x))u + \beta(t,(a,x)), \tag{5.58a}
$$

then the vector b_D can be represented, cf. (5.50a,b), (5.51), by

$$
b_D(t_D, h_D, y_0, u_D, (a,x)) = B_D(t_D, h_D, (a,x))u_D + \beta_D(t_D, h_D, y_0, (a,x)) \tag{5.58b}
$$

with a corresponding matrix B_D and vector β_D. With (5.58b), the assumptions of Theorem 5.2 and (5.51) the discretized state trajectory y_D reads

$$
y_D = A_D(\tilde{t}_D, h_D)^{-1} B_D(t_D, h_D, (a,x))u_D + A_D(\tilde{t}_D, h_D)^{-1}\beta_D(t_D, h_D, y_0, (a,x)). \tag{5.59a}
$$

The function $\beta_D = \beta_D(t_D, h_D, y_0, (a,x))$ can be represented by the sum

$$
\beta_D(t_D, h_D, y_0, (a,x)) = \beta_{D0}(t_0, h_0, y_0) + \tilde{\beta}_D(t_D, h_D, (a,x)) \tag{5.59b}
$$

of given functions $\beta_{D0} = \beta_{D0}(t_0, h_0, y_0)$ and $\tilde{\beta}_D = \tilde{\beta}_D(t_D, h_D, (a,x))$.

If A_N^{inv} denotes the submatrix of A_D^{-1} with the last n rows, then for the terminal subvector y_N of y_D we have

$$y_N = A_N^{inv}(\tilde{t}_D, h_D)B_D(t_D, h_D, (a,x))u_D + A_N^{inv}\beta_D(t_D, h_D, y_0, (a,x)). \tag{5.59c}$$

Inserting now (5.59a,b) into (5.57a,b) we get the following result:

Lemma 5.2 *For quadratic cost functions, the objective function of the minimization problem (5.55) can be represented by*

$$\begin{aligned} g_D(t_D, h_D, y_0, y_D, u_D, (a,x)) &= g_{ID} + \psi = \frac{h_0}{2}\big(y_0^T Q(t_0, (a,x))y_0 \\ &+ q(t_0, (a,x))^T y_0\big) + u_D^T \boldsymbol{Q}(t_D, h_D, (a,x))u_D + \boldsymbol{q}(t_D, h_D, y_0, (a,x))^T u_D \\ &+ \boldsymbol{d}(t_D, h_D, y_0, (a,x)), \end{aligned} \tag{5.60a}$$

where

$$\boldsymbol{Q}(t_D, h_D, (a,x)) := B_D^T\big({A_D^{-1}}^T Q_D A_D^{-1} + {A_N^{inv}}^T P A_N^{inv}\big)B_D + W_D, \tag{5.60b}$$

$$\begin{aligned} \boldsymbol{q}(t_D, h_D, y_0, (a,x)) &:= 2B_D^T\left({A_D^{-1}}^T Q_D A_D^{-1} + {A_N^{inv}}^T P A_N^{inv}\right)\beta_D \\ &+ B_D^T\left({A_D^{-1}}^T q_D + {A_N^{inv}}^T p\right), \end{aligned} \tag{5.60c}$$

$$\begin{aligned} \boldsymbol{d}(t_D, h_D, y_0, (a,x)) &:= \beta_D^T\left({A_D^{-1}}^T Q_D A_D^{-1} + {A_N^{inv}}^T P A_N^{inv}\right)\beta_D \\ &+ \beta_D^T\left({A_D^{-1}}^T q_D + {A_N^{inv}}^T p\right). \end{aligned} \tag{5.60d}$$

5.5.4.2 Tracking Problems

In case of tracking problems [11], with a prescribed trajectory $\overline{y} = \overline{y}(t)$, $t_0 \le t \le t_f$, the cost functions γ, ψ, see (5.56a,b), reads

$$\gamma(t, y, u, (a,x)) := (y - \overline{y})^T Q(t, (a,x))(y - \overline{y}) + u^T W(t, (a,x))u \tag{5.61a}$$

$$\psi(t_f, y, (a,x)) := 0. \tag{5.61b}$$

Thus, corresponding to (5.57a,b) the discretized total cost function of the tracking problem reads

$$\begin{aligned} g_D(t_D, h_D, y_0, y_D, u_D, (a,x)) &= \frac{h_0}{2}(y_0 - \overline{y}_0)^T Q(t_0, (a,x))(y_0 - \overline{y}_0) \\ &+ (y_D - \overline{y}_D)^T Q_D(\tilde{t}_D, h_D, (a,x))(y_D - \overline{y}_D) + u_D^T W(t_D, h_D, (a,x))u_D. \end{aligned} \tag{5.62a}$$

Inserting y_D, represented by (5.58a,b), (5.59a) into (5.62a), we get

$$g_D(t_D, h_D, y_0, y_D(u_D), u_D, (a,x)) = \frac{h_0}{2}(y_0 - \overline{y}_0)^T Q(t_0, (a,x))(y_0 - \overline{y}_0)$$
$$= u_D \mathbf{Q} u_D + u_D^T \mathbf{q} + \mathbf{d}, \tag{5.62b}$$

where

$$\mathbf{Q} := B_D^T A_D^{-1^T} Q_D A_D^{-1} B_D, \tag{5.62c}$$

$$\mathbf{q} := -2 B_D^T A_D^{-1^T} Q_D (A_D^{-1}\beta_D - \overline{y}_D), \tag{5.62d}$$

$$\mathbf{d} := (A_D^{-1}\beta_D - \overline{y}_D)^T Q_D (A_D^{-1}\beta_D - \overline{y}_D). \tag{5.62e}$$

Let us denote x_D the vector with the subvectors $y_0 - \overline{y}_0$, $A_D^{-1} B_D u_D + A_D^{-1}\beta_D - \overline{y}_D$, u_D let be $\mathbf{R}$ the positive (semi) definite block-diagonal matrix with the square matrices $\frac{h_0}{2} Q(t_0)$, Q_D, W_D. Then, g_D can be represented also by

$$g_D = x_D^T \mathbf{R} x_D = \|\Gamma x_D\|^2, \tag{5.62f}$$

where Γ is defined by $\mathbf{R} = \Gamma^T \Gamma$.

5.5.4.3 Endpoint Control

Given a certain end point y_f to be reached with minimum costs for the control input u, the cost functions γ, ψ, see (5.56a,b), can be selected as follows:

$$\gamma(t, y, u, (a,x)) := u^T W(t, (a,x)) u \tag{5.63a}$$

$$\psi(t_f, y, (a,x)) := (y - y_f)^T P(t_N, (a,x))(y_N - y_f). \tag{5.63b}$$

Here, corresponding to (5.57a,b) the discretized total cost function of the end point control problem reads

$$g_D(t_D, h_D, y_N, u_D, (a,x)) = (y_N - y_f)^T P(t_N, (a,x))(y_N - y_f)$$
$$+ u_D^T W(t_D, h_D, (a,x)) u_D. \tag{5.64}$$

5.5.4.4 Control Problems with Sublinear Objective Functions

Considering here tracking and end point control problem, the cost functions γ, ψ are given by

$$\gamma(t, y, u, (a,x)) := \max_l c_l(t, (a,x))^T (y - \overline{y}(t)) + c_{II}(t, u), \tag{5.65a}$$

$$\psi(t, y, (a, x)) := \max_l c_l(t_f, (a, x))^T (y - \overline{y}(t_f)), \tag{5.65b}$$

where $c_l = c_l(t, (a, x), 1 \leq l \leq l_{\max}$, are m_y-vectors of cost factors, $\overline{y} = \overline{y}(t)$ is the prescribed function to be tracked, and $c_{II} = c_{II}(t, u)$ denotes the costs for the control input u. Corresponding to (5.57a,b) we obtain then

$$\begin{aligned}
&g_{ID}(t_D, h_D, y_0, y_D, u_D, (a, x)) \\
&:= \frac{h_0}{2}\Big(\max_{l_0} c_{l_0}(t_0, (a, x))^T (y_0 - \overline{y}_0) + c_{II}(t_0, u_0)\Big) \\
&+ \sum_{j=0}^{N-2} \frac{h_j + h_{j+1}}{2}\Big(\max_{l_{j+1}} c_l(t_{j+1}, (a, x))^T (y_{j+1} - \overline{y}_{j+1}) + c_{II}(t_{j+1}, u_{j+1})\Big) \\
&+ \frac{h_{N-1}}{2}\Big(\max_l c_l(t_N, (a, x))^T (y_N - \overline{y}_N) + c_{II}(t_N, u_N)\Big).
\end{aligned} \tag{5.66a}$$

Moreover, the end point costs read

$$\psi(t_N, y_N, (a, x)) := \max_l c_l(t_N, (a, x))^T (y_N - \overline{y}_N). \tag{5.66b}$$

Define now the transformed state vectors

$$\tilde{y}_0 := \frac{h_0}{2} y_0, \tilde{y}_{j+1} := \frac{h_j + h_{j+1}}{2} y_{j+1}, 0 \leq j \leq N-2, \tilde{y}_N := \frac{h_{N-1}}{2} y_N, \tag{5.67a}$$

and in the same way the transformed states to be reached $\tilde{\overline{y}}_0, \tilde{\overline{y}}_j, 1 \leq j \leq N-1, \tilde{\overline{y}}_N$. Let us then denote $\tilde{y}_D, \tilde{\overline{y}}_D$, resp., the Nm_y-vector containing the subvectors $\tilde{y}_j, \tilde{\overline{y}}_j, resp., 1 \leq j \leq N$. Moreover, for an index set $l_D := \{l_1, l_2, \ldots, l_N\}, 1 \leq l_j \leq l_{max}, 1 \leq j \leq N$, we define the total cost vector

$$c_{l_D}(\tilde{t}_D) := \left(c_{l_1}^T, c_{l_2}^T, \ldots, c_{l_N}^T\right)^T. \tag{5.67b}$$

According to (5.66a,b) we have the following result:

Lemma 5.3 *For sublinear cost functions, the total approximate cost function $g_D = g_{ID} + \psi$ of the minimization problem (5.55) can be represented by*

$$\begin{aligned}
g_D(t_D, h_D, y_0, y_D, u_D, (a, x)) :=& \max_{1 \leq l_0 \leq l_{max}} c_{l_0}(t_0, (a, x))^T (\tilde{y}_0 - \tilde{\overline{y}}_0) \\
&+ \max_{1 \leq l_j \leq l_{max}, 1 \leq j \leq N} c_{l_D}(\tilde{t}_D, (a.x))^T (\tilde{y}_D - \tilde{\overline{y}}_D) \\
&+ \max_{1 \leq l_f \leq l_{max}} c_{l_f}(t_N, (a, x))^T (y_N - \overline{y}_N) + c_{IID}(t_D, h_D, u_D),
\end{aligned} \tag{5.68}$$

where the discretized trajectory y_D, the terminal state y_N, resp., are defined by (5.59a,b).

5.5.5 Reliability-Based Optimal Control

According to Sect. 5.5, see (5.38a–d,5.54a–d,5.55, 5.59a,b), the approximate performance function $s_D^* = s_D^*(t_D, h_D, y_0, (a, x))$ can be represented, assuming for simplification that $Y_{0D} := \mathbb{R}^{Nm_y}$, by

$$s_D^*(t_D, h_D, y_0, (a, x)) := \min_{u_D \in U_{0D}} g_D(t_D, h_D, y_0, y_D(u_D), u_D, (a, x)) - g_{max}, \tag{5.69a}$$

where $y_D(u_D) = y_D(t_D, h_D, y_0, u_D, (a, x))$, $y_N(u_D) = y_N(t_D, h_D, y_0, u_D, (a, x))$ is given by one of the above equations, as, e.g., (5.48d,5.51,5.59a,b).

In case that the control vector u_D is treated as an outer, selectable input variable, such as the design vector x, then, instead of (5.69a), we have the performance function—being a special case of (5.69a)

$$s_D^*(t_D, h_D, y_0, y_D(u_D), u_D, (a, x)) := g_D(t_D, h_D, y_0, y_D(u_D), u_D, (a, x)) - g_{max}. \tag{5.69b}$$

5.5.5.1 Computation of the Probability of Survival

Suppose now that the initial value $y_0 = y_0(\omega)$ and the parameter vector $a = a(\omega)$ are—without restrictions—stochastically independent random variables. In both cases (5.69a,b), the probability of survival p_s can be approximated as follows:

Theorem 5.7 *Let $h = h(t)$ denote a nonnegative, monotone increasing function on $\mathbb{R}$. Then,*

$$\begin{aligned} p_s(x) \approx p_{sD}(x) &= P\Big(\min_{u_D \in U_{0D}} g_D(t_D, h_D, y_0, y_D(u_D), u_D, (a, x)) \le g_{\max}\Big) \\ &\ge \max_{u_D \in U_{0D}} P\Big(g_D(t_D, h_D, y_0(\omega), y_D(u_D), u_D, (a(\omega), x)) \le g_{\max}, \Big) \\ &\ge 1 - \frac{1}{h(g_{\max})} \min_{u_D \in U_{0D}} Eh\Big(g_D(t_D, h_D, y_0(\omega), y_D(u_D), u_D, (a(\omega), x))\Big). \end{aligned} \tag{5.70}$$

Improvements of the second inequality can be obtained by minimizing the quotient in the second inequality subject to a certain class of nonnegative, monotone increasing functions $h(\cdot)$. If all measurable nonnegative monotone increasing functions are taken into account, then the minimum is attained at $h^* := 1_{[g_{max}, +\infty)}$, and the second inequality in (5.70) holds with "=".

Proof Since $g_D(.., \omega, ., u_D, \ldots) \le g_{max}$ implies that $\min_{u_D \in U_{0D}} g_D(.., \omega, ., u_D, \ldots) \le g_{max}$ for each given $u_D \in U_{0D}$, for $p_{sD}(x)$ we get the lower bound

$$p_{sD}(x) \ge \max_{u_D \in U_{0D}} P\Big(g_D(t_D, h_D, y_0(\omega), y_D(u_D), u_D, (a(\omega), x)) \le g_{\max}, \Big).$$

The second inequality in (5.70) follows from the Markov inequality. Improved lower bounds can be obtained by minimizing the quotient in this inequality subject to a certain class of nonnegative, monotone increasing functions $h(\cdot)$. If all measurable nonnegative monotone increasing functions are taken into account, then the minimum is attained at $h^* := 1_{[g_{max},+\infty)}$, and the second inequality in (5.70) holds with "=". □

From the above theorem we get the following consequences:

Corollary

(a) With a minimum probability of survival α, a sufficient condition for $p_{sD}(x) \geq \alpha$ is given by

$$\min_{u_D \in U_{0D}} Eh\big(g_D(t_D, h_D, y_0(\omega), y_D(u_D), u_D, (a(\omega), x))\big) \leq (1-\alpha)h(g_{\max}). \tag{5.71a}$$

(b) For nonnegative, monotone increasing concave functions on $\mathbb{R}$ we have

$$p_{sD}(x) \geq 1 - \frac{1}{h(g_{\max})} h\left(\min_{u_D \in U_{0D}} Eg_D(t_D, h_D, y_0(\omega), y_D(u_D), u_D, (a(\omega), x))\right). \tag{5.71b}$$

The optimal, but nonsmooth function $h^* := 1_{[g_{max},+\infty)}$ can be approximated by various smooth sigmoid functions, such as:

$$h_1(x) = \frac{1}{1+\exp(-k\,(x-g_{\max}))},\ h_2(x) = \frac{1}{2}\left(1+\mathrm{erf}\left(\frac{x-g_{\max}}{\sqrt{2}\sigma}\right)\right)$$
$$= \Phi\left(\frac{x-g_{\max}}{\sigma}\right), x \in \mathbb{R}, \tag{5.71c}$$

where erf(x) denotes the error function, Φ is the Gaussian distribution function, and $k > 0, \sigma > 0$ are positive parameters. For $k \to +\infty$, $\sigma \downarrow 0$, resp, the functions h_1, h_2, resp, approach the $0-1-$function h^*. Expanding the sigmoid functions h_1, h_2 in Taylor series at $x_0 = g_{\max}$, we get

$$h_1(x) = \frac{1}{2} + \frac{k}{4}(x-g_{\max}) - \frac{k^3}{48}(x-g_{\max})^3 + \frac{k^5}{480}(x-g_{\max})^5 - \ldots, \tag{5.71d}$$

$$h_2(x) = \frac{1}{2} + \frac{1}{\sqrt{2\pi}}\left(\frac{x-g_{\max}}{\sigma} - \frac{1}{2\cdot 3}\left(\frac{x-g_{\max}}{\sigma}\right)^3 + \frac{1}{2!2^2 5}\left(\frac{x-g_{\max}}{\sigma}\right)^5 - \ldots\right). \tag{5.71e}$$

The expectation $Eh(g_D) = Eh(g_D(\ldots, y_0(\omega), \ldots, a(\omega), \ldots))$ can be determined also by means of Taylor expansion of $\tilde{h}(b) := h(g_D)(b)$ with respect to $b := (y_0, a)$. The needed derivatives up to second order read, cf. [10, 11],

$$\nabla_b \tilde{h}(b) = h'(g_D(b))\nabla_b g_D(b)$$
$$\nabla_b^2 \tilde{h}(b) = h''(g_D(b))\nabla_b g_D(b)\nabla_b g_D(b)^T + h'(g_D(b))\nabla_b^2 g_D(b). \quad (5.71f)$$

For approximate performance functions g_D which are linear with respect to the parameter vector b, the second term in (5.71f) vanish.

5.5.5.2 Tracking Problems with Quadratic Cost Function

According to Corollary 5.2, (5.55), (5.62f) and Theorem 5.7, for the probability of survival we have

$$p_{sD}(x) \geq \max_{u_D \in U_{0D}} P\big(\|\Gamma x_D\| \leq \sqrt{g_{\max}}\big) \geq 1 - \frac{1}{h(\sqrt{g_{\max}})} \min_{u_D \in U_{0D}} Eh\big(\Gamma x_D\|\big), \quad (5.72a)$$

where

$$x_D(t_D, h_D, y_0, (a, x), u_D) := \begin{pmatrix} y_0 - \overline{y}_0 \\ A_D^{-1} B_D u_D + A_D^{-1} \beta_D - \overline{y}_D \\ u_D \end{pmatrix}, \quad (5.72b)$$

$$\mathbf{R} := \begin{pmatrix} \frac{h_0}{2} Q(t_0) & 0 & 0 \\ 0 & Q_D & 0 \\ 0 & 0 & W_D \end{pmatrix}, \quad (5.72c)$$

and $h = h(x)$ is a nonnegative, monotone increasing function on $\mathbb{R}$.

Taking $h(x) := x^2$, we get $h(\sqrt{g_{\max}}) = g_{\max}$ and

$$Eh\big(\Gamma x_D\|\big) = tr\mathbf{R}M_2(x_D(\cdot)), \quad (5.72d)$$

where $M_2(x_D(\cdot))$ denotes the matrix of second order moments of x_D.

5.5.5.3 Endpoint Control in Case of Sublinear Cost Function

With Lemma 5.3 and (5.59a,b), in case of endpoint control with a sublinear cost function the performance function reads

$$g_D(t_D, h_D, y_0, y_D, u_D, (a, x)) := \max_{1 \leq l_f \leq l_{\max}} c_{l_f}(t_N, (a, x))^T (y_N - \overline{y}_N)$$
$$+ c_{IID}(t_D, h_D, u_D), \quad (5.73)$$

where y_N is given by (5.59c). According to (5.70) and (5.72a–d) for the approximate probability $p_{sD}(x)$ we get the lower bound:

Theorem 5.8 *For optimal endpoint control problems with sublinear cost functions, lower bounds for the approximate probability of survival $p_{sD}(x)$ are given by*

$$p_{sD}(x) \geq \max_{u_D \in U_{0D}} P\big(g_D(t_D, h_D, y_0(\omega), y_D(u_D), u_D, (a(\omega), x)) \leq g_{\max},\big)$$
$$= \max_{u_D \in U_{0D}} P\big(c_{l_f}(t_N, (a, x))^T (y_N - \overline{y}_N)$$
$$+ c_{IID}(t_D, h_D, u_D) \leq g_{\max},\ \textit{for}\ 1 \leq l_f \leq l_{\max}\big), \tag{5.74}$$

where y_N is defined as in (5.59c). Corresponding lower bounds for $p_{sD}(x)$ hold also for general optimal control problems with sublinear cost functions.

With the representation (5.59a,c) of y_D, y_N and (5.59b) of the function β, from (5.74) we also get

$$p_{sD}(x) \geq \max_{u_D \in U_{0D}} P\big(c_{l_f}(t_N, (a, x))^T (A_N^{inv}(\tilde{t}_D, h_D) B_D(t_D, h_D, (a, x)) u_D$$
$$+ A_N^{inv}\big(b_{D0}(t_0, h_0, y_0) + \tilde{\beta}_D(t_D, h_D, (a, x))\big) - \overline{y}_N)$$
$$+ c_{IID}(t_D, h_D, u_D) \leq g_{\max},\ \text{for}\ 1 \leq l_f \leq l_{\max}\big). \tag{5.75a}$$

In many cases the function b_{D0} is linear in y_0, cf. (5.49), and $\tilde{\beta}_D$ is affine linear with respect to the parameter vector $a = a(\omega)$, hence

$$b_{D0}(t_0.h_0, y_0) = \Gamma_0(t_0, h_0) y_0,\ \tilde{\beta}_D(t_D, h_D, (a, x)) = \Gamma_1(t_D, h_D, x) a$$
$$+ \gamma_1(t_D, h_D, x). \tag{5.75b}$$

5.5.5.4 Further Lower Bound for p_{sD}

For deriving a further lower bound for p_{sD} we model the control input u_D by a random vector $u_D = \tilde{u}_D(\omega)$ with a certain distribution on the discretized control domain U_{0D}. We have the following result:

Theorem 5.9 *If $u_D = \tilde{u}_D(\omega)$, $\omega \in \Omega$, is a random vector taking values in U_{0D} with probability* 1, *then*

$$p_{sD}(x) \geq P\big(g_D(t_D, h_D, y_0(\omega), y_D(\tilde{u}_D(\omega)), \tilde{u}_D(\omega), (a(\omega), x)) \leq g_{\max}\big). \tag{5.76a}$$

Conversely, if $\tilde{u}_D^*(\omega)$ is defined by

$$\tilde{u}_D^*(\omega) := argmin_{u_D \in U_{0D}} g(t_D, h_D, y_0(\omega), y_D(u_D), u_D, (a(\omega), x)), \tag{5.76b}$$

then

$$p_{sD}(x) = P\big(g_D(t_D, h_D, y_0(\omega), y_D(\tilde{u}_D^*(\omega)), \tilde{u}_D^*(\omega), (a(\omega), x)) \leq g_{\max}\big). \tag{5.76c}$$

Proof For each $\omega \in \Omega$ such that

$$g_D(t_D, h_D, y_0(\omega), y_D(\tilde{u}_D(\omega)), \tilde{u}_D(\omega), (a(\omega), x)) \leq g_{\max}$$

we also get

$$\min_{u_D \in U_{0D}} g_D(t_D, h_D, y_0(\omega), y_D(u_D), u_D, (a(\omega), x)) \leq g_{\max},$$

which yields the first assertion.

If $\tilde{u}_D^*$ is defined by (5.76b), then $P(\tilde{u}_D^*(\omega) \in U_{0D}) = 1$ and

$$\begin{aligned} p_{sD}(x) =& P\left(\min_{u_D \in U_{0D}} g_D(t_D, h_D, y_0(\omega), y_D(u_D), u_D, (a(\omega), x)) \leq g_{\max}\right) \\ =& P\left(g_D(t_D, h_D, y_0(\omega), y_D(\tilde{u}_D^*(\omega)), \tilde{u}_D^*(\omega), (a(\omega), x)) \leq g_{\max}\right). \end{aligned}$$

□

Note 5.3 If $\tilde{u}_D = \tilde{u}_D(\omega)$ is independent of $(y_0(\omega), a(\omega))$, then the right hand side of (5.76a) can be represented by

$$\begin{aligned} & P\left(g_D(t_D, h_D, y_0(\omega), y_D(\tilde{u}_D(\omega)), \tilde{u}_D(\omega), (a(\omega), x)) \leq g_{\max}\right) \\ =& \int_{u_D \in U_{0D}} P\left(g_D(t_D, h_D, y_0(\omega), y_D(u_D), u_D, (a(\omega), x)) \leq g_{\max}\right) P_{\tilde{u}_D}(du_D), \end{aligned}$$

where $P_{\tilde{u}_D}$ is the probability distribution of $\tilde{u}_D$.

5.5.5.5 Reliability Computation by Using Copulas

Having a multiple cost function

$$\gamma_k = \gamma_k(t, y(t), u(t), (a, x)), k = 1, \ldots, r, \tag{5.77}$$

the objective function of the control problem may be defined, see (5.38a), by the maximum of the individual total costs along the time interval $[t_0, t_f]$:

$$g(y(\cdot), u(\cdot), (a, x)) := \max_{1 \leq k \leq r} g_k(y(\cdot), u(\cdot), (a, x))$$

$$= \max_{1\le k\le r}\left(\int_{t_0}^{t_f}\gamma_k\big(t, y(t), u(t), (a, x)\big)dt + \psi_k(t_f, y(t_f), (a, x))\right). \tag{5.78a}$$

Consequently, after discretization, the operating condition $g(y(\cdot), u(\cdot), (a, x)) \le g_{max}$ reads, cf. (5.4a–c),

$$\max_{1\le k\le r} g_{kD}(t_D, h_D, y_0, y_D, u_D, (a, x)) \le g_{\max}. \tag{5.78b}$$

In control problems, dynamic systems we can have certain integral constraints

$$\int_{t_0}^{t_f}\eta_k\big(t, y(t), u(t), (a, x)\big)dt \le \eta_{k,\max}, k = 1, \dots, r. \tag{5.78c}$$

After discretization we get again conditions of the type (5.78b):

$$\eta_{kD}(t_D, h_D, y_0, y_D, u_D, (a, x)) \le \eta_{k,\max}, k = 1, \dots, r. \tag{5.78d}$$

Furthermore, also pointwise constraints may occur:

$$\zeta_k\big(t_j, y(t_j), u(t_j), (a, x)\big) \le \zeta_{j,\max}, j \in J, \tag{5.78e}$$

with a certain set $J \subset \{1,2, \dots, N\}$, which are special constraints of the type (5.78b,d).

Hence, for simplification, suppressing the other arguments in (5.77–5.78e) in the following we represent the above conditions by systems of constraints involving a (joint) random vector $b = b(\omega)$ and a (joint) decision vector $w = (x, u_D)$:

$$h_k(b(\omega), w) \le h_{k,\max}, k = 1, \dots, r, \tag{5.78f}$$

with given functions $h_k = h_k(b, w)$, $k = 1, \dots, r$.

Note 5.4 If the approximate control u_D does not belong to the outer decision variables, then u_D is just an inner parameter, see (5.69a), or $u_D = u_D(\omega)$ belongs to the stochastic variables, see (5.76a–c).

In many applications one has then to determine and/or to maximize the joint probability of safety/survival:

$$p_s(w) := P\big(h_k(b(\omega), w)) \le h_{k,\max}, k = 1, \dots, r, \big) \tag{5.79a}$$

that the conditions (5.78f) hold. Consider the separated probabilities

$$p_{s,k}(w) = F(h_{k,\max}; w) := P\big(h_k(b(\omega), w) \le h_{k,\max}\big), k = 1, \dots, r, \tag{5.79b}$$

and the copula

$$C = C(p_1, p_2, \ldots, p_r), 0 \le p_k \le 1, k = 1, \ldots, r, \tag{5.79c}$$

representing the stochastic dependence or correlations of the one-dimensional random variables $h_k = h_k(b(\omega), w), k = 1, 2, \ldots, r$. The probability of safety $p_s = p_s(w)$ can be represented by

$$p_s(w) = C\big(F(h_{1,\max}; w), \ldots, F(h_{r,\max}; w)\big) \tag{5.79d}$$

with the marginal distribution functions $F = F(h_{k,\max}; w)$ of the $1 - dim$ random variables $h_k = h_k(b(\omega), w), k = 1, \ldots, r$. Consider now the set

$$PoS := \big\{PoS(w) : w \in W\big\} \tag{5.80a}$$

with

$$PoS(w) := \begin{pmatrix} F(h_{1,\max}; w) \\ F(h_{2,\max}; w) \\ \vdots \\ F(h_{r,\max}; w) \end{pmatrix} \tag{5.80b}$$

of marginal probabilities of safety, where W denotes the domain of w. For the maximization $\max_{w \in W} p_s(w)$ of the probability of safety $p_s = p_s(w)$ on W we have the following result:

Theorem 5.10 *Suppose that the copula $C = C(p)$ is strictly monotone increasing in each of its variables $p_k, k = 1, \ldots, r$. Moreover, assume that the copula $C(\cdot)$, describing the correlations of the $1 - dim$ random variables $h_k(\cdot), k = 1, \ldots, r$, is independent of the decision variable w. If w^* is an optimal solution of $\max_{w \in W} p_s(w)$, then $PoS(w^*)$ is a Pareto-optimal point of PoS.*

Proof Let w^* be an optimal solution of $\max_{w \in W} p_s(w)$. Assuming that $PoS(w^*)$ is not a Pareto-optimal point of PoS, then there is a point $\tilde{w} \in W$ such that

$$F(h_{j,\max}; w^*) \le F(h_{j,\max}; \tilde{w}), j = 1, \ldots, r,$$

$$F(h_{j,\max}; w^*) < F(h_{j,\max}; \tilde{w}) \text{ for an index } j = j_0.$$

According to (5.79d, 5.80a) and the assumption on C we get the contradiction $p_s(w^*) < p_s(\tilde{w})$. □

Sufficient conditions for the condition in the above theorem that the copula $C(\cdot)$ is independent if of the variables w are given in the following.

Lemma 5.4

(a) *A copula $C(\cdot)$ related to the performance or cost functions $h_k, k = 1, \ldots, r$, is independent of the variables w if this representation holds*

$$h_k = h_k(b, w) = h_k^{(I)}(b) + h_k^{(II)}(w), k = 1, \ldots, r, \tag{5.81a}$$

with certain functions $h_k^{(I)}, g_k^{(II)}, k = 1, \ldots, r$.

(b) *If the performance or cost functions $h_k, k = 1, \ldots, r$, are approximated, $h_k(b, w) \approx \tilde{h}_k(b, w)$ by the semi-linearizations*

$$\tilde{h}_k(b, w) := h_k(\bar{b}, w) + \nabla_b h_k(\bar{b}, w_0)^T (b - \bar{b}) \tag{5.81b}$$

of $h_k, k = 1, \ldots, r$, at $(\bar{b}, w_0)$, then the copula with respect to the approximates $\tilde{h}_k, k = 1, \ldots, r$, is independent of w.

Proof In case (a) the condition (5.78f) can be represented by

$$h_k^{(I)}(b(\omega) \leq h_{k,\max} - h_k^{(II)}(w), k = 1, \ldots, r. \tag{5.81c}$$

Hence, the $C(\cdot)$ is determined only by the random variables $h_k^{(I)}(b(\omega), k = 1, \ldots, r$, which are independent of w.

In case (b) for condition (5.78f) we get

$$\nabla_b h_k(\bar{b}, w_0)^T (b - \bar{b}) \leq h_{k,\max} - h_k(\bar{b}, w), k = 1, \ldots, r. \tag{5.81d}$$

Thus, also in this case the random variables on the left side of (5.81d) are independent of w. □

For the accuracy of the semi-linearization (5.81b) we have this result:

Lemma 5.5 *Let us denote $R_k = R_k(\bar{b}, w_0; \Delta b, \Delta w)$, $R_{k,\bar{b}} = R_{k,\bar{b}}(w_0; \Delta w)$, resp., denote the remainder term of the first order Taylor expansion of $h_k = h_k(b, w)$, $h_k = h_k(\bar{b}, w)$ at $(\bar{b}, w_0)$, w_0, respectively. Then,*

$$\begin{aligned} &h_k(\bar{b} + \Delta b, w_0 + \Delta w) - \tilde{h}_k(\bar{b} + \Delta b, w_0 + \Delta w) \\ &= R_k(\bar{b}, w_0; \Delta b, \Delta w) - R_{k,\bar{b}}(w_0; \Delta w), k = 1, \ldots, r. \end{aligned} \tag{5.82a}$$

The difference between the errors of the (full) linearization and the semi-linearization can be evaluated by the above lemma as follows:

Corollary *Denoting by $\Delta^L h_k, \Delta^{SL} h_k$, resp., the errors of the full, semi-linearization, resp., of h_k at $(\bar{b}, w_0)$, then for all (b, w) it holds*

$$\Delta^L h_k(b, w) - \Delta^{SL} h_k(b, w) = R_{k,\bar{b}}(w_0; \Delta w), k = 1, \ldots, r. \tag{5.82b}$$

Some special cases are shown next:

Example 5.3 (a) If h_k is linear in its argument (b, w), then $h_k = \tilde{h}_k$ and $\Delta^L h_k = \Delta^{SL} h_k$. (b) For a quadratic function h_k with the Hessian

$$\nabla^2 h_k(b, w) = \begin{pmatrix} Q_{bb} & Q_{wb} \\ Q_{bw} & Q_{ww} \end{pmatrix} \tag{5.82c}$$

we get $\Delta^L h_k(b, w) - \Delta^{SL} h_k(b, w) = \frac{1}{2}\Delta w^T Q_{ww} \Delta w$. c) If h_k is convex in w, then $\Delta^L h_k(b, w) \geq \Delta^{SL} h_k(b, w)$ for all (b, w)

5.5.5.6 Approximations of *PoS*

Approximations of the set PoS of marginal probabilities of safety can be obtained by using the Markov inequality: Given monotonous increasing functions $\phi_j = \phi_j(t), t \in \mathbb{R}, j = 1, \ldots, r$, for each index j we have

$$\begin{aligned} F(h_{j,\max}; w) &= P(h_j(b(\omega), w) \leq h_{j,\max}) \\ &\geq 1 - \frac{1}{\phi_j(h_{j,\max})} E\phi_j(h_j(b(\omega), w)). \end{aligned} \tag{5.83a}$$

Consequently, (5.83a) yields this result:

Theorem 5.11 *For given monotonous increasing functions $\phi_j = \phi_j(t), t \in \mathbb{R}, j = 1, \ldots, r$, PoS can be approximated from below as follows:*

$$PoS \geq \left\{ \begin{pmatrix} 1 - \frac{1}{\phi_1(h_{1,\max})} E\phi_1(h_1(b(\omega), w)) \\ 1 - \frac{1}{\phi_2(h_{2,\max})} E\phi_2(h_2(b(\omega), w)) \\ \vdots \\ 1 - \frac{1}{\phi_r(h_{r,\max})} E\phi_r(h_r(b(\omega), w)) \end{pmatrix}, w \in W \right\}, \tag{5.83b}$$

where the inequality corresponds to the representation of the two sets by means of the vectors $w \in W$.

Note 5.5 If the performance or cost functions $\phi_j = \phi_j(t), t \in \mathbb{R}, j = 1, \ldots, r$, are nonnegative, then also functions $\phi(t) = t, t^2, etc.$, can be taken.

References

1. Basudhar, A., Missoum, S., Harrison Sanchez, A.: Limit State Function Identification Using Support Vector Machines for Discontinuous Responses and Disjoint Failure Domains, pp. 1–11. Elsevier, Amsterdam (2008). https://doi.org/10.1016/j.probengmech.2007.08.004

2. van Brunt, B.: The Calculus of Variations. Springer, New York (2006)
3. Cocchetti, G., Maier, G.: Static shakedown theorems in piecewise linearized poroplasticity. Arch. Appl. Mech. **68**, 651–661 (1998). https://doi.org/10.1007/s004190050194
4. Gavin, H.P. Yan, S.: High-order limit state functions in the response surface method for structural reliability analysis. Struct. Saf. **30**(2), 162–179 (2008). https://doi.org/10.1016/j.strusafe.2006.10.003
5. Gerdts, M.: Optimal Control of ODES and ADES. De Gruyter, Berlin (2012)
6. Hardy, G., Littlewood, J., Pólya, G.: Inequalities. Cambridge Univ. Press, London (1973)
7. Kaliszky, S.: Plasticity: Theory and Engineering Applications. Elsevier, Amsterdam (1989)
8. Lewis, F.: Applied Optimal Control and Estimation. Prentice-Hall Inc., Englewood Cliffs (1992)
9. Marti, K.: Stochastic optimization methods in optimal engineering design under stochastic uncertainty. ZAMM **83**(11), 1–18 (2003)
10. Marti, K.: Stochastic Optimization Methods, 2nd edn. Springer, Berlin (2008). https://doi.org/10.1007/978-3-540-79458-5
11. Marti, K.: Stochastic Optimization Methods: Applications in Engineering and Operations Research, 3rd edn. Springer, Berlin (2015). https://doi.org/10.1007/978-3-662-46214-0
12. Neal, B.: The Plastic Methods of Structural Analysis. Chapman and Hall, London (1965)
13. Proppe, C.: Estimation of failure probabilities by local approximation of the limit state function. Struct. Saf. **30**, 277–290 (2008). https://doi.org/10.1016/j.strusafe.2007.04.001
14. Rozvany, G.: Structural Design via Optimality Criteria. Kluwer, Dordrecht (1989)
15. Schuëller, G., Gasser, M.: Some Basic Principles of Reliability-Based Optimization (RBO) of Structure and Mechanical Components. Lecture Notes in Economics and Mathematical Systems (LNEMS), vol. 458,pp. 80–103. Springer, Berlin (1998)
16. Schwarz, H., Köckler, N.: Numerische Mathematik, 8th edn. Vieweg und Teubner, Wiesbaden (2011)
17. Soong, T.: Active Structural Control: Theory and Practice. John Wiley, New York (1990)
18. Soong, T., Costantinou, M.: Passive and Active Structural Vibration Control in Civil Engineering. CISM Courses and Lectures, vol. 345. Springer, Wien (1994)
19. Straub, D. (ed.): Reliability and Optimization of Structural Systems. CRC Press, Taylor and Francis, London (2010)
20. Thoft-Christensen, P., Murotsu, Y.: Application of Structural Systems Reliability Theory. Springer, Berlin (1986)
21. Tin-Loi, F.: On the optimal plastic synthesis of frames. Eng. Optim. **16**, 91–108 (1990)
22. Valentine, F.: Convex Sets. McGraw-Hill, Inc., New York (1964)
23. Yang, J., Soong, T.: Recent advances in active control of civil engineering structures. Probab. Eng. Mech. **3**(4), 179–188 (1988). https://doi.org/10.1016/0266-8920(88)90010-0
24. Zabczyk, J.: Mathematical Control Theory, 2nd edn. Birkhäuser, Basel (1995)

Part II
Optimization by Stochastic Methods: Foundations and Optimal Control/Acceleration of Random Search Methods (RSM)

Chapter 6
Random Search Procedures for Global Optimization

6.1 Introduction

Solving optimization problems from engineering, as, e.g., parameter—or process—optimization problems

$$\min F(x) \text{ s.t. } x \in D, \tag{6.1}$$

where D is a subset of $\mathbb{R}^n$, one meets often the following situation:

(a) One should find the **global** optimum in (6.1), hence most of the deterministic programming procedures, which are based on local improvements of the performance index $F(x)$, will fail.
(b) Concerning the objective function F one has a **blackbox**—situation, i.e. there is only few a priori information about the structure of F, especially there is no knowledge about the direct functional relationship between the control or input vector $x \in D$ and its index of performance $F(x)$; hence—besides the more or less detailed a priori information about F—the only way of getting objective information about the structure of F is via evaluations of its values $F(x)$ by experiments or by means of a numerical procedure simulating the technical plant.

Consequently, engineers use in these situations often a certain search procedure for finding an optimal vector x, see, e.g., Box' EVOP method in [2] and the random search methods as first proposed by Anderson [1], Brooks [3], and Karnopp [6]. Obviously, deterministic search methods can be considered as special stochastic ones. Further classes of random search procedures are simulated annealing (SA) procedures [7], genetic algorithms (GA) [4], and further nature-oriented and evolution-type search methods [12]. For further information, see [11].

K. Marti, *Optimization Under Stochastic Uncertainty*, International Series in Operations Research & Management Science 296,
https://doi.org/10.1007/978-3-030-55662-4_6

An important property of (SA) methods is that also non-improving search steps are possible—with decreasing probability (cooling). In (GA) procedures, mutation, crossover and selection rules are applied to generate a sequence of sets (population) of points. Random changes occur during mutations. The optimal solution is then approximated by the sequence of best points in the current population.

In the basic random search routine considered in this section—allowing not only local improvements as in mathematical programming—a sequence of n-random vectors $X_0, X_1, \ldots, X_t, \ldots$ in D is constructed according to the following recurrence relation:

$$X_{t+1} := \begin{cases} z_{t+1}, z_{t+1} \in D \text{ and } F(z_{t+1}) < F(X_t) \\ X_t, \text{ if } z_{t+1} \notin D \text{ or } F(z_{t+1}) \geq F(X_t), \end{cases} \tag{6.2a}$$

$t = 0, 1, 2, \ldots$, where the starting point $X_0 := x_0$ is a realization x_0 of the random vector X_0 having the given distribution $P_{X_0} := \pi_{\text{start}}$ concentrated on the domain D_{start}. In many cases we have $D_{\text{start}} \subset D$. If the search process starts at a given, fixed point x_0, then $\pi_{\text{start}} = \varepsilon_{x_0}$, where ε_{x_0} denotes the one-point measure at the point x_0. Moreover, $z_1 = Z_1(\omega), z_2 = Z_2(\omega), \ldots$ are realizations of n-random vectors $Z_1, Z_2, \ldots$ such that

$$\begin{aligned} &P\big(Z_{t+1} \in B | X_0 = x_0, X_1 = x_1, \ldots, X_t = x_t, Z_1 = z_1, Z_2 = z_2, \ldots, Z_t = z_t\big) \\ &\quad = P\big(Z_{t+1} \in B | X_0 = x_0, X_1 = x_1, \ldots, X_t = x_t\big) = \pi_t(x^t, B), \end{aligned} \tag{6.2b}$$

where $x^t := (x_0, x_1, \ldots, x_t)$ and $\pi_t(x^t, \cdot)$ is a given transition probability distribution, as, e.g., a joint normal distribution with mean x_t and covariance matrix $Q = Q_t$.

According to Definition (6.2a), given the states $X^t = x^t$, first an n-vector z_{t+1} is generated randomly according to the distribution $\pi_t(x^t, \cdot)$. Then, if z_{t+1} drops into the area of success $G_F(x_t)$, where

$$G_F(x) := \big\{y \in D : F(y) < F(x)\big\}, \tag{6.3}$$

we move to $X_{t+1} = z_{t+1}$, otherwise we stay at $X_{t+1} = X_t$. Thus, the whole search process X_t stays within the union $D_0 := D_{\text{start}} \cup D$ of the domain of the starting points and the feasible domain of the basic optimization problem (6.1). If the set of starting points $D_{\text{start}} \subset D$ is contained in the feasible domain D, then $D_0 = D$.

Moreover, we observe that if $X_{t+1} \in G_F(x_t)$, then also $X_s \in G_F(x_t)$ for all $s > t$. Furthermore, if $F^* = \inf\big\{F(x) : x \in D\big\}$ and, for given levels $\varepsilon > 0$, $M < 0$, resp., the set of $\varepsilon-$, $M-$optimal solutions of (6.1) is defined by

$$B_{\varepsilon,M} := \big\{y \in D : F(y) \leq F^* + \varepsilon, \text{if } F^* \in \mathbb{R}, F(y) \leq M, \text{if } F^* = -\infty, \text{resp.}\big\}, \tag{6.4a}$$

then

$$X_s \in B_{\varepsilon,M} \Rightarrow X_{s+1} \in B_{\varepsilon,M}, s = 0, 1, 2, \ldots. \tag{6.4b}$$

Hence,

$$P(X_s \in B_{\varepsilon,M}) \leq P(X_{s+1} \in B_{\varepsilon,M}), s = 0, 1, 2, \ldots. \tag{6.4c}$$

In the following we assume that the objective function F of (6.1) is a measurable function on $\mathbb{R}^n$.

6.2 The Convergence of the Basic Random Search Procedure

For considering the convergence behavior of the search method (6.2a), we examine the probability

$$P(X_t \in B_{\varepsilon,M}), t = 0, 1, 2, \ldots,$$

that the t-th iterate X_t is an ε-, M-optimal solution, resp., of (6.1), where $\varepsilon > 0$, $M < 0$ are given numbers. According to the considerations at the end of Sect. 6.1 these probabilities form a nondecreasing, convergent sequence, and due to (6.4b) we have that

$$X_t \notin B_{\varepsilon,M} \Leftrightarrow X_0 \notin B_{\varepsilon,M}, X_1 \notin B_{\varepsilon,M}, \ldots, X_t \notin B_{\varepsilon,M}, \tag{6.5a}$$

hence,

$$\begin{aligned} P(X_t \in B_{\varepsilon,M}) &= 1 - P(X_0 \notin B_{\varepsilon,M}, X_1 \notin B_{\varepsilon,M}, \ldots, X_t \notin B_{\varepsilon,M}) \\ &= 1 - \int_{x_0 \notin B_{\varepsilon,M}} P(X_1 \notin B_{\varepsilon,M}, \ldots, X_t \notin B_{\varepsilon,M} | X_0 = x_0) \pi_t(dx_0). \end{aligned} \tag{6.5b}$$

Denoting by $K_t(x^t, \cdots)$ the conditional distribution of X_{t+1} given $X_0 = x_0, X_1 = x_1, \ldots, X_t = x_t$, we have

$$K_t(x^t, B) = \pi_t\big(x^t, B \cap G_F(x_t)\big) + \Big(1 - \pi_t(\big(x^t, G_F(x_t)\big)\Big)\varepsilon_{x_t}(B), \tag{6.6a}$$

where ε_x is the one-point-measure at x. Thus, with $\overline{B}_{\varepsilon,M} := D_0 \setminus B_{\varepsilon,M}$, we get

$$P(X_1 \notin B_{\varepsilon,M}, \ldots, X_t \notin B_{\varepsilon,M} | X_0 = x_0) = \int_{x_1 \in \overline{B}_{\varepsilon,M}} K_0(x_0, dx_1) \ldots$$

$$\times \cdot \int\limits_{x_{t-1}\in\overline{B}_{\varepsilon,M}} K_{t-2}(x^{t-2}, dx_{t-1}) \cdot \int\limits_{x_t\in\overline{B}_{\varepsilon,M}} K_{t-1}(x^{t-1}, dx_t). \tag{6.6b}$$

Considering first the t-th integral in the above equation, we obtain

$$\int\limits_{x_t\in\overline{B}_{\varepsilon,M}} K_{t-1}(x^{t-1}, dx_t) = K_{t-1}(x^{t-1}, D_0 \setminus B_{\varepsilon,M}) \tag{6.6c}$$

$$= K_{t-1}(x^{t-1}, D_0) - K_{t-1}(x^{t-1}, B_{\varepsilon,M})$$
$$= 1 - K_{t-1}(x^{t-1}, B_{\varepsilon,M}).$$

Suppose now that $D_0 = D$. Having $x_{t-1} \notin B_{\varepsilon,M}$, we get $\varepsilon_{x_{t-1}}(B_{\varepsilon,M}) = 0$ and $B_{\varepsilon,M} \subset G_F(x_{t-1})$, see the definitions (6.3), (6.4a). Hence, (6.6a,c) yield

$$\int\limits_{x_t\in\overline{B}_{\varepsilon,M}} K_{t-1}(x^{t-1}, dx_t) \le 1 - \pi_{t-1}(x^{t-1}, B_{\varepsilon,M}) \tag{6.7}$$

$$\le 1 - \inf\left\{\pi_{t-1}(x^{t-1}, B_{\varepsilon,M}) : x_s \in D_0 \setminus B_{\varepsilon,M}, 0 \le s \le t-1\right\}$$

for all $x_s \in D_0 \setminus B_{\varepsilon,M}, s = 0.1, \ldots, t-1$.

Defining now $\alpha_t, t = 0, 1, \ldots,$ by

$$\alpha_t := \alpha_t(B_{\varepsilon,M}) = \inf\left\{\pi_t(x^t, B_{\varepsilon,M}) : x_s \in D_0 \setminus B_{\varepsilon,M}, 0 \le s \le t\right\}, \tag{6.8}$$

from (6.6a,b) we obtain now

$$P(X_1 \notin B_{\varepsilon,M}, \ldots, X_t \notin B_{\varepsilon,M} | X_0 = x_0) \le \prod_{s=0}^{t-1}(1 - \alpha_s(B_{\varepsilon,M})) \tag{6.9a}$$

Hence, by (6.5b) and (6.9a) it is

$$P(X_t \in B_{\varepsilon,M})$$
$$= 1 - \int\limits_{x_0\notin B_{\varepsilon,M}} P(X_1 \notin B_{\varepsilon,M}, \ldots, X_t \notin B_{\varepsilon,M} | X_0 = x_0)\pi_{\text{start}}(dx_0)$$

$$\ge 1 - (1 - \pi_{\text{start}}(B_{\varepsilon,M})) \prod_{s=0}^{t-1}(1 - \alpha_s(B_{\varepsilon,M})). \tag{6.9b}$$

Since $\log u \le u - 1$, for $u > 0$ we have

$$(1 - \pi_{\text{start}}(B_{\varepsilon,M})) \prod_{s=0}^{t-1}(1 - \alpha_t) \leq \exp\left(-\pi_{\text{start}}(B_{\varepsilon,M}) - \sum_{s=0}^{t-1} \alpha_s\right) \tag{6.9c}$$

and therefore also

$$P(X_t \in B_{\varepsilon,M}) \geq 1 - \exp\left(-\pi_{\text{start}}(B_{\varepsilon,M}) - \sum_{s=0}^{t-1} \alpha_s\right). \tag{6.9d}$$

Thus, from (6.9d) we get the following convergence result:

Theorem 6.1 *Suppose now that* $D_0 = D$. *The search process (6.2a) has the following convergence properties:*

(a) *If for an* $\varepsilon > 0$, $M < 0$, *resp.,*

$$\sum_{s=0}^{\infty} \alpha_s(B_{\varepsilon,M}) = +\infty, \tag{6.10}$$

then $\lim_{t\to\infty} P(X_t \in B_{\varepsilon,M}) = 1$.

(b) *Suppose that* $F^* \in \mathbb{R}$ *and*

$$\lim_{n\to\infty} P(X_n \in B_\varepsilon) = 1 \textit{ for every } \varepsilon > 0. \tag{6.11}$$

Then $\lim_{n\to\infty} F(X_n) = F^*$ *w.p.1 (with probability one),*

(c) *Assume that* $F^* \in \mathbb{R}$ *and* F *is continuous and that the level sets* D_ε *are nonempty and compact for each* $\varepsilon > 0$. *Then* $\lim_{t\to\infty} F(X_t) = F^*$ *implies that also* $\lim_{t\to\infty}$ *dist* $(X_t, D^*) = 0$, *where dist* (X_t, D^*) *denotes the distance between* X_t *and the set* D^* *of global minimum points of (6.1).*

Proof For a proof of assertions (b) and (c), see [9]. □

Note 6.1

(a) For the case that the distribution π_t of Z_{t+1} does not depend on the states x^t, preliminary versions of the decisive inequality (6.9a) may already be found in the early Random Search literature, see, e.g., [3].
(b) Comparing the above theorem with the 0-1-laws of probability theory we observe that this result is essentially a consequence from the Borel–Cantelli-type laws, see, e.g., [10, p. 400] and [5, p.1–6, 51–52].
(c) If the feasible domain D can be reached also from starting points x_0 outside D, then the above result holds also.

Working with random search procedures, one observes that the rate of convergence—especially near to the optimum—may be very poor. Hence, in the following we consider a modified random search procedures with an improved convergence behavior.

6.2.1 Discrete Optimization Problems

Consider now the case that D contains a finite number r of elements $d_i \in \mathbb{R}^n$, thus,

$$D = \{d_1, d_2, \dots, d_r\}. \tag{6.12a}$$

Furthermore, assume that $D_{\text{start}} \subset D$, hence $D_0 = D$, and let

$$P\big(Z_{t+1} \in B | X_0, X_1, \dots, X_t\big) = P\big(Z_{t+1} \in B | X_t\big). \tag{6.12b}$$

Hence, (Z_t) and (X_t) are discrete-time stochastic processes. Therefore (Z_t) is described by a transition matrix (π^t_{ij}) from $X_t = i$ to $Z_{t+1} = j$, and the iterates (X_t) are described by the transition matrix (p^t_{ij}) from $X_t = i$ to $X_{t+1} = j, t = 0, 1, \dots$. For $X_t = d_i$ we have $G_F(d_i) = G_F(i) = \big\{j : F(d_j) < F(d_i)\big\}$. The relationship between (π^t_{ij}) and (p^t_{ij}) reads then:

$$p^t_{ij} = p^{(t,t+1)}_{ij} = \begin{cases} 0, & \text{if } j \notin G_F(i) \text{ and } j \neq i \\ 1 - \sum\limits_{l \in G_F(i)} \pi^t_{il}, & \text{if } j = i \\ \pi^t_{ij}, & \text{if } j \in G_F(i). \end{cases} \tag{6.12c}$$

Assuming now stationary search variables $Z_t(\omega)$, i.e. in case $\pi^t_{ij} = \pi_{ij}$ for all $t = 0, 1, \dots$, then also (X_t) is stationary and by searching for stationary distributions of (p_{ij}) we get this result:

Theorem 6.2 *Let $\pi_{ij} > 0$ for all $i, j = 1, \dots, r$ or suppose that $\sum\limits_{j \in G_F(i)} \pi_{ij} > 0$ for all $1 \leq i \leq r$ such that d_i is not a solution of the optimization problem (6.1). Then $\big(X_t(\omega)\big)$ converges with probability one to a solution of problem (6.1).*

Proof Without limitation we may assume here that $0 < \varepsilon < F_{max} - F^*$. According to (6.8) and the above assumptions, for the minimum probabilities $\alpha_t = \alpha_t(\varepsilon)$ we have

$$\alpha_t(\varepsilon) = \inf\big\{\pi_t(x^t, B_\varepsilon) : x_s \in D \setminus B_\varepsilon, 0 \leq s \leq t\big\} = \inf\big\{\pi(x_t, B_\varepsilon) : x_t \in D \setminus B_\varepsilon\big\}$$

$$= \inf\left\{\sum_{d_j \in B_\varepsilon} \pi_{ij} : d_i \notin B_\varepsilon\right\} =: \alpha_0 > 0, \tag{6.12d}$$

provided that $\pi_{ij} > 0$ for all indices i, j. Since $\alpha_t(\varepsilon) = \alpha_0 > 0$ for all $t = 0, 1, \dots,$ and each ε, $0 < \varepsilon < F_{max} - F^*$, the assertion follows now from Theorem 6.1. Since in the present case there are a finite number elements of feasible points $d_i, i \dots, r$, and the sets $G_F(i)$, B_ε are contained in each other for corresponding values of ε, $F(d_i)$, resp., the proof for the second case follows then also from (6.12d). □

Example 6.1 For illustration we may assume—without limitation—that the elements $d_1, d_2, \ldots, d_r$ of the feasible domain D are arranged such that

$$F(d_1) < F(d_2) < \ldots < Fd_r). \tag{6.13a}$$

Hence, d_1 is the unique minimum point, and the remaining points d_j are arranged in strictly increasing order of the function values $F(d_j)$. With the stationary transition probabilities $\pi_{ij}^{(t,t+1)} = \pi_{ij}$ from X_t to Z_{t+1}, the stationary transition matrix $P^t = P := (p_{ij})$ from X_t to X_{t+1} reads

$$P = \begin{pmatrix} 1 & 0 & 0 & \ldots 0 \\ \pi_{21}^t & 1 - \pi_{21}^t & 0 & \ldots 0 \\ \pi_{31}^t & \pi_{32}^t & 1 - (\pi_{31}^t + \pi_{32}^t) & \ldots 0 \\ \vdots & \vdots & \vdots & \ddots \vdots \\ \pi_{r1}^t & \pi_{r2}^t & \pi_{r3}^t & \ldots \pi_{rr}^t \end{pmatrix}. \tag{6.13b}$$

Corresponding to Theorem 6.2 we find that $q^T := (1, 0, \ldots, 0)$ is a left fixed point of P and $\lim_{t\to\infty} X_t = q$ with probability 1.

6.3 Adaptive Random Search Methods

In this section we describe a general method how to find search variables (Z_t) such that the convergence of (X_t) towards a solution of our basic problem (6.1) is accelerated. This can be achieved by an adaptive selection of the probability distribution of the search variates $Z_1, Z_2, \ldots$. In order to control the sequence (Z_t) we assume that the probability distribution

$$\pi_t(x_0, x_1, \ldots x_t, \cdot) = \pi_t(a_t, x_0, x_1, \ldots, x_t, \cdot) \tag{6.14a}$$

of Z_{t+1} depends on a control parameter vector $a_t \in A_t(x_0, x_1, \ldots, x_t)$, where $A_t \subset A$ is the set of admissible controls at time t and given state-history $x^t := (x_0, x_1, \ldots, x_t)$. Moreover A_t is assumed to be contained in a fixed set A. By

$$\delta = (\delta_t)_{t\geq 0}, \delta_t : \mathbb{R}^{n(1+t)} \to A, t = 0, 1, \ldots \tag{6.14b}$$

we denote a decision rule, composed of the control functions or strategies $\delta_t, t = 0, 1, \ldots$, such that the control parameter vectors a_t are given by

$$a_t := \delta_t(x^t) \in A_t(x^t) \text{ for } x_s \in D, 0 \leq s \leq t, t = 0, 1, \ldots. \tag{6.14c}$$

The set Δ of admissible decision rules δ is defined then by

$$\Delta := \{\delta : \delta = (\delta_t)_{t\geq 0}, \delta_t(x_0, x_1, \ldots, x_t) \in A_t(x_0, x_1, \ldots, x_t)$$
$$\text{for } x_s \in D, 0 \leq s \leq, t = 0, 1, \ldots\}. \tag{6.14d}$$

Note 6.2 Since the transition probabilities $\pi_t(a_t, x_0, x_1, \ldots, x_t, \cdot)$ depend on the controls a_t, the expectation operator $E = E^\delta$ depends on the decision rule δ.

Looking for an *optimal* decision rule δ^*, clearly we have to guarantee that the process (X_t) generated by δ^* converges actually to a solution of (6.1).

Note that the reachability property in Theorem 6.1 holds, e.g. if the decision rules satisfies the condition, see (6.8),

$$\sum_{t=0}^{\infty} \inf\Big\{\pi_t\big(\delta_t(x^t), x^t, B_\varepsilon\big) : x^t = (x_0, x_1, \ldots, x_t), x_s \in D \setminus B_\varepsilon, 0 \leq s \leq t\Big\} = +\infty, \tag{6.15a}$$

where cf. (6.4a)

$$B_\varepsilon := \{y \in D : F(y) \leq F^* + \varepsilon\}. \tag{6.15b}$$

In the stationary case $\pi_t(a_t, x^t, \cdot) = \pi(a_t, x_t, \cdot)$ and $\delta_t(x^t) = \delta(x_t), t = 0, 1, \ldots$ (6.15a) is reduced to the much simpler condition

$$\inf\Big\{\pi\big(\delta(x), x, B_\varepsilon\big) : x \in D \setminus B_\varepsilon\Big\} > 0. \tag{6.15c}$$

Appropriate utility- or reward-criterion for the evaluation of the individual steps $X_t \to X_{t+1}$ of the search process (X_t) are, e.g.

(a) Probability of success

$$u_t(x_t, x_{t+1}) = \begin{cases} 1, & x_{t+1} \in G_F(x_t) \\ 0, & \text{otherwise,} \end{cases} \tag{6.16a}$$

hence $E^\delta\big(u_t(a_t, X_t, X_{t+1})|X_t\big) = P\big(X_{t+1} \in G(X_t)|X_t\big)$ is the (conditional) probability of a success in the state X_t.

(b) Step length

$$u_t(X_t, X_{t+1}) = \begin{cases} \|X_{t+1} - X_t\|^p, & X_{t+1} \in G_F(X_t) \\ 0, & \text{otherwise,} \end{cases} \tag{6.16b}$$

where $p \geq 1$ is a fixed number. Here $E\big(u_t(a_t, X_t, X_{t+1}|X_t\big)$ is the average step length of X_{t+1} into the area of success $G(X_t)$.

A modification of the above case is

(c) Relative step length

$$u_t(X_t, X_{t+1}) = \begin{cases} \left(\frac{\|X_{t+1}-X_t\|}{\|X_t\|}\right)^p & , X_{t+1} \in G_F(X_t) \\ 0 & , \text{otherwise.} \end{cases} \tag{6.16c}$$

Obviously, any linear combination of the above three criterion yields a further criterion. In the following we suppose

$$F^* = \inf\{F(x) : x \in D\} > -\infty.$$

Search procedures with an improved performance can be constructed now by maximizing [8] the expected (total) reward function

$$U_\infty(x_0, \delta) := E^\delta \sum_{s=0}^{\infty} \varrho^s u_s(\delta_s(x^s), x_s, x_{s+1}), \tag{6.17a}$$

with respect to the decision rule $\delta = (\delta_s)$ involving the control functions δ_s satisfying the constraints (6.14c). Here, $\varrho, 0 < \varrho < 1$, denotes still a certain *discount factor*.

For the maximization of the expectation $U_\infty(x_0, \delta)$ next to we consider the $(T-t)$-stage search processes (X_s) starting at time t and running then up to time $T > t$. Hence, with $X^s = (X_0, X_1, \ldots, X_s), X_0 := x_0$, and $a_s = \delta(X^s), s = t, t+1, \ldots, T-1$, let

$$\begin{aligned} &U_T(t, x^t; \delta_t, \ldots, \delta_{T-1}) \\ &:= E^\delta\Big(\sum_{s=t}^{T-1} \varrho^{(s-t)} u_s(\delta_s(X^s), X_s, X_{s+1}) | X_0 = x_0, X_1 = x_1, \ldots, X_t = x_t\Big) \end{aligned} \tag{6.17b}$$

denote the conditional expected reward of this $(T-t)-$stage process, given the time history $X_s = x_s, s = 0, 1, \ldots, t$, and the control functions $\delta_t, \ldots, \delta_{T-1}$.

Denote by $K_t(a_t, x^t, \cdot)$ the transition probabilities for $X_t \to X_{t+1}$

$$K_t(a_t, x^t, B) = P(X_{t+1} \in B | X^t = x^t) \tag{6.18a}$$

of the process (X_t) based on search variates (Z_t) controlled by control inputs $a_t, t = 0, 1, \ldots$, where $X^t = (X_0, X_1, \ldots, X_t)$ and B is any Borel subset of $\mathbb{R}^n$. According to the basic definition (6.2a,b) of X_t and (6.14a-d) it holds

$$K_t(x^t, B) = K_t(a_t, x^t, B) = \pi_t\big(a_t, x^t, B \cap G(x_t)\big) + \Big(1 - \pi_t\big(a_t, x^t, G(x_t)\big)\Big)\varepsilon_{x_t}(B), \tag{6.18b}$$

cf. (6.6a), where $a_t = \delta_t(x^t)$, and ε_x denotes again the one-point measure at the point $x \in \mathbb{R}^n$.

Due to the above definitions, the reward functions $U_T(t, x^t; \delta_t, \ldots, \delta_{T-1})$, $t = 0, 1 \ldots, T-2, T-1$, see (6.17b), satisfy the recurrence rations

$$U_T(t, x^t; \delta_t, \ldots, \delta_{T-1}) = \int \Big(u_t\big(\delta_t(x^t), x_t, x_{t+1}\big) + \varrho U_T\big(t+1, (x^t, x_{t+1}); \delta_{t+1}, \ldots, \delta_{T-1})\big)\Big) K_t(\delta_t(x^t), dx_{t+1}) = \bar{u}_t(\delta_t(x^t), x^t) + \varrho \int U_T\big(t+1, (x^t, y); \delta_{t+1}, \ldots, \delta_{T-1})\big) K_t(\delta_t(x^t), x^t, dy), \tag{6.19a}$$

where

$$\bar{u}_t(a_t, x^t) = E\big(u_t(a_t, X_t, X_{t+1}) | X^t = x^t\big) = \int u_t(a_t, x_t, y) K_t(a_t, x^t, dy). \tag{6.19b}$$

With the set Δ of admissible decision rules δ, cf. (6.14d), the value function of the $(T-t)$-stage process $X_t, \ldots, X_T$ with given state-history x^t is now defined by

$$V_t^T(x^t) := \sup_{\delta_t, \ldots, \delta_{T-1}} \big\{U_T(t, x^t; \delta_t, \ldots, \delta_{T-1}) : \delta_s(x^s) \in A_s(x^s), x^s \in D^{(s+1)}, t \le s \le T-1\big\}, \tag{6.20}$$

where $D^{(s+1)}$ denotes the $(s+1)$-fold Cartesian product of D.

As mentioned already above, the set of restrictions in (6.20) should also include a condition guaranteeing that the whole search process (X_t) controlled by the decision rule $\delta = (\delta_t)$ satisfies a reachability condition according to Theorem 6.1. However, in many practical problems this condition may be deleted since the optimal decision functions δ_t^* defined by the optimization problem (6.20) can be shown to generate a search process (X_t^*) fulfilling a sufficient reachability condition.

From (6.19a,b), for the value functions $V_t^T(x^t)$ we get then the following recurrence relation:

Theorem 6.3 *Let $V_T^T(x^T) = 0$ for all $x^T \in \mathbb{R}^{(T+1)n}$. If for all steps t under consideration the maximum is attained in (6.20), then the following backwards recurrence relation holds*

$$V_t^T(x^t) = \sup_{a \in A_t(x^t)} \int \Big(u_t(a, x_t, y) + V_{t+1}^T\big((x^t, y)\big)\Big) K_t(a, x^t, dy) \tag{6.21}$$

$$= \sup_{a \in A_t(x^t)} \Big(\bar{u}_t(a, x^t) + \int V_{t+1}^T(x^t, y) K_t(a, x^t, dy) \Big),$$

$t = T-1, T-2, \ldots, 1, 0,$ *where* $a = \delta_t(x^t)$.

Proof Omitting for simplification the constraint set in (6.20), from (6.19a,b) we get

$$\begin{aligned} V_t^T(x^t) :=& \sup_{\delta_t, \ldots, \delta_{T-1}} U_T(t, x^t; \delta_t, \ldots, \delta_{T-1}) \\ =& \sup_{\delta_t} \sup_{\delta_{t+1}, \ldots, \delta_{T-1}} U_T(t, x^t; \delta_t, \ldots, \delta_{T-1}) = \sup_{\delta_t} \sup_{\delta_{t+1}, \ldots, \delta_{T-1}} \Big(\bar{u}_t(\delta_t(x^t), x^t) \\ &+ \varrho \int U_T\big(t+1, (x^t, y); \delta_{t+1}, \ldots, \delta_{T-1}\big) K_t(\delta(x^t), x^t, dy) \Big) \\ =& \sup_{a_t \in A_t(x^t)} \Big(\bar{u}_t(a_t, x^t) + \varrho \sup_{\delta_{t+1}, \ldots, \delta_{T-1}} \\ &\times \int U_T\big(t+1, (x^t, y); \delta_{t+1}, \ldots, \delta_{T-1}\big) K_t(\delta(x^t), x^t, dy) \Big). \end{aligned} \tag{6.22a}$$

Now, according to (6.17b) we have

$$\begin{aligned} U_T(t+1, (x^t, y); \delta_{t+1}, \ldots, \delta_{T-1}) =& E_{x^t} U_T(t+1, (x^t, y), \delta_{t+1}(x^t, y), \\ & \delta_{t+2}(x^t, y, X_{t+2}), \ldots, \delta_{T-1}(x^t, y, X_{t+2}, \ldots, X_{T-1})), \end{aligned} \tag{6.22b}$$

where $E_{x^t, y}$ denotes the conditional expectation given $X^{t+1} = (x^t, y)$ and X_j are random vectors defined by (6.2a,b) and U_T is the expected sum in (6.17b). Taking now, cf. (6.22a), the integral in (6.22b) with respect to y and then the supremum with respect to $\delta_{t+1}, \ldots, \delta_{T-1}$ under the constraints $\delta_s(x^s) \in A_s(x^s), x^s \in D^{(s+1)}, t+1 \le s \le T-1$, see (6.20), the question is whether the integral and the supremum can be interchanged. Assuming that the suprema in (6.22b) are attained at $a_s^* = \delta_s^*(x^s), x^s \in D^{(s+1)}, t+1 \le s \le T-1$, with the conditional expectation operator E_{x^t} with respect to $X^t = x^t$, from (6.22b) we get

$$\begin{aligned} & E_{x^t} U_T\big(t+1, (x^t, y); \delta_{t+1}^*, \ldots, \delta_{T-1}^*\big) \\ \le & \sup_{\delta_{t+1}, \ldots, \delta_{T-1}} E_{x^t} U_T\big(t+1, (x^t, y); \delta_{t+1}, \ldots, \delta_{T-1})\big) \\ \le & E_{x^t} \sup_{\delta_{t+1}, \ldots, \delta_{T-1}} U_T\big(t+1, (x^t, y); \delta_{t+1}, \ldots, \delta_{T-1})\big) \\ = & E_{x^t} U_T\big(t+1, (x^t, y); \delta_{t+1}^*, \ldots, \delta_{T-1}^*)\big). \end{aligned} \tag{6.22c}$$

Thus, (6.22c) yields

$$\sup_{\delta_{t+1},\ldots,\delta_{T-1}} E_{x^t} U_T\big(t+1,(x^t,y);\delta_{t+1},\ldots,\delta_{T-1})\big)$$
$$= E_{x^t} \sup_{\delta_{t+1},\ldots,\delta_{T-1}} U_T\big(t+1,(x^t,y);\delta_{t+1},\ldots,\delta_{T-1})\big) = V_{t+1}^T(x^t,y). \tag{6.22d}$$

The assertion follows now from Eq. (6.22a) and (6.22d). □

Remark 6.1 According to the definition (6.18b) of $K_t(a_t, x^t, \cdot)$ we have

$$\int V_{t+1}^T(x^t,y)K_t(a,x^t,dy) = \int_{y\in G(x_t)} V_{t+1}^T(x^t,y)\pi_t(a,x^t,dy)$$
$$+ V_{t+1}^T(x^t,x_t)\Big(1-\pi_t\big(a,x^t,G(x_t)\big)\Big). \tag{6.23a}$$

Furthermore, assuming $u_t(a,x,x) = 0$ for all $t = 0, 1, \ldots,$ and $x \in \mathbb{R}^n$, we have, cf. (6.19b),

$$\bar{u}_t(a,x^t) = \int_{y\in G(x_t)} u_t(a,x_t,y)\pi_t(a,x^t,dy). \tag{6.23b}$$

In the important Markovian case, i.e. if

$$\pi_t(a,x^t,\cdot) = \pi_t(a,x_t,\cdot) \text{ and } A_t(x^t) = A_t(x_t), \tag{6.23c}$$

the value function V_t^T depends only on x_t, see (6.14a-d), (6.17b), (6.18a,b), and (6.21) has the form

$$V_t^T(x_t) = \sup_{a\in A_t(x_t)} \left(\bar{u}_t(a,x_t) + \int V_{t+1}^T(y)K_t(a,x_t,dy)\right). \tag{6.23d}$$

In the one-stage case $t = T-1$ Eq. (6.21) has the simple form

$$V_{T-1}^T\left(x^{T-1}\right) = \sup\left\{\bar{u}_{T-1}\left(a,x^{T-1}\right) : a \in A_{T-1}\left(x^{T-1}\right)\right\}. \tag{6.23e}$$

6.3.1 *Infinite-Stage Search Processes*

The decision process defined by (6.21) is called the **sequential stochastic decision process associated with the random search procedure (6.2a,b)**. An important variant of this decision process results in the infinite-stage stationary Markovian case:

Let $\pi_t(a_t, x^t, \cdot) = \pi(a_t, x_t, \cdot)$, $A_t(x^t) = A(x_t)$, $u_t(a_t, x_t, x_{t+1}) = u(a_t, x_t, x_{t+1})$, $\delta_t(x^t) = \delta(x_t), t = 0, 1, \ldots$. Moreover, let $0 < \varrho < 1$ be a certain discount factor. According to Theorem 6.3, the value function $V_t^T = V_t^T(x)$ of the (T-t)-stage process depends only on the state $x_t = x$ and fulfills the recurrence relation:

$$V_t^T(x) = \sup_{a \in A(x)} \left(\bar{u}(a, x) + \int V_{t+1}^T(y) K(a, x, dy) \right), \tag{6.24a}$$

$t = T-1, T-2, \ldots, 1, 0$, where $a = \delta(x_t)$. Introducing the stage transformation $(T-t) \to t$, the transformed value function

$$W_t(x) := V_{T-t}^T(x), t = 0, 1, \ldots, \tag{6.24b}$$

satisfies (insert $s := T - t$ and replace then again $s \to t$) the forward recurrence relations

$$W_t(x) = \sup_{a \in A(x)} \left(\bar{u}(a, x) + \varrho \int W_{t-1}(y) K(a, x, dy) \right), t = 0, 1, \ldots, \tag{6.24c}$$

where the functional equation (6.24c) holds for each integer T, and we have, cf. Theorem 6.3, $W_0(x) = 0$.

Under certain conditions the sequence $\big(W_t(x)\big)$ is convergent to the function $W^*(x)$ satisfying the asymptotic functional equation

$$W^*(x) = \sup_{a \in A(x)} \left(\bar{u}(a, x) + \varrho \int W^*(y) K(a, x, dy) \right). \tag{6.24d}$$

Moreover, an optimal decision rule δ^* is then given by $\delta^*(x) = a^* \in A(x)$, where a^* is a solution of the maximization problem in (6.24d).

6.4 Convex Problems

For simplicity we consider here only the minimization of a real valued convex function $F : \mathbb{R} \to \mathbb{R}$ with respect to $D = \mathbb{R}$, where we assume that the second derivative F'' exists and $F''(x) > 0$ for all $x \in \mathbb{R}$. The interval $G(x) = \big\{y \in \mathbb{R} : F(y) < F(x)\big\}$ may now be approximated by the interval

$$H(x) = \left\{ y \in K : F'(x)(y - x) + \frac{F''(x)}{2}(y - x)^2 < 0 \right\}. \tag{6.25a}$$

It is easy to see that

$$H(x) = \left\{ y \in \mathbb{R} : x < y < x - 2\frac{F'(x)}{F''(x)} \right\}, \text{ if } F'(x) < 0, \tag{6.25b}$$

$$H(x) = G(x) = \emptyset \qquad , \text{ if } F'(x) = 0, \tag{6.25c}$$

$$H(x) = \left\{ y \in \mathbb{R} : x - 2\frac{F'(x)}{F''(x)} < y < x \right\}, \text{ if } F'(x) > 0. \tag{6.25d}$$

For the conditional distribution $\pi(a, x, \cdot)$ of the search variables (Z_t) given $X_t = x$ we choose now a normal distribution with mean $\mu = x$ and variance $\sigma^2 = a^2$. Hence, in this case our decision parameter a is then the standard deviation σ. Furthermore, according to the above approximation of $G(x)$ by $H(x)$, we approximate the utility function $u(a, x, y)$ of Sect. 6.3, by

$$\tilde{u}(a, x, y) = \begin{cases} |y - x| \,, & y \in H(x) \\ 0 & , \text{otherwise} \end{cases}. \tag{6.26a}$$

Obviously, the stochastic decision process associated with the random search procedure (6.2a,b) is stationary and $\bar{u}(a, x)$ may be approximated by

$$\bar{\tilde{u}}(a, x) := \int_{y \in H(x)} \tilde{u}(a, x, y)\pi(a, x, dy) = \frac{\sigma}{\sqrt{2\pi}} \left(1 - \exp\left(-\frac{1}{2} \left(\frac{2F'(x)}{\sigma F''(x)} \right)^2 \right) \right). \tag{6.26b}$$

Starting from $\widetilde{W}_0(x) = W_0(x) = 0$, the approximate $\widetilde{W}_1(x) = \sup_{\sigma > 0} \bar{\tilde{u}}(\sigma, x)$ to the value function $W_1(x)$, see (6.24c) and the approximative decision function $\tilde{\sigma}_1 = \tilde{\sigma}_1(x)$, defined by $\widetilde{W}_1(x) = \bar{\tilde{u}}(\tilde{\sigma}_1, x)$, are given by the following theorem:

Theorem 6.4 *Let g be the function $g(t) = \frac{1}{\sqrt{2\pi}t}\left(1 - \exp(-\frac{1}{2}t^2)\right)$, and let us denote $t^* > 0$ the number where g attains its maximum g^*. Then,*

$$\widetilde{W}_1(x) = g^* \left| \frac{2F'(x)}{F''(x)} \right| \textit{ and } \tilde{\sigma}_1(x) = \frac{1}{t^*} \left| \frac{2F'(x)}{F''(x)} \right|. \tag{6.27}$$

Proof Using the transformation $\sigma \to t := \frac{1}{\sigma} \left| \frac{2F'(x)}{F''(x)} \right|$, according to (6.26b) and the above definition of the function $g = g(t)$, we have $\bar{\tilde{u}}(\sigma, x) = \left| \frac{2F'(x)}{F''(x)} \right| g(t)$. This yields the assertion. □

Obviously, according to (6.26a), $\widetilde{W}_1(x) = g^* \left| \frac{2F'(x)}{F''(x)} \right|$ is an approximate to the average step length $s_1(x)$ of the first step $X_0 \to X_1$ of the search process (X_t). Comparing this result with Newton's method $x \to y = x - \alpha(x)\frac{F'(x)}{F''(x)}, \alpha(x) > 0$

for the minimization of F, we observe that in Newton's method the step length $s_N(x) = |y - x| = \alpha(x)\left|\frac{F'(x)}{F''(x)}\right|$ has—up to a normalizing factor—the same form as $s_1(x)$.

Similar results are obtained from comparisons of Theorem 6.4 with deterministic and stochastic gradient procedures.

In general, the computation of the further iterates $\widetilde{W}_t$ and $\tilde{\delta}_t, t = 2, 3, \ldots$ will be in general hardly carried out in practice, because of its difficulty and because $\tilde{\sigma}(x) = \tau(x)\left|\frac{2F'(x)}{F''(x)}\right|$ with a normalizing factor $\tau(x) > 0$ is a reasonable approximate to the optimal decision rule. This is also confirmed by numerical experiments. On the other hand, for the quadratic case

$$F(x) = x^2$$

we can obtain the exact results. In fact, then we have that $H(x) = G(x)$ and $W_1(x) = \widetilde{W}_1(x) = 2g^*|x|$ as also $\sigma_1(x) = \tilde{\sigma}_1(x) = \frac{2}{t^*}|x|$. For solving now the functional equation (6.24b) we work therefore with the assumptions

$$W^*(x) = C|x| \text{ and } \sigma^*(x) = c|x| \tag{6.28}$$

where C, c are positive constants.

Theorem 6.5 *The optimal value W^* and the optimal decision rule $\delta^*(x) = \sigma^*(x)$ of the infinite-stage stationary stochastic decision process associated with the random search procedure for the minimization of $F(x) = x^2$ has the form (6.28), where $c \approx \frac{8\sqrt{\pi}}{4\sqrt{\pi} - \varrho\sqrt{2}}$ and $C \approx \frac{1}{\sqrt{2\pi}} - \frac{\varrho}{4\pi}$.*

Note 6.3 As was mentioned in Sect. 6.1, often an analytic expression for F is not known and only the function values $F(x)$ may be obtained. Hence the derivatives $F'(x), F''(x)$ in the "optimal" decision rule $\tilde{\sigma}(x) = \tau(x)\left|\frac{2F'(x)}{F''(x)}\right|$ must be estimated from observations of F.

References

1. Anderson, R.L.: Recent advances in finding best operating conditions. J. Am. Stat. Assoc. **48**(264), 789–798 (1953). http://www.jstor.org/stable/2281072
2. Box, G.: Evolutionary operation: a method for increasing industrial productivity. J. R. Stat. Soc. Ser. C **6**(2), 81–101 (1957). https://doi.org/10.2307/2985505
3. Brooks, S.: A discussion of random methods for seeking maxima. Oper. Res. **6**(2), 244–251 (1958). https://doi.org/10.1287/opre.6.2.244

4. Goldberg, D.: The Design of Innovation: Lessons from and for Competent Genetic Algorithms. Kluwer Acad, Publ., Boston (2002)
5. Iosifescu, M., Theodorescu, R.: Random Process and Learning. Springer, Berlin (1969)
6. Karnopp, D.C.: Random search techniques for optimization problems. Automatica **1**(2–3), 111–121 (1963). https://doi.org/10.1016/0005-1098(63)90018-9
7. Kirkpatrick, S., Gelatt, C.D., Vecchi, M.P.: Optimization by simulated annealing. Science **220**(4598), 671–680 (1983)
8. Neumann, K.: Dynamische Optimierung. Bibliographisches Institut, Mannheim (1969)
9. Rappl, G.: Konvergenzraten von Random-Search-Verfahren zur globalen Optimierung. Ph.D. Thesis, UniBw München (1984)
10. Richter, H.: Wahrscheinlichkeitstheorie. Springer, Berlin (1966)
11. Spall, J.: Introduction to stochastic search and optimization. J. Wiley, Hoboken (2003)
12. Zabinsky, Z.: Stochastic Adaptation Search for Global Optimization. Springer, New York (2003)

Chapter 7
Controlled Random Search Under Uncertainty

7.1 The Controlled (or Adaptive) Random Search Method

In order to increase the rate of convergence of the basic search method (6.2a), according to Sect. 6.3 we consider the following procedure, cf. [2–4]. Based on the basic random search method (6.2a), by means of the definitions (I)–(III) we describe first an (infinite-stage) **sequential stochastic decision process associated with (6.2a)**.

(I) We use next to the fact that the transition probabilities $\pi_t(x^t, \cdot)$ depend

$$\pi_t(x^t, \cdot) = \pi_t(a, x^t, \cdot)$$

usually on certain parameters $a = (a_j)_{j\varepsilon J} \in A$, as, e.g., on certain (mixed) moments of the random vector Z_{t+1}. Let

$$\hat{h}^t = (x_0, x_1, \ldots, x_t, z_1, \ldots, z_t)$$

be the process history of X_t, Z_t up to time t. The idea, developed first in [2, 3], is now to run the random search not with a fixed parameter a, but to use an "optimal" control

$$a = a_t^*(x^t) \text{ or } a = a_t^*(\hat{h}^t)$$

of the parameter a such that a certain criterion measuring the progress of the search, as, e.g., the probability of a search success or the mean step length into the area of success at each step $X_t \to X_{t+1}$, is as large as possible. In the following,

K. Marti, *Optimization Under Stochastic Uncertainty*, International Series in Operations Research & Management Science 296,
https://doi.org/10.1007/978-3-030-55662-4_7

$$h^t = (x_0, x_1, \ldots, x_t, z_1, z_2, \ldots, z_t, a_0, a_1, \ldots, a_{t-1})$$

denotes the total process history up to time t.

(II) To each step $x_t \to X_{t+1}$ there is associated a conditional mean (search-) gain

$$E\Big(u_t(a_t, x_t, X_{t+1})|h^t\Big) = E\big(u_t(a_t, x_t, Z_{t+1})|h^t\big). \tag{7.1a}$$

Working, e.g., with the probability of a search success resp. the mean improvement of F resp. the mean (relative) step length into the area of success, u_t is given by

$$\begin{aligned} u_t(a_t, x_t, z_{t+1}) &= 1, = F(x_t) - F(z_{t+1}), = \|x_t - z_{t+1}\| \\ &= \frac{\|x_t - z_{t+1}\|}{\|x_t\|}, \text{ resp., if } z_{t+1} \in G_F(x_t), \\ u_t &= 0, \text{ if } z_{t+1} \notin G_F(x_t). \end{aligned} \tag{7.1b}$$

Calculating the conditional mean again in (7.1a) we have to solve next the integrals of the type

$$J(x^t, F) = \int\limits_{\substack{F(z_{t+1}) < F(x_t) \\ z_{t+1} \in D}} u_t(a_t, x_t, z_{t+1}) \pi_t(a_t, x^t, dz_{t+1}). \tag{7.2}$$

However, because of the black-box-situation concerning the objective function F of (6.1), i.e. having available at stage t only the discrete F-values $F(x_0), F(z_1), \ldots, F(z_t)$ as also the given some a priori information on F, the inequality $F(z_{t+1}) < F(x_t)$ in (7.2) cannot be evaluated in general. Consequently, the integral $J(x^t, F)$ cannot be computed in general because of the missing knowledge about F.

(III) In order to cope with this uncertainty, we may proceed in the following two different ways:

(III.1) Approximation of the area of success $G_F(x_t)$, see (6.3), by a set $\widetilde{G}_F(x_t)$ which may be described by the information available on F up to the current time t. As an example we mention here the following **Random-Search-Newton-Method** [3]: Similar to the (deterministic) Newton method in optimization, at state x_t we approximate first the value $F(z_{t+1})$ by the second order Taylor polynomial

$$\begin{aligned} F(z_{t+1}) \approx F(x_t) &+ \nabla F(x_t)^T (z_{t+1} - x_t) \\ &+ \frac{1}{2}(z_{t+1} - x_t)^T \nabla^2 F(x_t)(z_{t+1} - x_t) \end{aligned} \tag{7.3a}$$

at x_t, where the derivatives $\nabla F, \nabla^2 F$ of F may be obtained approximately by a numerical differentiation procedure using the process history h^t.

Then, $G_F(x_t)$ can be approximated by the set

$$\widetilde{G}_F(x_t;\hat{h}^t) := \big\{y \in D : \widetilde{\nabla F}(x_t;\hat{h}^t)^T(y - x_t) + \frac{1}{2}(y - x_t)^T \widetilde{\nabla^2 F}(x_t;\hat{h}^t)(y - x_t) < 0\big\}, \tag{7.3b}$$

where $\widetilde{\nabla F}(x_t;\hat{h}^t), \widetilde{\nabla^2 F}(x_t;\hat{h}^t)$ are approximations to the gradient, Hessian, resp., $\nabla F(x_t), \nabla^2 F(x_t)$ of F at x_t based on the process history h^t. The conditional mean gain (7.1a,b) be approximated now by

$$\begin{aligned} E\big(u_t(a_t, x_t, Z_{t+1})|h^t\big) &\approx \\ &\approx E\big(\tilde{u}_t(a_t, x_t, Z_{t+1})|\hat{h}^t\big) \\ &:= \int\limits_{z_{t+1} \in \widetilde{G}_F(x_t;\hat{h}^t)} u_t(a_t, x_t, z_{t+1})\pi_t(a_t, x^t, dz_{t+1}). \end{aligned} \tag{7.3c}$$

By (I), (II), (III.1) we have now an infinite-stage sequential stochastic decision process for the definition of an optimal parameter control $a_t^* = a_t^*(h^t)$ speeding up the random search according to the chosen searching-gain criteria u_t. Since in practice our aim is to speed up to some extent the convergence of the basic random search procedure (6.2a) for solving (6.1), the computational effort for finding an *optimal* decision rule $a_t^*, t = 0, 1, \ldots$, should remain in realistic bounds. Hence, for practical purposes we are not interested in the exact solution $a_t^* = a_t^*(h^t)$ of the associated decision process, but we are looking for a sub-optimal control a_t^* obtainable by a reasonable computational effort. Approximating therefore the infinite-stage decision process by the family of 1-stage decision processes, a_t^* may be defined by

$$a_t^* = a_t^*(\hat{h}^t) \in \arg\max_{a \in A_t(x_t)} E\big(\tilde{u}_t(a, x_t, Z_{t+1})|\hat{h}^t\big), \tag{7.4}$$

where $A_t = A_t(x_t)$ denotes still the set of parameters available at (t, x_t). Having by the application of a_t^* a local improvement of the convergence behavior of the random search, we will show later on also the convergence of the controlled process X_t^* towards $B_{\varepsilon,M}$.

Now the second method for handling the uncertainty concerning F is described.

(III.2) Because of the missing information about the objective function F, actually we have a **sequential stochastic decision process under uncertainty**. Hence, the conditional mean search gain (7.1a,b) has to be defined by

$$E\big(u_t(a_t, x_t, X_{t+1})|h^t\big) = \int J(x^t, F)\mu_{h^t}(dF), \tag{7.5}$$

where $J(x^t, F)$ is given by (7.2), and μ_{h^t} is the **conditional distribution of the unknown F, given the process history $\mathbf{h^t}$**. For the proper definition of μ_{h^t} we need first a mathematical representation of the given a priori information about F, as, e.g., "*F is an unknown polynomial in n variables*" or "*$a(x) \le F(x) \le b(x)$ for all $x \in D$ with given functions $a(\cdot), b(\cdot)$.*"

We use a **Bayesian model** for the unknown F: We assume that there is a measurable space $(\Theta, \mathcal{A})$ of parameters θ and an a priori distribution μ^0 of the parameters θ on the σ-algebra $\mathcal{A}$ such that the objective function F of (6.1) is a realization

$$F(x) = f(x, \theta^0), x \in \mathbb{R}^n, \tag{7.6a}$$

of a stochastic function $y = f(x, \theta), x \in \mathbb{R}^n, \theta \in \Theta$, on $\mathbb{R}^n$, where θ^0 is the true, but unknown parameter. We assume that each realization $f(\cdot, \theta)$ of f is a measurable function on $\mathbb{R}^n$.

Note 7.1 We observe that similar models for unknown functions in engineering and optimization have been used also in [1] and [5].

In the next Sect. 7.2 we will show that

$$\mu_{h^t} = \mu_{\hat{h}^t}, \tag{7.6b}$$

i.e. the posterior distribution μ_{h^t} of F depends only on $\hat{h}^t = (x_0, x_1, \dots, x_t, z_1, \dots, z_t)$, but not on $a_0, a_1, \dots, a_{t-1}$. Hence, due to (7.5) it is

$$E\big(u_t(a_t, x_t, X_{t+1})|h^t\big) = \int J(x^t, F)\mu_{h^t}(dF)$$
$$= \int J(x^t, F)\mu_{\hat{h}^t}(dF) = E\big(u_t(a_t, x_t, X_{t+1})|\hat{h}^t\big). \tag{7.7a}$$

Approximating as in approach (III.1) the associated infinite-stage stochastic decision process by the family of 1-stage decision processes, a (sub-) optimal control a_t^* may be defined by

$$a_t^* = a_t^*(\hat{h}^t) = a_t^*(x_t, \mu_{\hat{h}^t}) \in \arg \max_{a \in A_t(x_t)} E\big(u_t(a_t, x_t, X_{t+1})|\hat{h}^t\big). \tag{7.7b}$$

Hence in both cases (7.4), (7.7b), the π_t-parameter control a_t^* depends on the (X_t, Z_t)-history $\hat{h}^t$ only.

7.1.1 The Convergence of the Controlled Random Search Procedure

As mentioned at the end of (III.1) we have now to consider the convergence of the process $X_0^*, X_1^*, \ldots$ controlled by $a_t^*, t = 0, 1, \ldots$, towards the set $B_{\varepsilon,M}$ of (ε, M)-optimal solutions of (6.1), cf. (6.4a).

For this controlled procedure, denoted by Z_t^*, X_t^*, we consider similar to Sect. 6.3 first the conditional distribution $\widetilde{K}_t(\hat{h}^t, \cdot)$ of the tuple (Z_{t+1}^*, X_{t+1}^*), given $Z_t^* = z^t, X_t^* = x^t$ as also given the unknown F. Denoting by $T_{F,x}$ the mapping

$$T_{F,x}(y) = \begin{cases} y, \text{ if } y \in G_F(x), \\ x, \text{ else} \end{cases}, \tag{7.8}$$

obviously we have that $X_{t+1} = T_{F,x_t}(Z_{t+1})$ and therefore

$$\begin{aligned}
\widetilde{K}_t(\hat{h}^t, A \times B) &:= P(Z_{t+1}^* \in A, X_{t+1}^* \in B | Z^{*t} = z^t, X^{*t} = x^t) \\
&= P\left(Z_{t+1}^* \in A, T_{F,x_t}(Z_{t+1}^*) \in B | Z^{*t} = z^t, X^{*t} = x^t\right) \\
&= \int\limits_{\substack{z_{t+1} \in A \\ T_{F,x_t}(z_{t+1}) \in B}} \pi_t\left(a_t^*(\hat{h}^t), x^t, dz_{t+1}\right) \\
&= \int\limits_{\substack{z_{t+1} \in A \cap B \\ z_{t+1} \in G_F(x_t)}} \pi_t(a_t^*, x^t, dz_{t+1}) + \left(\int\limits_{\substack{z_{t+1} \in A \\ z_{t+1} \notin G_F(x_t)}} \pi_t(a_t^*, x^t, dz_{t+1}) \right) \varepsilon_{x_t}(B),
\end{aligned} \tag{7.9}$$

where the last equality follows from $z_{t+1} \in G_F(x_t) \Rightarrow T_{F,x_t}(z_{t+1}) = z_{t+1}$ and $z_{t+1} \notin G_F(x_t) \Rightarrow T_{F,x_t}(z_{t+1}) = x_t$.

Corresponding to (6.5a,b), (6.6a,b), for the unknown, but fixed objective function F it holds then that

$$P(X_t^* \in B_{\varepsilon,M}|F) = 1 - \big(1 - 1_{B_{\varepsilon,M}}(x_0)\big)P(X_1^* \notin B_{\varepsilon,M}, \ldots, X_t^* \notin B_{\varepsilon,M}|F) \tag{7.10}$$

Putting $\overline{B}_{\varepsilon,M} := D \setminus B_{\varepsilon,M}$, we get

$$\begin{aligned}
&P(X_1^* \notin B_{\varepsilon,M}, \ldots, X_t^* \notin B_{\varepsilon,M} | F) = \\
&\quad = P\left((Z_1^*, X_1^*) \in \mathbb{R}^n \times \overline{B}_{\varepsilon,M}, (Z_2^*, X_2^*) \in \mathbb{R}^n \times \overline{B}_{\varepsilon,M}, \ldots, (Z_t^*, X_t^*)\right. \\
&\quad \left.\in \mathbb{R}^n \times \overline{B}_{\varepsilon,M} | F\right) \\
&\quad = \int\limits_{\substack{z_1 \in \mathbb{R}^n \\ x_1 \in \overline{B}_{\varepsilon,M}}} \widetilde{K}_0(x_0, dz_1, dx_1) \ldots \int\limits_{\substack{z_{t-1} \in \mathbb{R}^n \\ x_{t-1} \in \overline{B}_{\varepsilon,M}}} \widetilde{K}_{t-2}(\hat{h}^{t-2}, dz_{t-1}, dx_{t-1}), \\
&\quad \times \int\limits_{\substack{z_t \in \mathbb{R}^n \\ x_t \in \overline{B}_{\varepsilon,M}}} \widetilde{K}_{t-1}(\hat{h}^{t-1}, dz_t, dx_t). \qquad (7.11)
\end{aligned}$$

Proceeding now as in (6.7–6.9d), we obtain next to by means of $x_{t-1} \notin B_{\varepsilon,M}$ and (7.8)

$$\begin{aligned}
\int\limits_{\substack{z_t \in \mathbb{R}^n \\ x_t \in \overline{B}_{\varepsilon,M}}} \widetilde{K}_{t-1}(\hat{h}^{t-1}, dz_t, dx_t) &= \widetilde{K}_{t-1}\left(\hat{h}^{t-1}, \mathbb{R}^n \times (D \setminus B_{\varepsilon,M})\right) \\
&= 1 - \widetilde{K}_{t-1}(\hat{h}^{t-1}, \mathbb{R}^n \times B_{\varepsilon,M}) \\
&= 1 - \int\limits_{T_{F,x_{t-1}}(z_t) \in B_{\varepsilon,M}} \pi_{t-1}(a_{t-1}^*, x^{t-1}, dz_t) \\
&= 1 - \pi_{t-1}\left(a_{t-1}^*, x^{t-1}, B_{\varepsilon,M} \cap G_F(x_{t-1})\right) \\
&= 1 - \pi_{t-1}(a_{t-1}^*, x^{t-1}, B_{\varepsilon,M}), \qquad (7.12a)
\end{aligned}$$

where the last two equalities hold because of $x_{t-1} \notin B_{\varepsilon,M}$. Denoting by U a subset of $\mathbb{R}^n$ containing all supports of the conditional distributions $\pi_t(a_t, x_t, \cdot)$, $t = 0, 1, \ldots$, corresponding to α_t in Sect. 6.2.1

$$\alpha_t^* = \alpha_t^*(\varepsilon, M, F) = \inf\left\{\pi_t\left(\alpha_t^*(\hat{h}^t), x^t, B_{\varepsilon,M}\right) : z_s \in \mathbb{R}^n, x_s \in D \setminus B_{\varepsilon,M}, 1 \le s \le t\right\} \qquad (7.12b)$$

is the minimal probability for finding an (ε, M)-optimal solution of (6.1) at stage $t = 1, 2, \ldots$. Because of

$$\int\limits_{\substack{z_t \in \mathbb{R}^n \\ x_t \in B_{\varepsilon,M}}} \widetilde{K}_{t-1}(\hat{h}^{t-1}, dz_t, dx_t) \le 1 - \alpha_{t-1}^*, t = 1, 2, \ldots, \qquad (7.12c)$$

from (7.11) we obtain with $\alpha_0^* = \pi_0\big(a_0^*(x_0), x_0, B_{\varepsilon,M}\big)$ that

$$\begin{aligned} &P(X_1^* \notin B_{\varepsilon,M}, \ldots, X_t^* \notin B_{\varepsilon,M}|F) \\ &\leq \Big(1 - \pi_0\big(a_0^*(x_0), x_0, B_{\varepsilon,M}\big)\Big) \sum_{s=1}^{t-1}(1-\alpha_s^*) = \prod_{s=0}^{t-1}(1-\alpha_s^*). \end{aligned} \tag{7.13}$$

Thus, by (7.10) and (7.13) yield, cf. (6.9a–d),

$$P(X_t^* \in B_{\varepsilon,M}|F) \geq 1 - \prod_{s=0}^{t-1}(1-\alpha_s^*) \geq 1 - \exp\left(-\sum_{s=0}^{t-1}\alpha_s^*\right). \tag{7.14}$$

Hence, we proved the following result, see Theorem 6.1.

Theorem 7.1 (Convergence of the Controlled Random Search Method) *If* $\sum_{s=1}^{\infty} \alpha_s^*(\varepsilon, M, F) = +\infty$, *then* $\lim_{t\to\infty} P(X_t^* \in B_{\varepsilon,M}|F) = 1$.

Example 7.1 Let $\pi_t(a_t, x^t, \cdot)$ be a normal distribution with mean x_t and covariance matrix $Q = (\sigma_i^2\delta_{ik})$, where

$$a_t = a_t(\hat{h}^t) = \big(\sigma_1(\hat{h}^1), \sigma_2(\hat{h}^t), \ldots, \sigma_n(\hat{h}^t)\big)^T$$

and $\delta_{ij} = 1$ if $i = j$, $\delta_{ij} = 0$ if $i \neq j$. If we know about F at least that $B_{\varepsilon,M}$ is bounded and has non-zero measure, then from the above theorem we get immediately this consequence.

Corollary 7.1 *Let $B_{\varepsilon,M}$ be bounded and have non-zero Lebesgue-measure. If the variance control $a_t^* = \big(\sigma_1^*(\hat{h}^t), \ldots \sigma_n^*(\hat{h}^t)\big)^T$ fulfills $\sigma_t^*(\hat{h}^t) \geq \sigma_0 > 0$ for all $t = 1, 2, \ldots$ and all values of $\hat{h}^t$, then $\sum_{s=1}^{\infty} \alpha_s^* = \infty$ and therefore $\lim_{t\to\infty} P(X_t^* \in B_{\varepsilon,M}|F) = 1$.*

For practical purposes—besides the convergence of X_t^* towards an (ε, M)-optimal solution of (6.1)—of great importance is a stopping criterion for the searching procedure.

7.1.2 A Stopping Rule

A suitable criterion for terminating the search at stage $t = T$ would be

$$P(X_T^* \notin B_{\varepsilon,M}) \leq p, \tag{7.15}$$

i.e. the probability that X_T^* is not an (ε, M)-optimal solution of (6.1) is not greater than a prescribed small value p. By (7.11) and (7.13) for this probability we have the estimate

$$P(X_T^* \notin B_{\varepsilon,M}) = P(X_T^* \notin B_{\varepsilon,M}|F) \leq \prod_{s=0}^{T-1} \big(1 - \alpha_s^*(\varepsilon, M, F)\big). \tag{7.16}$$

If the minimal probabilities $\alpha_s^*(\varepsilon, M, F)$ have a lower bound $\tilde{\alpha} > 0$, then from (7.15) we get obviously this next result.

Lemma 7.1 *Let $\alpha_t^*(\varepsilon, M, F) > \tilde{\alpha} > 0$ for all $t = 0, 1, 2, \ldots$. Then the stopping criterion (7.15) is satisfied if*

$$T \geq \frac{\log p}{\log(1 - \tilde{\alpha})}. \tag{7.17}$$

Due to the uncertainty concerning F we cannot work in general with the expression in (7.15) and (7.16) directly. However, replacing the unknown function F by its conditional distribution $\mu_{h^{T-1}}$ known at stage $T - 1$, we may approximate (7.15) by the stopping rules

$$\int P\big(X_T^* \notin B_{\varepsilon,M}(F)|F\big)\mu_{h^{T-1}}(dF) \leq p, \tag{7.18a}$$

$$\int \prod_{s=0}^{T-1} \big(1 - \alpha_s^*(\varepsilon, M, F)\big)\mu_{h^{T-1}}(dF) \leq p, \tag{7.18b}$$

7.2 Computation of the Conditional Distribution of F Given the Process History: Information Processing

According to the uncertainty-model for the unknown objective function F of (7.6a) described in Sect. 7.1, (III.2), we assume that F is a realization $F(x) = f(x, \theta^0)$ of a stochastic function $y = f(x, \theta), \theta \in \Theta$. Moreover, μ^0 denotes the a priori distribution (on a σ-algebra $\mathcal{A}$ on Θ) of the stochastic parameter θ, and by $f = f(x)$ we denote any realization $f(\cdot, \theta)$ of the stochastic function $y = f(x, \theta)$. Instead of $\mu^0(d\theta)$, $\mu_{h^t}(d\theta)$, we also write $\mu^0(df)$, $\mu_{h^t}(df)$, resp., interpreting then μ^0, μ_{h^t} as conditional probability distributions on a certain σ-algebra $\mathcal{A}_{\mathcal{F}}$ on an appropriate space $\mathcal{F}$ of possible objective functions f defined (at least) on the admissible domain D, $D \subset \mathbb{R}^n$.

Given the process history $h^t = (x_0, x_1, \ldots, x_t, z_1, z_2, \ldots, z_t, a_0, a_1, \ldots, a_{t-1})$ obtained from the search process $(Z_t, X_t, t = 1, 2, \ldots$, to find the minimum of the function $F = F(x)$, according to the definition of the random search process we have the following relations:

$$T_{F,x_s}(z_{s+1}) = x_{s+1}, t_0, 1, \ldots, t-1, \tag{7.19}$$

where $T_{F,x_s}(z_{s+1})$ is defined by (6.3), (7.8).

At each stage $s = 0, 1, \ldots, t-1$, there are then three different possibilities:

(i) $z_{s+1} \notin D$ *(search-failure I)*. In this case it is $z_{s+1} \notin G_F(x_s)$ and therefore

$$x_{s+1} = T_{F,x_s}(z_{s+1}) = x_s.$$

But in the same way, for each realization f of $f(x,\theta)$ because of $z_{s+1} \notin D$ we find that

$$T_{f,x_s}(z_{s+1}) = x_s.$$

Hence, because of $x_{s+1} = x_s$, in this case (i) the constraint

$$T_{f,x_s}(z_{s+1}) = x_{s+1}$$

is satisfied automatically and can therefore be omitted.

(ii) $z_{s+1} \in D$ and $F(z_{s+1}) \geq F(x_s)$ *(search-failure II)*. In this case we have again that

$$x_{s+1} = T_{F,x_s}(z_{s+1}) = x_s.$$

Now we have for f the constraint

$$T_{f,x_s}(z_{s+1}) = x_{s+1} = x_s$$

which holds if and only if

$$f(z_{s+1}) \geq f(x_s).$$

Indeed, assume that $f(z_{s+1}) < f(x_s)$. But this implies that $z_{s+1} \neq x_s$ and $x_{s+1} = T_{f,x_s}(z_{s+1}) = z_{s+1} \neq x_s$ which is a contradiction to the constraint $T_{f,x_s}(z_{s+1}) = x_{s+1} = x_s$.

(iii) $z_{s+1} \in D$ and $F(z_{s+1}) < F(x_s)$ *(search success)*. Here it is

$$x_{s+1} = F_{F,x_s}(z_{s+1}) = z_{s+1} \neq x_s,$$

hence for f we have the constraint

$$T_{f,x_s}(z_{s+1}) = x_{s+1} = z_{s+1} \neq x_s$$

which is possible if and only if

$$f(z_{s+1}) < f(x_s).$$

Indeed, assume that $f(z_{s+1}) \geq f(x_s)$. This implies $T_{f,x_s}(z_{s+1}) = x_s$, which is a contradiction.

Based on the segment x_s, z_{s+1}, z_{s+1} of the time history $\hat{h}^t$ and the observed values of the true, but unknown function $F(x) = f(x, \theta^0)$, we find that the following constraints for the parameter θ of the unknown model function $f = f(x, \theta)$:

$$T_{f,x_s}(z_{s+1}) = x_{s+1} \text{ is equivalent to } \begin{cases} \text{no constraint, if } z_{s+1} \notin D \\ f(z_{s+1}) \geq f(x_s), \text{ if } z_{s+1} \in D \\ \text{and } F(z_{s+1}) \geq F(x_s) \\ f(z_{s+1}) < f(x_s), \text{ if } z_{s+1} \in D \\ \text{and } F(z_{s+1}) < F(x_s). \end{cases} \tag{7.20a}$$

Consequently, given the time history $\Theta(h^t)$, $\hat{h}^t$, the set $\Theta h^t = \Theta(\hat{h}^t)$ of admissible parameters θ up to stage t is defined recursively by

$$\Theta(h^0) = \Theta(h^0) := \Theta \tag{7.20b}$$

$$\Theta(\hat{h}^{t+1}) = \begin{cases} \Theta(\hat{h}^t), \text{ if } z_{t+1} \notin D \\ \{\theta \in \Theta(\hat{h}^t) : f(z_{t+1}, \theta) \geq f(x_t, \theta)\}, \text{ if } z_t \in D \\ \text{and } F(z_{t+1}) \geq F(x_t) \\ \{\theta \in \Theta(\hat{h}^t) : f(z_{t+1}, \theta) < f(x_t, \theta)\}, \text{ if } z_t \in D \\ \text{and } F(z_{t+1} < F(x_t). \end{cases} \tag{7.20c}$$

Note 7.2 Obviously we have

$$\theta^0 \in \Theta(h^t) \subset \Theta(h^{t-1}), t = 0, 1, \ldots.$$

From the above consideration we get the following result:

Theorem 7.2 (Representation of the Admissible Set of Parameters)

(a) Given the time history h^t, the recursion for the admissible parameter domains $\Theta(\hat{h}^t)$ can be given by

$$\Theta(h^0) = \Theta(h^0) := \Theta, \tag{7.21a}$$

$$\Theta(\hat{h}^{t+1}) = \{\theta \in \Theta(\hat{h}^t) : T_{f(\cdot,\theta),x_t}(z_{t+1}) = x_{t+1}\}, t = 0, 1, \ldots \tag{7.21b}$$

(b) Let then the index sets $I_1(\hat{h}^t)$, $I_2(\hat{h}^t)$ be defined by

$$I_1(\hat{h}^t) = \{s : 0 \leq s \leq t-1, z_{s+1} \in D \text{ and } F(z_{s+1}) \geq F(x_s)\}, \quad (7.21c)$$

$$I_2(\hat{h}^t) = \{s : 0 \leq s \leq t-1, z_{s+1} \in D \text{ and } F(z_{s+1}) < F(x_s)\}. \quad (7.21d)$$

By these definitions the set $\Theta(\hat{h}^t)$ of admissible parameters at stage t can be represented in the following form:

$$\Theta(\hat{h}^t) = \{\theta \in \Theta : f(z_{s+1}, \theta) \geq f(x_s, \theta) \text{ for } s \in I_1(\hat{h}^t), \quad (7.21e)$$
$$f(z_{s+1}, \theta) < f(x_s, \theta) \text{ for } s \in I_2(\hat{h}^t)\}.$$

Note 7.3 The inequality relations of the sample functions f among their values at the arising points $x_s, z_{s+1}, s = 0, 1, \dots, t-1$, corresponds to the observed inequality relations of true, but unknown objective function F.

Having the set $\Theta(\hat{h}^t)$ of admissible parameters θ of the model $f = f(x, \theta)$ for the analytically not given objective function F, given the process history $\hat{h}^t$, we get this result:

Theorem 7.3 (Conditional Probability of F**)** *With the a priori distribution μ^0 of the model parameters θ of the unknown objective function F, the conditional probability of F, given the process history $\hat{h}^t$, reads*

$$\mu_{\hat{h}^t}(W) = \frac{\mu^0(W \cap \Theta(\hat{h}^t))}{\mu^0(\Theta(\hat{h}^t))}, \quad (7.22)$$

for each measurable set $W \subset \Theta$.

Example 7.2

$$f(x, \theta) = \sum_{i=1}^{r} f_i(x)\theta_i \quad (7.23a)$$

for F, where $f_1, f_2, \dots, f_r$ are known functions and $\theta_i, i = 1, \dots, r$, are unknown real parameters with an a priori distribution μ^0 on $(\Theta, \mathcal{A}) = (\mathbb{R}^r, \mathcal{B}^r)$, then $\Theta(\hat{h}^t)$ is described by a finite number ($\leq t$) of linear inequalities.

Here, the conditional distribution $\mu_{\hat{h}^t}$ of F, given the search history $\hat{h}^t$, is the restriction of the a priori distribution μ^0 to set $\Theta(\hat{h}^t)$. Thus, if μ^0 has a probability density $\phi_0 = \phi_0(\theta)$, then the probability density $\phi_{\hat{h}^t} = \phi_{\hat{h}^t}(\theta)$ of $\mu_{\hat{h}^t}$ reads

$$\phi_{\hat{h}^t}(\theta) = \mathbf{1}_{\Theta(\hat{h}^t)}(\theta) \frac{\phi_0(\theta)}{\mu^0(\Theta(\hat{h}^t))}. \quad (7.23b)$$

References

1. Kushner, A.: A versatile stochastic model of a function of unknown and time varying form. J. Math. Anal. Appl. **9**, 379–388 (1962)
2. Marti, K.: Random search in optimization problems as a stochastic decision process (adaptive random search). Math. Meth. Oper. Res. **36**, 223–234 (1979)
3. Marti, K.: Adaptive zufallssuche in der optimierung. ZAMM **60**, 357–359 (1980)
4. Marti, K.: Controlled Random Search Procedures For Global Optimization. Lecture Notes in Control and Information Sciences, vol. 81, pp. 457–474. Springer, Berlin (1986)
5. Mockus, J.: Bayesian approach to global optimization: theory and applications In: Mathematics and Its Applications, vol. 37. Kluwer Academic Publishers, Dordrecht (1989). https://doi.org/10.1007/978-94-009-0909-0

Chapter 8
Controlled Random Search Procedures for Global Optimization

8.1 Introduction

Solving optimization problems arising from engineering and economics, as, e.g., parameter- or process-optimization problems,

$$\min F(x) \text{ s.t. } x \in D, \tag{8.1}$$

where D is a measurable subset of $\mathbb{R}^d$ and F is a measurable real function defined (at least) on D, one meets often the following situation:

(I) One should find the **global** minimum F^* and/or a **global** minimum point x^* of (8.1). Hence, most of the deterministic programming procedures, which are based on local improvements of the objective function $F(x)$, will fail.
(II) Concerning the objective function $F(x)$ one has a **black-box**-situation, i.e. there is only few a priori information about F especially there is no (complete) knowledge about the direct functional relationship between the control or input vector $x \in D$ and its function value $y = F(x)$. Hence, besides the limited a priori information about F, only by evaluating F numerically or by experiments at certain points $z_1, z_2, \ldots$ of $\mathbb{R}^d$ one gets further information on F.

Consequently, engineers use in these situations usually a certain search procedure for finding the global minimum F and an optimal solution x^x of (8.1), see, e.g., Box' EVOP-method [2] and the random search methods as first proposed by Anderson [1], Brooks [3], and Karnopp [5]. More recent descriptions of random search procedures were given by [6–9, 12–14].

In the random search method considered here the sequence $X_0(\omega), X_1(\omega), \ldots, X_n(\omega), \ldots$ of random iterates is constructed according to the following recurrence schema:

K. Marti, *Optimization Under Stochastic Uncertainty*, International Series in Operations Research & Management Science 296,
https://doi.org/10.1007/978-3-030-55662-4_8

$$X_{n+1}(\omega) = \begin{cases} z_{n+1}, & \text{if } z_{n+1} \in D \text{ and } F(z_{n+1}) < F\big(X_n(\omega)\big) \\ X_n(\omega), & \text{else}, \end{cases} \tag{8.2}$$

$n = 0, 1, \ldots,$ where $x_0(\omega) = x_0 \in D$ is a given starting point in D and $z_1, z_2, \ldots, z_n, \ldots$ are realizations of a sequence of random d-vectors

$$Z_1(\omega), Z_2(\omega), \ldots, Z_n(\omega), \ldots$$

having conditional distributions

$$P\big(Z_{n+1}(\omega) \in B | X_0 = x_0, X_1 = x_1, \ldots, X_n = x_n, Z_1 = z_1, \ldots, Z_n = z_n\big)$$
$$= P\big(Z_{n+1}(\omega) \in B | X_n = x_n\big) = \pi_n(x_n, B) \tag{8.3a}$$

for each Borel subset B of $\mathbb{R}^d$. Here $\pi_n(x_n, \cdot), n = 0, 1, \ldots,$ is a sequence of transition probability measures to be selected by the user of the search procedure. In many concrete cases Z_{n+1} has a d-dimensional normal distribution with mean vector μ_n and covariance matrix Λ_n, i.e.

$$\pi_n(x_n, \cdot) = N(\mu_n, \Lambda_n), n = 0, 1, \ldots, \tag{8.3b}$$

where $\mu_n = \mu_n(x_n)$ and $\Lambda_n = \Lambda_n(x_n)$ are certain functions of the last state (n, x_n).

Let the area of success $G_F(x)$ at a point $x \in \mathbb{R}^d$ be defined by

$$G_F(x) = \big\{y \in D : F(y) < F(x)\big\}. \tag{8.4}$$

At an iteration point x_n by the random search procedure (8.2) a d-vector z_{n+1} is generated randomly according to the transition probability distribution $\pi_n(x_n, \cdot)$ and from x_n we move to $x_{n+1} = z_{n+1}$ provided that $z_{n+1} \in G_F(x_n)$. Otherwise we stay at $x_{n+1} = x_n$ and generate a new random point z_{n+2} according to the distribution $\pi_{n+1}(x_{n+1}, \cdot) = \pi_{n+1}(x_n, \cdot)$.

We observe that $X_{n+1} \in G_F(x_n)$ implies $X_t \in G_F(x_n)$ for all $t > n$. Let the set D_ε of ε-optimal solutions of our global minimization problem (8.1) be defined by

$$D_\varepsilon = \big\{y \leq D : F(y) \leq F^* + c\big\}, \tag{8.5}$$

where $c > 0$ and F^* is given by

$$F^* = \inf\big\{F(x) : x \in D\big\};$$

let $F^* > -\infty$. Note that $X_n \in D_\varepsilon$ implies $X_t \in D_\varepsilon$ for all $t > n$, hence

$$P(X_n \in D_\varepsilon), n = 1, 1, \ldots$$

is a non decreasing sequence for each fixed $\varepsilon > 0$.

8.2 Convergence of the Random Search Procedure

Let $\alpha_n(D_\varepsilon)$ denote the minimal probability that at the nth iteration step $X_n \rightarrow S_{n+1}$ we reach the set D_ε from any point $X_n = x_n$ outside this set, i.e.

$$\alpha_n(D_\varepsilon) = \inf\left\{\pi_n(x_n, D_\varepsilon) : x_n \in D \setminus D_\varepsilon\right\}. \tag{8.6}$$

According to [6] we have this

Theorem 8.1

(a) If for an $\varepsilon > 0$

$$\sum_{n=0}^{\infty} \alpha_n(D_\varepsilon) = +\infty, \tag{8.7}$$

then $\lim_{n\to\infty} P(X_n \in D_\varepsilon) = 1$ *for every* $\varepsilon > 0$.

(b) Suppose that

$$\lim_{n\to\infty} P(X_n \in D_\varepsilon) = 1 \text{ for every } \varepsilon > 0. \tag{8.8}$$

Then $\lim_{n\to\infty} F(X_n) = F^*$ *w.p.1 (with probability one) for every starting point* $x_0 \in D$.

(c) Assume that F *is continuous and that the level sets* D_ε *are nonempty and compact for each* $\varepsilon > 0$. *Then* $\lim_{n\to\infty} F(X_n) = F^*$ *implies that also* $\lim_{n\to\infty}$ *dist* $(X_n, D^*) = 0$, *where dist* (X_n, D^*) *denotes the distance between* X_n *and the set* $D^* = D_0$ *of global minimum points* x^* *of (8.1).*

Example 8.1 If $\pi_n(x_n, \cdot) = \pi(\cdot)$ is a fixed probability measure, then $\lim_{n\to\infty} F(X_n) = F^*$ w.p.1 holds, provided that

$$\pi\left(\left\{y \in D : F(y) \leq F^* + \varepsilon\right\}\right) > 0 \text{ for each } \varepsilon > 0.$$

This if true e.g. if D_ε has a nonzero Lebesgue measure for all $\varepsilon > 0$ and π has a probability density ϕ with $\phi(x) > 0$ almost everywhere.

Note 8.1 Further convergence results of this type were given by Oppel [11], Rappl [13].

Knowing several (weak) conditions which guarantee the convergence w.p.1 of (X_n) to the global minimum F^*, to the set of global minimum points D^*, resp., one should also have some information concerning the rate of convergence of $F(X_n)$, (X_n) to F^*, D^*, respectively.

By Rappl [13] we have now the following result. Of course, as in the deterministic optimization, in order to prove theorems about the speed of convergence, the optimization problem (8.1) must fulfill some additional regularity conditions.

Theorem 8.2 *Suppose that $D^* \neq \emptyset$ and the transition probability measure $\pi(x_n, \cdot)$ is a d-dimensional normal distribution $N\big(\mu(x_n), \Lambda\big)$ with a fixed covariance matrix Λ.*

(a) Let D_ε be bonded for some $\varepsilon = \varepsilon_0 > 0$ and assume that F is convex in a certain neighborhood of D^*. Then

$$\lim_{n\to\infty} n^\gamma \big(F(X_n) - F^*\big) = 0 \text{ w.p.1} \tag{8.9}$$

for each constant γ such that $0 < \gamma < \frac{1}{d}$ and every starting point x_0.

(b) Let D_ε be compact, $D^* = \{x^*\}$, where $x^* \in int(D)$ (= interior of D), and suppose that F is continuous and twice continuously differentiable in a certain neighborhood of x^*. Moreover, assume that F has a positive definite Hessian matrix at x^*. Then for each starting point $x_0 \in D$ it is

$$\lim_{n\to\infty} n^\gamma \big(F(X_n) - F^*\big) = 0 \text{ w.p. 1 for each } 0 < \gamma < \frac{2}{d},$$

$$\lim_{n\to\infty} n^\gamma \|X_n - x^*\| = 0 \text{ w.p. 1 for each } 0 < \gamma < \frac{1}{d}, \tag{8.10}$$

$$\lim_{n\to\infty} \sup n^{\frac{2}{d}} E\big(F(X_n) - F^*\big) \leq \tau(x_0) < +\infty,$$

where $\tau(x_0)$ is a nonnegative finite constant depending on the starting point $x_0 \in D$ and E denotes the expectation operator.

(c) Under the same assumptions as in (b) we also have for each starting point $x_0 \in D$, $x_0 \neq x^*$,

$$\lim_{n\to\infty} \inf n^{\frac{2}{d}} E\big(F(X_n) - F^*\big) \geq h(x_0), \tag{8.11}$$

where $h(x_0)$ is a nonnegative constant depending on the starting point x_0. Furthermore for each $x_0 \in D$, $x_0 \neq x^*$, it is

$$\lim_{n\to\infty} \inf n^\gamma E\|X_n - x^*\| = +\infty \text{ for each } \gamma > \frac{2}{d}. \tag{8.12}$$

Note 8.2

(a) Theorem 8.2 holds also for many non-normal classes of transition probability measures $\pi_n(x_n, \cdot)$), see [13].

(b) It turns out that under the assumptions of Theorem 8.2b the speed of convergence of (8.2) to the global minimum of (8.1) is **exactly** given by

$$E\big(F(X_n) - F^* = O(n^{-\frac{2}{d}}). \tag{8.13}$$

(c) The above convergence rates reflect the fact that in practice one observes that the speed of convergence may be very poor—especially near to the optimum of (8.1).

Hence, using random search procedures, a main problem is the control of the basic random search algorithm (8.2) such that the speed of convergence of (X_n) of F^*, D^*, resp., is increased.

8.3 Controlled Random Search Methods

A general procedure how to speed up the search routine (8.2) is described in [6–8]. The idea is to **associate with the random search routine (8.2) a sequential stochastic decision process** defined by the following items (I)–(III):

(I) We observe that the conditional probability distribution $\pi_n(x_n, \cdot)$ of Z_{n+1} given $X_n = x_n$ depends in general on a certain (vector valued) parameter a, i.e.

$$\pi_n(x_n, \cdot) = \pi_n(a, x_n, \cdot), a \in A, \tag{8.14}$$

where A is the set of admissible parameters a. The method, developed first in [6–8], is now to run the algorithm (8.2) not with a fixed parameter a, but to use an *optimal* control

$$a = a_n^*(x_n) \tag{8.15}$$

of a such that a certain criterion—to be explained in (II)—is maximized.

Exemplary, $\pi_n(x_n, \cdot)$ is assumed to be a d-dimensional normal distribution with mean μ_n and covariance matrix Λ_n . Hence, in this case we have

$$a = (\mu, \Lambda) \in A := M \times Q, \tag{8.16}$$

where $M \subset \mathbb{R}^d$ and Q is the set containing all symmetric, positive definite $d \times d$ matrices and the zero matrix.

(II) To search step $X_n \to X_{n+1}$ there is associated a mean search gain

$$U_n(a_n, x_n) = E\big(u(x_n, X_{n+1})|X_n = x_n\big), \tag{8.17}$$

where the gain function $u(x_n, x_{n+1})$ is defined, e.g., by

$$u(x_n, x_{n+1}) = \begin{cases} 1, \text{ if } x_{n+1} \in G_F(x_n) \\ 0, \text{ else} \end{cases} \tag{8.18a}$$

$$u(x_n, x_{n+1}) = \begin{cases} F(x_n) - F(x_{n+1}), & \text{if } x_{n+1} \in G_F(x_n) \\ 0, & \text{else} \end{cases} \tag{8.18b}$$

$$u(x_n, x_{n+1}) = \begin{cases} \|x_n - x_{n+1}\|, & \text{if } x_{n+1} \in G_F(x_n) \\ 0, & \text{else} \end{cases} \tag{8.18c}$$

Hence, in the first case $U_n(a_n, x_n)$ is the probability of a search success , in the second case $U_n(a_n, x_n)$ is the mean improvement of the value of the objective function and in case (8.18c) $U_n(a_n, x_n)$ is the mean step length of a successful iteration step $X_n \to X_{n+1}$.

(III) Obviously, the convergence behavior of the random search process (S_n) can be improved now by maximizing the mean total search gain

$$U_\infty = U_\infty(a_0, a_1, \ldots) := E \sum_{n=0}^{\infty} \rho^n u(X_n, X_{n+1})$$

subject to the controls $a_n = a_n(x_n) \in A, n = 0, 1, \ldots$, where $\rho > 0$ is a certain discount factor. This maximization can be done in principle by the methods of stochastic dynamic programming, see e.g. [10].

8.4 Computation of Optimal Controls

In order to weaken the computational complexity, the infinite-stage stochastic decision process defined in Sect. 8.3 is replaced by the sequence of 1-stage decision problems

$$\max\ U_n(a_n, x_n) \quad \text{s.t. } a_n \in A,$$

$n = 0, 1, 2, \ldots$. Hence, the *optimal* control $a_n^* = a^*(x_n)$ is defined as a solution of

$$\max_{a \in A} \int_{y \in G_F(x)} u(x, y)\pi(a, x, dy). \tag{8.19}$$

In the following we consider the gain function (8.18b), i.e.

$$u(x, y) = F(x) - F(y).$$

Since an exact analytical solution of (8.19) is not possible in general, we have to apply some approximations. Firstly, the area of success $G_F(x)$ is approximated according to

$$G_F(x) \approx \left(y \in \mathbb{R}^d : \nabla F(x)^T (y - x) + \frac{1}{2}(y - x)^T \nabla^2 F(x)(y - x) < 0 \right), \tag{8.20}$$

where $\nabla F(x)$ denotes the gradient of F and $\nabla^2 F$ is the Hessian matrix of F at x. We assume that $\nabla^2 F(x)$ is regular and $\nabla F(x) \neq 0$. Defining then the vector $w \in \mathbb{R}^d$ by

$$y - x = w - \nabla^2 F(x)^{-1} \nabla F(x), \tag{8.21}$$

the quadratic inequality contained in (8.20) has the form

$$w^T \frac{\nabla^2 F(x)}{r} w < 1,$$

where $r > 0$ is defined by

$$r = \nabla F(x)^T \nabla^2 F(x)^{-1} \nabla F(x).$$

By the Cholesky-decomposition of $\frac{\nabla^2 \Gamma(x)}{r}$ we can determine a matrix Γ such that

$$\frac{\nabla^2 F(x)}{r} = \Gamma \Gamma^T. \tag{8.22}$$

Defining

$$v = \Gamma^T w, \tag{8.23}$$

the approximation (8.20) of $G_F(x)$ can be represented according to (8.21) and (8.22) by

$$G_F(x) \approx \left\{ x_N + \Gamma^{-1^T} v : \|v\| < 1 \right\}, \tag{8.24}$$

where $\| \cdot \|$ is the Euclidean norm and x_N is given by

$$x_N = x - \nabla^2 F(x)^{-1} \nabla F(x).$$

It is then easy to verify that by the same transformation

$$v = v(\omega) := \Gamma^T (y(\omega) - x_N)$$

the search gain $u(x, y) = F(x) - F(y)$ can be approximated by

$$u(x, y) \approx \frac{r}{2}\left(1 - \|v\|^2\right). \tag{8.25}$$

By means of (8.24) and (8.25) the objective function $U(a, x)$, $a = (\mu, \Lambda)$, of (8.19) can be approximated by

$$\widetilde{U}(a, Q) = \frac{r}{2} \int\limits_{\|v\|<1} \left(1 - \|v\|^2\right) f(q, Q, v) dv, \tag{8.26}$$

where the probability density $f = f(q, Q, v)$ of the transformation $v = v(\omega)$ of $y = y(\omega)$ is given by

$$f(q, Q, v) = \frac{1}{(2\pi)^{d/2}(\det Q)^{1/2}} \exp\left(-\frac{1}{2}(v - q)^T Q^{-1}(v - q)\right).$$

Here the d-vector q and the positive definite $d \times d$ matrix Q are given by

$$q = Ev(\omega) = \Gamma^T(\mu - x_n), \tag{8.27}$$

$$Q = cov(v(\cdot)) = \Gamma^T \Lambda \Gamma. \tag{8.28}$$

By the 1-1-transformations (8.27) and (8.28), the maximization problem (8.19) can be approximated by

$$\max_{q \in K, Q \in \mathcal{Q}} \widetilde{U}(q, Q), \tag{8.29}$$

where K and $\mathcal{Q}$ are defined by

$$K = K(M) = \left\{\Gamma^T(\mu - x_N) : \mu \in M\right\},$$

$$\mathcal{Q} = \{0\} \cup \{\Lambda : \Lambda \text{ positive definite } d \times d \text{ matrix}\}$$

and M is a certain subset of $\mathbb{R}^d$.

By the preceding considerations we obtain now this result.

Theorem 8.3 *Let q^*, Q^* be an optimal solution of (8.29) and define μ^*, Λ^* by*

$$\mu^* = x_N + \Gamma^{T^{-1}} q^*,$$

$$\Lambda^* = (\Gamma Q^{*-1} \Gamma^T)^{-1}. \tag{8.30}$$

Then the 1-stage optimal control $a^*(x) = (\mu^*, \Lambda^*)$ is given approximately by (8.30).

In order to determine q^* and Q^*, we suppose now that the feasible set M for the mean value μ is defined by

$$M = \{\mu \in \mathbb{R}^d : \gamma_1^2 r < (\mu - x_N)^T \nabla^2 F(x)(\mu - x_N) \le \gamma_2^2 r\}, \tag{8.31}$$

where $0 < \gamma_1 \le \gamma_2$ are arbitrary, but fixed constants. In this case $K = K(M)$ is given by

$$K = \{q \in \mathbb{R}^d : \gamma_1 \le \|q\| \le \gamma_2\},$$

where $\|\cdot\|$ denotes the Euclidean norm. Note that the important case $M = \{x\}$, i.e. $\mu = x(=$ last iteration point), corresponds to the case $\gamma_1 = \gamma_2 = 1$.

Assume now that M is given by (8.31). Since each Q has the form $Q = T\Delta T^T$, where T is an orthogonal matrix and Δ is a diagonal matrix, the minimization problem (8.29) is equivalent to

$$\begin{aligned} \max \widetilde{U}(q, \Delta) \text{ s.t. } \ & \gamma_1 \le \|q\| \le \gamma_2, \\ & \Delta \in \mathcal{Q}, \Lambda \text{ diagonal.} \end{aligned} \tag{8.32}$$

By a further approximation, we find then that an optimal solution q^*, Q^* of (8.32) is given approximately by this equation

$$\begin{aligned} q^* &= k^* 1, 1 = (1, \ldots, 1)^T, k^* \in \mathbb{R} \\ Q^* &= c^* I, I = \text{ identity matrix}, c^* > 0. \end{aligned} \tag{8.33}$$

Now (8.30), (8.33), and (8.22) yield

$$\begin{aligned} \mu^* &= x_N + k^* \Gamma'^{-1} 1, \\ \Lambda^* &= \left(\Gamma \frac{1}{c^*} \Gamma'\right)^{-1} = c^*(\Gamma\Gamma')^{-1} = c^* \left(\frac{\nabla^2 F(x)}{r}\right)^{-1} \\ &= c^* r \nabla^2 F(x)^{-1}. \end{aligned}$$

Hence, we have this result.

Theorem 8.4 *The 1-stage optimal control $a^*(x) = (\mu^*, \Lambda^*)$ of the random search procedure (8.2) is given approximately by*

$$\begin{aligned} \mu^* &= x_N + k^* \Gamma'^{-1}, \\ \Lambda^* &= c^* \Big(\nabla F(x)' \nabla^2 F(x)^{-1} \nabla F(x)\Big) \nabla^2 F(x)^{-1}, \end{aligned} \tag{8.34}$$

where $k^* \in \mathbb{R}$, $c^* > 0$ are certain fixed parameters.

8.5 Convergence Rates of Controlled Random Search Procedures

Assume that the random search procedure (8.2) has normal distributed search variates $Z_1(\omega), Z_2(\omega), \ldots, Z_n(\omega), \ldots$ controlled by means of the following control law

$$\begin{aligned}\mu^0(x) &= x \\ \Lambda^0(x) &= c\Big(\nabla F(x)'\nabla^2 F(x)^{-1}\nabla F(x)\Big)\nabla^2 F(x)^{-1},\end{aligned} \tag{8.35}$$

where $c > 0$ is a fixed parameter. For control (8.35) we obtain, see the later considerations, this result:

Theorem 8.5 *Suppose that D is a compact, convex subset of $\mathbb{R}^d$ and let x^* be the unique optimal solution of (8.1). Let $x^* \in int(D)$ (= interior of D) and assume that F is twice continuously differentiable in a certain neighborhood of x^*. Moreover, suppose that $\nabla^2 F$ is positive definite at x^*. Then there is a constant $\kappa > 1$ such that*

$$\kappa^n E\big(F(X_n) - F^*\big) \to 0 \textit{ as } n \to \infty$$

and

$$\kappa^n\big(F(X_n) - F^*\big) \to 0 \textit{ as } n \to \infty, \textit{ w.p. 1} \tag{8.36}$$

for all starting points contained in a certain neighborhood of x^*.

Note 8.3

(a) Comparing Theorems 8.2 and 8.5, we find that—at least locally—the convergence rate of (8.2) is increased very much by applying a suitable control, as e.g. the control (8.35).
(b) However, the high convergence rate (8.36) holds only if the starting point x_0 is sufficiently close to x^*, while the low convergence rate found in Theorem 8.2 holds for arbitrary starting points $x_0 \in D$.

Hence, the question arises whether by a certain combination of a controlled random search procedure we also can guarantee a linear convergence rate for all starting points $x_0 \in D$.

Given an increasing sequence N of integers

$$n_1 < n_2 < \ldots < n_k < n_{k+1} < \ldots,$$

let the controls $a_n = (\mu_n, \Lambda_n)$ of the normal distributed search variates $Z_{n+1}(\omega), n = 0, 1, 2, \ldots$, be defined by

$$\mu_n = x_n$$

and

$$\Lambda_n = \begin{cases} \Lambda^0(x_n) \ , & \text{if } n \in N \\ R & , \text{if } n \notin N, \end{cases} \tag{8.37}$$

where $\Lambda^0(x)$ is defined by (8.35) and R is a fixed positive definite $d \times d$ matrix.

Hence, according to (8.37), the search procedure is controlled only at the times $n_1, n_2, \ldots$.

Now, we have this result.

Theorem 8.6 *Suppose that D is a compact, convex subset of $\mathbb{R}^d$ and let $x^* \in int(D)$ be the unique optimal solution of (8.1). Assume that F is twice continuously differentiable in a certain neighborhood of x^* and let $\nabla^2 F(x^*)$ be positive definite. Define then*

$$h_n = \max\{k : n_k < n\}.$$

Then for every starting point $x_0 \in D$ there is a constant $\beta > 1$ such that

$$\beta^{h_n} E\big(F(X_n) - F^*\big) \to 0 \text{ as } n \to \infty$$

and

$$\beta^{h_n}\big(F(X_n) - F^*\big) \to 0 \text{ as } n \to \infty, \text{ w.p. 1}, \tag{8.38}$$

provided that $\lim\sup_{n\to\infty} \frac{h_n}{n} < 1.$

Note 8.4

(a) Hence, the linear convergence rate (8.38) can be obtained by a suitable control of the type (8.37) for each starting points $x_0 \in D$.
(b) If $n_k = \frac{k}{p}$ for some $p \in \mathbb{N}$, then $\beta^{h_n} = (\sqrt[p]{\beta})^n$.
(c) The assertion of Theorem 8.6 follows by using results of [13].

8.6 Numerical Realizations of Optimal Control Laws

In order to realize the control laws obtained in (8.34), (8.35), (8.37), one has to compute the gradient $\nabla F(x)$ and the inverse Hessian Matrix $\nabla^2 F(x)^{-1}$ of F at x. However, since the derivatives ∇F and $\nabla^2 F$ of F are not given in analytical form in practice, the gradient and the Hessian matrix of F must be approximated by means of the information obtained about F during the search process. Hence, for an approximate computation of F and $\nabla^2 F$ we may use the sequence of sample points, iteration points, and function values

$$x_0, F(x_0), z_1, F(z_1), x_1, z_2, F(z_2), x_2, \ldots.$$

In order to define a recursive approximation procedure, for $n = 0, 1, 2, \ldots$ let denote

$$\begin{aligned} g_n &\quad \text{the approximation of } \nabla F(x_n), \\ B_n &\quad \text{the approximation of } \nabla^2 F(x_n), \\ H_n &\quad \text{the approximation of } \nabla^2 F(x_n)^{-1}. \end{aligned}$$

Proceeding recursively, we suppose at the n-th stage of the search process we know the approximations g_n, B_n, and H_n of $\nabla F(x_n)$, $\nabla^2 F(x_n)$, and $\nabla^2 F(x_n)^{-1}$, respectively. Hence, we may compute—approximately—the control $a_n = (\mu_n, \Lambda_n)$ according to one of the formulas (8.34), (8.35), or (8.37) by replacing $\nabla F(x_n)$ and $\nabla^2 F(x_n)^{-1}$ by g_n, H_n, respectively. The search process (8.2) yields then the sample point z_{n+1}, its function value $F(z_{n+1})$ and the next iteration point x_{n+1}. Now we have to perform the update

$$g_n \to g_{n+1}, B_n, \to B_{n+1} \text{ and } H_n \to H_{n+1} \tag{8.39}$$

by using the information $s_n = x_{n+1} - x_n$, $F(x_n)$, z_{n+1}, $F(z_{n+1})$, x_{n+1} about F.

(a) **Search failure at x_n**
If $z_{n+1} \notin D$ or $F(z_{n+1}) \geq F(x_n)$, then $x_{n+1} = x_n$. Since in this case we stay at x_n, we may define the update (8.39) by

$$\begin{aligned} g_{n+1} &= g_n, \\ B_{n+1} &= B_n, \\ H_{n+1} &= H_n. \end{aligned}$$

(b) **Search success at x_n**
In this case it is $z_{n+1} \in D$ and $F(z_{n+1}) < F(x_n)$, hence $x_{n+1} = z_{n+1} \neq x_n$. By a quadratic approximation of F at x_{n+1} we find then

$$\begin{aligned} F(x_n) \approx\ & F(x_{n+1}) + \nabla F(x_{n+1})^T (x_n - x_{n+1}) \\ & + \frac{1}{2}(x_n - x_{n+1})^T \nabla^2 F(x_{n+1})(x_n - x_{n+1}) \end{aligned}$$

and therefore

$$\nabla F(x_{n+1}^T)s_n - \frac{1}{2}s_n^T \nabla^2 F(x_{n+1})s_n \approx \Delta F_n, \tag{8.40}$$

where

$$s_n = x_{n+1} - x_n = z_{n+1} - x_n,$$
$$\Delta F_n = F(x_{n+1}) - F(x_n) = F(z_{n+1}) - F(x_n).$$

Now we have to define the new approximations g_{n+1} and B_{n+1} of $\nabla F(x_n n + 1)$ and $\nabla^2 F(x_{n+1})$, respectively.

Because of (8.40), in order to define the update (8.39), we demand next to the following

Modified Quasi–Newton Condition

$$g_{n+1}^T s_n - \frac{1}{2}s_n^T B_{n+1}s_n = \Delta F_n \tag{8.41}$$

or

$$g_{n+1}^T s_n - \frac{1}{2}s_n^T B_{n+1}s_n < 0. \tag{8.42}$$

Note 8.5

(i) In contrary to (8.41), the modified Quasi-Newton condition (8.42) uses only the information that the function value of F at x_{n+1} is less than that at x_n.

(ii) If $\Delta F_n = F(x_{n+1}) \cdot F(x_n) < 0$, then $-s_n = x_n - x_{n+1}$ is an ascent direction of F at x_{n+1}. Hence, since $\nabla F(x_{n+1})$ is the best ascent direction of F at x_{n+1}, $-s_n$ may be used to define the approximation g_{n+1} of $\nabla F(x_{n+1})$.

Since g_{n+1} is not completely determined by the modified Quasi-Newton condition (8.41) or (8.42), resp., there are still many possibilities to define the update formulas (8.39). Clearly, since B_n is an approximation to a symmetric matrix, we suppose that B_n is a symmetric matrix.

(A) **Additive rank-one-updates**

In order to select a particular tuple (g_{n+1}, B_{n+1}) we may require that (g_{n+1}, B_{n+1}) is an optimal solution $(\overline{g}, \overline{B})$ of the distance minimization problem

$$\min d_1(\overline{B}, B) + d_2(\overline{g}, g) \tag{8.43}$$
$$\text{s.t. } \overline{g}^T s - \frac{1}{2}s^T \overline{B}s = \Delta F,$$

where $B = B_n$, $g = g_n$, $\Delta F = \Delta F_n$ and d_1, d_2 are certain distance measures. We suppose here that d_1, d_2 are defined by

$$d_1(\overline{B}, B) = \frac{1}{2} \sum_{i,j-1}^{d} (\overline{b}_{ij} - b_{ij})^2, \tag{8.44}$$

$$d_2(\overline{g}, g) = \frac{1}{2} \sum_{j-1}^{d} (\overline{g}_j - g_j)^2,$$

where $\overline{b}_{ij}, b_{ij}$ are the elements of $\overline{B}$ and B, resp., and $\overline{g}_j, g_j$ denote the components of $\overline{g}$, g, respectively.

Note 8.6 The minimization (8.43) is a generalization of the minimality principles characterizing some of the well known Quasi-Newton update formulas, see e.g. [4].

Solving (8.43), (8.44), we find that $\overline{g}$, $\overline{B}$ are given by

$$\overline{g} = g - \lambda s \tag{8.45}$$

$$\overline{B} = B + \frac{\lambda}{2} s s^T, \tag{8.46}$$

where the Lagrange multiplier λ is given by

$$\lambda = \frac{g^T s - \frac{1}{2} s^T B s - \Delta F}{s^T s \left(1 + \frac{1}{4} s^T s\right)}. \tag{8.47}$$

If the distance functions d_1, d_2 are changed, then other update formulas may be generated. If e.g. d_2 is replaced by $d_2(\overline{g}, g) = \frac{1}{2}(\overline{g} - g)B^{-1}(\overline{g} - g)$, then $\overline{g} = g - \lambda B s$.

Supposing now that B is positive definite, it is known that the matrix $\overline{B}$ defined by (8.46) is positive definite if and only if

$$1 + \frac{\lambda}{2} s^T H s > 0, \tag{8.48}$$

where $H = B^{-1}$ is our approximation to the inverse Hessian matrix $\nabla^2 F(x)^{-1}$ of F at $x = x_n$. Hence, if $\overline{H} = \overline{B}^{-1}$ denotes the approximation of the inverse Hessian matrix of F at x_{n+1}, then by (8.46) and (8.48) the following update formula $H \to \overline{H}$ for the inverse Hessian matrix of Γ may be established:

$$\overline{H} = \begin{cases} H - \frac{1}{2} \frac{\lambda}{1 + \frac{\lambda}{2} s^T H s} H s s^T H, & \text{if (8.48) holds} \\ H, & \text{else,} \end{cases} \tag{8.49}$$

Updates in the Case of a Search Failure
If $z_{n+1} \notin D$ or $F(z_{n+1}) \geq F(x_n)$, then we stay at $x_{n+1} = x_n$ and we may define therefore $\overline{g} = g$, $\overline{B} = B$, and $\overline{H} = H$. However, also in the case of a search failure the tuple $\big(z_{n+1}, F(z_{n+1})\big)$ yields new information about F, provided only that $z_{n+1} \neq x_n$. Hence, replacing the modified Quasi-Newton condition (8.41) by

$$\overline{g}^T s + \frac{1}{2} s^T \overline{B} s = \Delta F,$$

where now $s = z_{n+1} - x_n$, $\Delta F = F(z_{n+1}) - F(x_n)$, we may derive by the above procedure also update formulas $g \to \overline{g}$, $B \to \overline{B}$, $H \to \overline{H}$ for defining improved approximation $\overline{g}, \overline{B}, \overline{H}$ of $\nabla F, \nabla^2$ and $\nabla^2 F^{-1}$, respectively, at $x_{n+1} = x_n$.

(B) **Multiplicative rank-one-updates**
By (8.45)–(8.49) we have given a first concrete procedure for the realization of the optimal control laws (8.34), (8.35), and (8.37), respectively. Indeed, having e.g. the mean $\mu_n = x_n$ and the covariance matrix

$$\Lambda_n = c^*(g_n^T H_n g_n) H_n, \tag{8.50}$$

the random variable Z_{n+1} may be defined by

$$Z_{n+1} = \mu_n + \Gamma_n Z_{n+1}^0,$$

where Z_{n+1}^0 is a normal distributed with mean zero and covariance matrix equal to the identity matrix, and Γ_n is a $d \times d$ matrix such that

$$\Gamma_n \Gamma_n^T = \Lambda_n. \tag{8.51}$$

Hence, at each iteration point x_n the (Cholesky-)decomposition (8.51) of Λ_n has to be computed.

In order to omit this time-consuming step, we still ask whether update formulas $\Gamma_n \to \Gamma_{n+1}$ for the Cholesky-factors Γ_n may be obtained.

Since $H_n = B_n^{-1}$ we suppose that B_n may be represented by

$$T_n T_n^T = B_n.$$

Then Λ_n is given by

$$\Lambda_n = c^*\big((T_n^{-1} g_n)^T T_n^{-1} g_n\big) T_n^{-1T} T_n^{-1}$$

and the factor Γ_n may be defined, cf. (8.50) by

$$\Gamma_n = \sqrt{c^*} \|T_n^{-1} g_n\| T_n^{-1T}. \tag{8.52}$$

In order to define the update $T \to \overline{T}$, where $T = T_n$ and $\overline{T} = T_{n+1}$ with $T_{n+1}T_{n+1}^T = B_{n+1}$, we require that T is changed only in the direction of $s = x_{n+1} - x_n$. Hence, we assume that

$$\overline{T} = \left(I + \frac{\gamma - 1}{s^T s}, ss^T \right) T,$$

where γ is real parameter to be determined. Furthermore, the distance minimization problem (8.43) is then replaced by

$$\begin{aligned} &\min d_1(\overline{T}, T) + d_2(\overline{g}, g) \\ &\text{s.t. } \overline{g}^T s - \frac{1}{2} s^T \overline{B} s = \Delta F, \end{aligned} \tag{8.53}$$

where now

$$\overline{B} = \left(I + \frac{\gamma - 1}{s^T s} ss^T \right) B \left(I + \frac{\gamma - 1}{s^T s} ss^T \right), \text{ with } B = TT^T. \tag{8.54}$$

If the distance functions d_1, d_2 are again defined corresponding to (8.44), then

$$d_1(\overline{T}, T) = \frac{1}{2} \left(\frac{\gamma - 1}{s's} \right)^2 \|ss^T T\|_E^2, \tag{8.55}$$

where $\|T\|_E$ denotes the Euclidian norm of T. Hence, by (8.53) a particular tuple $(\overline{g}, \gamma)$ is selected. Because of (8.54) and (8.55), the minimization problem (8.53) has the form

$$\begin{aligned} &\min \frac{\|ss^T T\|_E^2}{2} \left(\frac{\gamma - 1}{s's} \right)^2 + \frac{1}{2} \|\overline{g} - g\|_E^2 \\ &\text{s.t. } \overline{g}^T s - \frac{\gamma^2}{2} s^T B s - \Delta F, \end{aligned} \tag{8.56}$$

hence, the tuple $(\overline{g}, \gamma)$ is projected onto the parabola in $\mathbb{R}^{d+1}$ defined by the constraint in (8.56).

Note 8.7

(a) Other update formulas may be gained by changing the objective function of (8.56).
(b) Also for multiplicative updates, similar update rules may be derived in case of search failures.

References

1. Anderson, R.L.: Recent advances in finding best operating conditions. J. Am. Stat. Assoc. **48**(264), 789–798 (1953). http://www.jstor.org/stable/2281072
2. Box, G.: Evolutionary operation: a method for increasing industrial productivity. J. R. Stat. Soc. Ser. C **6**(2), 81–101 (1957). https://doi.org/10.2307/2985505
3. Brooks, S.: A discussion of random methods for seeking maxima. Oper. Res. **6**(2), 244–251 (1958). https://doi.org/10.1287/opre.6.2.244
4. Dennis, J., Schnabel, R.: Numerical Methods for Unconstrained Optimization and Nonlinear Equations. Englewood Cliffs, Prentice-Hall (1983)
5. Karnopp, D.C.: Random search techniques for optimization problems. Automatica **1**(2-3), 111–121 (1963). https://doi.org/10.1016/0005-1098(63)90018-9
6. Marti, K.: Random search in optimization problems as a stochastic decision process (adaptive random search). Meth. Oper. Res. **36**, 223–234 (1979)
7. Marti, K.: Adaptive zufallssuche in der optimierung. ZAMM **60**, 357–359 (1980)
8. Marti, K.: On accelerations of the convergence in random search methods. Meth. Oper. Res. **38**, 391–406 (1980)
9. Müller, P.E.A.: Optimierung mittels stochastischer suchverfahren. Wissenschaftliche Zeitschrift der TU Dresden **21**(1), 69–75 (1983)
10. Müller, P., Nollau, V.: Steuerung Stochastischer Prozesse. Akademie-Verlag, Berlin (1984)
11. Oppel, U.: Random search and evolution. In: Transactions Symposium Applied Mathematics. Thessaloniki (1976)
12. Rappl, G.: Optimierung durch Zufallsirchungsverfahren, Simulation und Optimierung det. und stoch. dynam. Systeme (1980)
13. Rappl, G.: Konvergenzraten von Random-Search-Verfahren zur globalen Optimierung. Ph.D. Thesis, UniBw München (1984)
14. Zielinski, R., Neumann, P.: Stochastische Verfahren zur Suche nach dem Minimum einer Funktion. Akademie-Verlag, Berlin (1983)

Part III
Random Search Methods (RSM): Convergence and Convergence Rates

Chapter 9
Mathematical Model of Random Search Methods and Elementary Properties

Random search methods [2–4] are special stochastic optimization methods to solve the following problem:

Minimize $f(x)$ with regard to $x \in D$, where
$f : M \to \mathbb{R}$ and $D \subset M \subset \mathbb{R}^d$ $(d \in \mathbb{N})$.

The principle on which this method is based is extremely simple and clear. It is similar to the mutation and selection of the biological evolution.

The search is started at an arbitrary starting point $x_0 \in D$. Now a distribution $p_1(x_0, \cdot)$, e.g. a d-dimensional normal distribution with the mean value x_0 and the covariance matrix $C(x_0)$, is used to generate a point y_1. This process is equivalent to mutation. If y_1 lies in D, f is evaluated at y_1 and the two functional values $f(x_0)$ and $f(y_1)$ are compared with each other. If y_1 is better than x_0 (i.e. $f(y_1) \leq f(x_0)$), then $x_1 := y_1$. If y_1 does not lie in D or y_1 is not better than x_0, i.e. $f(y_1) \geq f(x_0)$, then x_0 is retained, i.e. $x_1 := x_0$. This process is equivalent to the selection. Thus the first step of the method is completed. In the second step a distribution $p_2(x_0, x_1, \cdot)$ is used to generate an additional mutation point y_2 which is compared with x_1. Analogously to the first step, this provides the selection point x_2. If this process is continued, it leads to a sequence of random points $x_1, x_2, \ldots$ from D with $f(x_0) \geq f(x_1) \geq f(x_2) \geq \ldots$.

The description also shows that in order to use such a method it only has to be assumed that the function f can be evaluated at any value $x \in D$. Furthermore, it can be seen that one and the same method being applied to two functions f_1 and f_2, which are identical on D, gives the same selection points $x_1, x_2, \ldots$ etc. In other words: A method is not influenced by the way f behaves outside of D. Therefore we may assume without loss of generality (and we will do this in the further course of the discussion) that the domain set M is identical with the set of admissible points D. Therefore we will have to deal with the following basic problem:

K. Marti, *Optimization Under Stochastic Uncertainty*, International Series in Operations Research & Management Science 296,
https://doi.org/10.1007/978-3-030-55662-4_9

$$\begin{aligned}&\text{Minimize } f(x) \text{ with regard to } x \in D, \text{ where}\\ &f : D \to \mathbb{R} \text{ and } D \subset \mathbb{R}^d \text{ not empty .}\end{aligned} \tag{9.1}$$

For technical reasons we will furthermore assume that D and f are measurable, i.e. $D \in \mathcal{B}^d$ and f $\mathcal{B}^d_{|D} - \mathcal{B}^1$ measurable ($\mathcal{B}^d$ = Borel-σ-algebra on $\mathbb{R}^d$). In addition, we will always suppose that f is not constant. This does not represent a restriction, since the minimization problem 9.1 is trivial if f is constant.

Let us now once more have a look at the initial description of the search algorithm. We notice that, if D and f are given, the starting point $x_0 \in D$ and the distributions p_n are the decisive quantities for the practical execution of a random search method. With these quantities the mutation points $y_1, y_2, \ldots$ are generated, and the selection rule

$$\begin{aligned}x_n &= \begin{cases} y_n \text{ if } y_n \in D \text{ and } f(y_n) < f(x_{n-1}) \\ x_{n-1} \text{ otherwise}\end{cases} \quad \text{, therefore}\\ x_n &= y_n \cdot 1_{\{x \in D : f(x) < f(x_{n-1})\}}(y_n) \\ &\quad + x_{n-1} \cdot 1_{\mathbb{R}^d \setminus \{x \in D : f(x) < f(x_{n-1})\}}(y_n) \qquad (n \in \mathbb{N})\end{aligned} \tag{9.2}$$

yields the selection points $x_0, x_1, x_2, \ldots$

We furthermore notice that $x_n = x_{n+1} = x_{n+2} = \ldots$ is satisfied if x_n lies in the minimal set $D^* := \{x \in D : f(x) \leq f(y) \; \forall \; y \in D\}$, due to the fact that from time n on an improvement is no longer possible. It is therefore of no importance how the distributions p_n on D^* are defined. We will use this later on repeatedly without expressly mentioning it.

Using the theorem of Ionescu-Tulcea (cf. [1], p. 48 ff.), it can now be demonstrated that for every sequence of transition probabilities p_n from $(\mathbb{R}^d)^n$ to $\mathbb{R}^d$ and every starting point $x_0 \in D$ there is a probability space $(\Omega, \mathcal{A}, P)$ and sequences of d-dimensional random vectors $(Y_n)_{n \in \mathbb{N}})$, $(X_n)_{n \in \mathbb{N}}$ with the properties

$$\begin{aligned}X_0 &= x_0 \qquad P-\text{a.s.}\\ X_n &= Y_n \cdot 1_{\{x \in D : f(x) < f(X_{n-1})\}}(Y_n)\\ &\quad + X_{n-1} \cdot 1_{\mathbb{R}^d \setminus \{x \in D : f(x) < f(X_{n-1})\}}(Y_n) \qquad P-\text{a.s.}\end{aligned} \tag{9.3}$$

and

$$\begin{aligned}&p_n(x_0, \ldots, x_{n-1}, dy_n) = P_{Y_n}(dy_n \mid X_{n-1} = x_{n-1}, \ldots, X_0 = x_0)\\ &\quad \text{for all } n \in \mathbb{N}\\ &\text{(i.e. } p_n(x_0, \ldots, x_{n-1}, dy_n) \text{ is a version of the}\\ &\text{conditional distribution of } Y_n \text{ under the hypothesis that}\\ &X_{n-1} = x_{n-1}, \ldots, X_0 = x_0).\end{aligned} \tag{9.4}$$

This statement, which we do not want to prove at this point because it has only motivating character, demonstrates that the following definition makes sense.

Definition 9.1 Let $x_0 \in D$. For $n \in \mathbb{N}$ let p_n be a transition probability from $(\mathbb{R}^d)^n$ to $\mathbb{R}^d$ (more exactly: from $(\mathbb{R}^d)^n \times (\mathcal{B}^d)^n$ to $\mathbb{R}^d \times \mathcal{B}^d$). A sequence of d-dimensional random vectors $(X_n : n \in \mathbb{N}_0)$, which is defined on a probability space $(\Omega, \mathcal{A}, P)$, is called *random search method* with *starting point* x_0 and *mutation sequence* $(p_n : n \in \mathbb{N})$ if a sequence of d-dimensional random distributions $(Y_n : n \in \mathbb{N})$ exists on $(\Omega, \mathcal{A}, P)$ such that (9.3) and (9.4) are satisfied.

$(X_n : n \in \mathbb{N}_0)$ is called a random search method with the mutation sequence $(p_n : n \in \mathbb{N})$ if this is a random search method with this mutation sequence and an arbitrary starting point x_0.

In the following the abbreviation R-S-M will be used for the word random search method.

Definition 9.2 Let $(X_n : n \in \mathbb{N}_0)$ be an R-S-M with the mutation sequence $(p_n : n \in \mathbb{N})$.

If $n \in \mathbb{N}$, then

$$\begin{aligned}
&q_n(x_0, \ldots, x_{n-1}, B) \\
&\quad := p_n\left(x_0, \ldots, x_{n-1}, B \cap \{x \in D : f(x) < f(x_{n-1})\}\right) \\
&\quad + \epsilon_{x_{n-1}}(B) \cdot p_n\left(x_0, \ldots, x_{n-1}, \mathbb{R}^d \setminus \{x \in D : f(x) < f(x_{n-1})\}\right) \qquad (9.5) \\
&\quad \left(x_0, \ldots, x_{n-1} \in D, B \in \mathcal{B}^d\right) \\
&\quad (\epsilon_x = \text{ singular distribution }).
\end{aligned}$$

Then for all $n \in \mathbb{N}$

$$q_n(x_0, \ldots, x_{n-1}, dx_n) = P_{X_n}(dx_n \mid X_{n-1} = x_{n-1}, \ldots, X_0 = x_0). \qquad (9.6)$$

The sequence $(q_n : n \in \mathbb{N})$, as defined according to (9.5), is called the *selection sequence* belonging to $(p_n : n \in \mathbb{N})$.

Proof Because of 9.3 we have for all $B \in \mathcal{B}^d$

$$\begin{aligned}
\{X_n \in B\} &= \{X_n \in B\} \cap \{Y_n \in \{x \in D : f(x) < f(X_{n-1})\}\} \quad \cup \\
&\quad \left(\{X_n \in B\} \cap \left\{Y_n \in \left(\mathbb{R}^d \setminus \{x \in D : f(x) < f(X_{n-1})\}\right)\right\}\right) \\
&= \{Y_n \in B \cap \{x \in D : f(x) < f(X_{n-1})\}\} \quad \cup \\
&\quad \left(\{X_{n-1} \in B\} \cap \left\{Y_n \in \left(\mathbb{R}^d \setminus \{x \in D : f(x) < f(X_{n-1})\}\right)\right\}\right)
\end{aligned}$$

From this it follows that

$$
\begin{aligned}
&P\left(X_n \in B \mid X_{n-1}, \ldots, X_0\right) \\
&\quad = P\left(Y_n \in B \cap \{x \in D : f(x) < f(X_{n-1})\} \mid X_{n-1}, \ldots, X_0\right) + \\
&\quad\quad + P(\{X_{n-1} \in B\} \cap \left\{Y_n \in (\mathbb{R}^d \setminus \{x \in D : f(x) < f(X_{n-1})\}\right\} \\
&\quad\quad \mid X_{n-1}, \ldots, X_0) \\
&\quad = p_n(X_0, \ldots, X_{n-1}, B \cap \{x \in D : f(x) < f(X_{n-1})\}) + \\
&\quad\quad + 1_{\{X_{n-1} \in B\}} \cdot p_n(X_0, \ldots, X_{n-1}, \mathbb{R}^d \setminus \{x \in D : f(x) < f(X_{n-1})\}).
\end{aligned}
$$

With that 9.6 is demonstrated. □

Remark 9.1 Let $(X_n : n \in \mathbb{N}_0)$ and $(\hat{X}_n : n \in \mathbb{N}_0)$ be R-S-Ms defined on $(\Omega, \mathcal{A}, P)$ resp. $(\hat{\Omega}, \hat{\mathcal{A}}, \hat{P})$ with the same starting point x_0 and the same mutation sequence $(p_n : n \in \mathbb{N})$. Then we have for all $n \in \mathbb{N}_0$:

$$
P_{(X_0, \ldots, X_n)} = \hat{P}_{(\hat{X}_0, \ldots, \hat{X}_n)}. \tag{9.7}
$$

Proof Because of $P_{X_0} = \epsilon_{x_0} = \hat{P}_{\hat{X}_0}$ (9.7) is valid for $n = 0$.

Let $h : (\mathbb{R}^d)^{k+2} \to \mathbb{R}$ be a measurable nonnegative function and (9.7) be satisfied $n = k$.

Then the following holds

$$
\begin{aligned}
&\int h(x_0, \ldots, x_{k+1}) dP_{(X_0, \ldots, X_{k+1})} = \\
&\int \int h(x_0, \ldots, x_k, x_{k+1}) P_{X_{k+1}}(dx_{k+1} \mid X_k = x_k, \ldots, X_0 = x_0) dP_{(X_0, \ldots, X_k)} = \\
&\int \int h(x_0, \ldots, x_k, x_{k+1}) q_{k+1}(x_0, \ldots, x_k, dx_{k+1}) d\hat{P}_{(\hat{X}_0, \ldots, \hat{X}_k)} = \\
&\int \int h(x_0, \ldots, x_k, x_{k+1}) \hat{P}_{\hat{X}_{k+1}}(dx_{k+1} \mid \hat{X}_k = x_k, \ldots, \hat{X}_0 = x_0) d\hat{P}_{(\hat{X}_0, \ldots, \hat{X}_k)} = \\
&\int h(x_0, \ldots, x_{k+1}) d\hat{P}_{(\hat{X}_0, \ldots, \hat{X}_{k+1})}.
\end{aligned}
$$

Therefore (9.7) is satisfied for $n = k + 1$

Now the assertion follows from complete induction. □

The statement of Note 9.1 allows it to speak of *that* R-S-M with the starting point x_0 that belongs to the mutation sequence $(p_n : n \in \mathbb{N})$. When this is subsequently done, it should be always seen in the sense of (9.7).

The next lemma contains some elementary statements about R-S-Ms. In the following sections these statements will be needed continually.

Lemma 9.1 *Let* $(X_n : n \in \mathbb{N}_0)$ *be an R-S-M with the mutation sequence* $(p_n : n \in \mathbb{N})$ *and the associated* $(q_n : n \in \mathbb{N})$.

Then we have the following elementary properties:

(1) $$q_n(x_0, \ldots, x_{n-1}, \{x \in D : f(x) < f(x_{n-1})\}) = $$
$$= p_n(x_0, \ldots, x_{n-1}, \{x \in D : f(x) < f(x_{n-1})\})$$
for all $n \in \mathbb{N}$ *and any* $x_0, \ldots, x_{n-1} \in D$

(2) $$q_n(x_0, \ldots, x_{n-1}, \{x \in D : f(x) < f(x_{n-1})\} \cup \{x_{n-1}\}) =$$
$$= q_n(x_0, \ldots, x_{n-1}, \{x \in D : f(x) \leq f(x_{n-1})\}) = 1$$
for all $n \in \mathbb{N}$ *and any* $x_0, \ldots, x_{n-1} \in D$

(3) *If* $g : \mathbb{R}^d \to \mathbb{R}$ *is integrable with respect to* $p_n(x_0, \ldots, x_{n-1}, \cdot)$, *then* g *is also integrable with respect to* $q_n(x_0, \ldots, x_{n-1}, \cdot)$ *and we have for all* $n \in \mathbb{N}$ *and* $x_0, \ldots, x_{n-1} \in D$:

$$\int (g(x_{n-1}) - g(x_n)) q_n(x_0, \ldots, x_{n-1}, dx_n) =$$
$$= \int_{\{x \in D : f(x) < f(x_{n-1})\}} (g(x_{n-1}) - g(x_n)) p_n \cdot$$
$$\cdot (x_0, \ldots, x_{n-1}, dx_n)$$

(4) *For every nonnegative, measurable function* $h : D^k \to \mathbb{R}$

$$\int \cdots \int \int h(x_{n+1}, \ldots, x_{n+k}) q_{n+k} \cdot$$
$$\cdot (x_0, \ldots, x_n, x_{n+1}, \ldots, x_{n+k-1}, dx_{n+k}) \cdot$$
$$\ldots q_{n+2}(x_0, \ldots, x_n, x_{n+1}, dx_{n+2}) q_{n+1}(x_0, \ldots, x_n, dx_{n+1})$$

is one version of the conditional expectation

$$E(h(X_{n+1}, \ldots, X_{n+k}) \mid X_n = x_n, \ldots, X_0 = x_0)$$

(5) $$P\left(\bigcap_{i=0}^{\infty} \{X_i \in D\}\right) = P\left(\bigcap_{i=0}^{\infty} \{f(X_{i+1} \leq f(X_i)\}\right) = 1$$

for every starting point $x_0 \in D$

Proof properties (1) and (2) follow directly from the definition of q_n.

This definition also implies property (3) for indicator functions $g(x) = l_B(x) \quad (B \in \mathcal{B}^d)$ and with that also for step functions $g_n(x) = \sum_{i=1}^{n} c_i l_{B_i}(x) \quad (B_i \in \mathcal{B}^d, c_i \geq 0)$.

Since any measurable nonnegative function $g : \mathbb{R}^d \to \mathbb{R}$ is an isotonic limit of step functions g_n, property (3) holds for the positive/negative part of each measurable function $g : \mathbb{R}^d \to \mathbb{R}$ using the monotone convergence theorem. Then the assertion from Part (3) follows from the linearity of the integral.

To (4):

$$
\begin{aligned}
&E(h(X_{n+1}, \ldots, X_{n+k}) \mid X_n = x_n, \ldots, X_0 = x_0) \\
&\quad = \int h(x_{n+1}, \ldots, x_{n+k}) P_{(X_{n+1},\ldots,X_{n+k})} \cdot \\
&\qquad \cdot (d(x_{n+1}, \ldots, x_{n+k}) \mid X_n = x_n, \ldots, X_0 = x_0) \\
&\quad = \int \int h(x_{n+1}, \ldots, x_{n+k-1}, x_{n+k}) P_{X_{n+k}} \cdot \\
&\qquad \cdot (dx_{n+k} \mid X_{n+k-1} = x_{n+k-1}, \ldots, X_{n+1} = x_{n+1}, X_n = x_n, \ldots \\
&\qquad \ldots, X_0 = x_0) P_{(X_{n+1},\ldots,X_{n+k-1})} \cdot \\
&\qquad \cdot (d(x_{n+1}, \ldots, x_{n+k-1}) \mid X_n = x_n, \ldots, X_0 = x_0) = \ldots = \\
&\quad = \int \ldots \int h(x_{n+1}, \ldots, x_{n+k}) P_{X_{n+k}} (dx_{n+k} \mid X_{n+k-1} = x_{n+k-1}, \ldots \\
&\qquad \ldots, X_0 = x_0) \ldots P_{X_{n+2}} (dx_{n+2} \mid X_{n+1} = x_{n+1}, \ldots, X_0 = x_0) \cdot \\
&\qquad \cdot P_{X_{n+1}} (dx_{n+1} \mid X_n = x_n, \ldots, X_0 = x_0).
\end{aligned}
$$

Concerning (9.6) now the assertion follows.

Property (5) follows directly from (9.3). □

Definition 9.3 An R-S-M with the mutation sequence $(p_n : n \in \mathbb{N})$ is called *Markov* method if the transition probability p_n depends only on the last condition, i.e.:
for all $n \in \mathbb{N}$ a transition probability $\hat{p}_n$ from $\mathbb{R}^d$ to $\mathbb{R}^d$ exists with

$$p_n(x_0, \ldots, x_{n-1}, \cdot) = \hat{p}_n(x_{n-1}, \cdot) \quad (x_0, \ldots, x_{n-1} \in \mathbb{R}^d).$$

Additionally, if $\hat{p}_n$ does not depend on n, i.e. if there is a transition probability p from $\mathbb{R}^d$ to $\mathbb{R}^d$ so that

$$p_n(x_0, \ldots, x_{n-1}, \cdot) = p(x_{n-1}, \cdot) \quad (x_0, \ldots, x_{n-1} \in \mathbb{R}^d)$$

is satisfied, then the method is called *stationary* with an associated mutation transition probability p.

The notations

$$p_n(x_0, \ldots, x_{n-1}, \cdot) = p_n(x_{n-1}, \cdot)$$

and

$$p_n(x_0, \ldots, x_{n-1}, \cdot) = p(x_{n-1}, \cdot)$$

may indicate that this is the Markov or stationary case.

References

1. Gänssler, P., Stute, W.: Wahrscheinlichkeitstheorie. Springer, Berlin (1977)
2. Marti, K.: Controlled Random Search Procedures For Global Optimization. Lecture Notes in Control and Information Sciences, vol. 81, pp. 457–474. Springer, Berlin (1986)
3. Spall, J.: Introduction to Stochastic Search and Optimization. J. Wiley, Hoboken (2003)
4. Zabinsky, Z.: Stochastic Adaptation Search for Global Optimization. Springer, New York (2003)

Chapter 10
Special Random Search Methods

10.1 R-S-M with Absolutely Continuous Mutation Sequence

An important subclass is formed by those methods whose mutation transition probabilities p_n have Lebesgue densities.

Definition 10.1 If for all $n \in \mathbb{N}$ $(\mathcal{B}^d)^{n+1} - \mathcal{B}^1$ measurable functions $g_n : (\mathbb{R}^d)^{n+1} \to \mathbb{R}_+$ exist with

$$p_n(x_0, \ldots, x_{n-1}, B) = \int_B g_n(x_0, \ldots, x_{n-1}, y_n)\lambda^d(dy_n)$$
$$\times \left(x_0, \ldots, x_{n-1} \in \mathbb{R}^d, B \in \mathcal{B}^d\right),$$

then the mutation sequence $(p_n : n \in \mathbb{N})$ is called absolutely continuous with the associated density sequence $(g_n : n \in \mathbb{N})$.

If $g_n(x_0, \ldots, x_{n-1}, y_n)$ depends only on the variables x_{n-1} and y_n, then the associated method $(X_n : n \in \mathbb{N}_0)$ is obviously a Markov method. If on top of that $g_n = g$ applies to any $n \in \mathbb{N}$, we have the stationary case, and g is called the density belonging to the R-S-M $(X_n : n \in \mathbb{N}_0)$.

Example 10.1

$$g_n(x_0, \ldots, x_{n-1}, y_n) = g(x_{n-1}, y_n) = \left(2\pi(\sigma(x_{n-1})^2)\right)^{-\frac{d}{2}} \exp\left[-\frac{| y_n - x_{n-1} |^2}{2(\sigma(x_{n-1}))^2}\right]$$

with $\sigma : \mathbb{R}^d \to \mathbb{R}_+$ measurable.

This is the stationary case, and $p(x_{n-1}, \cdot)$ is the d-dimensional normal distribution with the mean x_{n-1} and $(\sigma(x_{n-1}))^2 I$ as covariance matrix (I = unity matrix).

K. Marti, *Optimization Under Stochastic Uncertainty*, International Series in Operations Research & Management Science 296,
https://doi.org/10.1007/978-3-030-55662-4_10

10.2 Random Direction Methods

A variety of random search methods used in practice is characterized by the fact that at each step a random direction is generated independently of past history, and that subsequently a random step length is realized independently of the selected direction (cf. [1–6]). If at the times $0, 1, \ldots, n-1$ the search method is at the selection points $x_0, x_1, \ldots, x_{n-1}$, then the next mutation point y_n will be generated as follows:

(i) First a point s_n and thus a random direction on the unit sphere is selected at random in accordance with the uniform distribution.
(ii) Starting from x_{n-1} in the direction s_n point y_n is now generated on the unlimited line. This is done by realizing a distribution, $\lambda_n(x_0, \ldots, x_{n-1}, \cdot)$ on $\mathbb{R}_+$ (step length distribution) and setting $y_n = x_{n-1} + s_n r_n$.

Notation and Conditions

Let $d \in \mathbb{N}$ and $S := \{x \in \mathbb{R}^d :| x |= 1\}$. Let γ be the normalized uniform distribution on $(S, \mathcal{B}^d \mid S)$ (cf. Appendix A).

Definition 10.2

(1) p_n is said to be of the random direction type if there is a transition probability λ_n from $(\mathbb{R}^d)^n$ to $\mathbb{R}_+$ with

$$\begin{aligned} &p_n(x_0, \ldots, x_{n-1}, B) = \\ &\quad = (\gamma \otimes \lambda_n(x_0, \ldots, x_{n-1}, \cdot)) (\{(s, r) \in S \times \mathbb{R}_+ : x_{n-1} + rs \in B\}) \\ &\quad \times \left(x_0, \ldots, x_{n-1} \in \mathbb{R}^d, B \in \mathcal{B}^d\right). \end{aligned} \tag{10.1}$$

(2) An R-S-M $(X_n : n \in \mathbb{N}_0)$ with the mutation sequence $(p_n : n \in \mathbb{N})$ is called random direction method (in short: R-D-M) if p_n is of the random direction type for all $n \in \mathbb{N}$. The transition probabilities λ_n are called step length distributions.

10.3 Relationships Between Random Direction Methods and Methods with an Absolutely Continuous Mutation Sequence

The following theorems include sufficient conditions for the fact that an R-D-M is a method with an absolutely continuous mutation sequence and vice versa.

Theorem 10.1 *If p_n is of the random direction type and if the transition probability λ_n has a Lebesgue density h_n, i.e.:*
There is a $(\mathcal{B}^d)^n \otimes \mathcal{B}^1 - \mathcal{B}^1$ - measurable function $h_n : (\mathbb{R}^d)^n \times \mathbb{R}_+ \to \mathbb{R}_+$ with

$$\lambda_n(x_0, \dots, x_{n-1}, B) = \int_B h_n(x_0, \dots, x_{n-1}, r)\lambda^1(dr)$$

$$\left(B \in \mathcal{B}^1 \mid \mathbb{R}_+\right).$$

Then the transition probability p_n is absolutely continuous with the density

$$g_n(x_0, \dots, x_{n-1}, y_n) = \frac{\Gamma(\frac{d}{2})}{2\pi^{\frac{d}{2}}} \cdot \frac{h_n(x_0, \dots, x_{n-1}, \mid y_n - x_{n-1} \mid)}{\mid y_n - x_{n-1} \mid^{d-1}} \quad 1_{\mathbb{R}^d \setminus \{x_{n-1}\}}(y)$$

with $\Gamma(x)$ denoting the gamma function

Proof For $x_0, \dots, x_{n-1} \in \mathbb{R}^d$ and $B \in \mathcal{B}^d$ we have

$$\begin{aligned}
&p_n(x_0, \dots, x_{n-1}, B) = \\
&\quad (\gamma \otimes \lambda_n(x_0, \dots, x_{n-1}, \cdot))(\{(s, r) : x_{n-1} + rs \in B\}) = \\
&\quad \int_{S \times \mathbb{R}_+} 1_B(x_{n-1} + rs)(\gamma \otimes \lambda_n(x_0, \dots, x_{n-1}, \cdot))(d(s, r)) = \\
&\quad \int_{\mathbb{R}_+} \int_S 1_B(x_{n-1} + rs)\gamma(ds)\lambda_n(x_0, \dots, x_{n-1}, dr) = \\
&\quad \int_{\mathbb{R}_+} \int_S r^{d-1} \frac{1_B(x_{n-1} + rs)h_n(x_0, \dots, x_{n-1}, \mid rs \mid)}{\mid rs \mid^{d-1}} \gamma(ds)\lambda^1(dr) = \\
&\quad \frac{\Gamma(\frac{d}{2})}{2\pi^{\frac{d}{2}}} \int_{\mathbb{R}^d} \frac{1_B(x_{n-1} + y_n)h_n(x_0, \dots, x_{n-1}, \mid y_n \mid)}{\mid y_n \mid^{d-1}} \lambda^d(dy_n) = \\
&\quad \frac{\Gamma(\frac{d}{2})}{2\pi^{\frac{d}{2}}} \int_B \frac{h_n(x_0, \dots, x_{n-1}, \mid y_n - x_{n-1} \mid)}{\mid y_n - x_{n-1} \mid^{d-1}} \lambda^d(dy_n).
\end{aligned}$$

The last equation follows from the integral transformation theorem and the translation invariance of λ^d. The last but one equation follows from Lemma A.1. □

Theorem 10.2 *Let the transition probability p_n be absolutely continuous with the density g_n. If there is a $(\mathcal{B}^d)^n \otimes \mathcal{B}^1 - \mathcal{B}^1$—measurable function $h_n : (\mathbb{R}^d)^n \times \mathbb{R}_+ \to \mathbb{R}_+$ with*

$$g_n(x_0, \dots, x_{n-1}, y_n) = h_n(x_0, \dots, x_{n-1}, \mid y_n - x_{n-1} \mid),$$

then p_n is of the random direction type with the step length distribution λ_n, where

$$\lambda_n(x_0, \ldots, x_{n-1}, A) = \frac{2\pi^{\frac{d}{2}}}{\Gamma(\frac{d}{2})} \int_A r^{d-1} h_n(x_0, \ldots, x_{n-1}, r) \lambda^1(dr)$$

$$(A \in \mathcal{B}^1 \mid \mathbb{R}_+).$$

Proof Analogously to the proof of Theorem 10.1, Lemma A.1 is used to show that for every $x_0, \ldots, x_{n-1} \in \mathbb{R}^d$ and $B \in \mathcal{B}^d$ holds

$$\begin{aligned}
&(\gamma \otimes \lambda_n(x_0, \ldots, x_{n-1}, \cdot))(\{(s, r) : x_{n-1} + rs \in B\}) = \\
&\qquad \int_{\mathbb{R}^d} 1_B(x_{n-1} + y_n) h_n(x_0, \ldots, x_{n-1}, \mid y_n \mid) \lambda^d(dy_n) = \\
&\qquad \int_{\mathbb{R}^d} 1_B(y_n) h_n(x_0, \ldots, x_{n-1}, \mid y_n - x_{n-1} \mid) \lambda^d(dy_n) = \\
&\qquad \int_B g_n(x_0, \ldots, x_{n-1}, y_n) \lambda^d(dy_n) = \\
&\qquad p_n(x_0, \ldots, x_{n-1}, B).
\end{aligned}$$

□

Remark 10.1

(1) Theorem 10.2 shows that in the situation of Example 10.1 the method is an R-D-M. Furthermore, for the step length distributions λ_n we have

$$\begin{aligned}
\lambda_n(x_0, \ldots, x_{n-1}, A) &= \lambda(x_{n-1}, A) = \\
&= \frac{2^{1-\frac{d}{2}}}{\Gamma(-\frac{d}{2})(\sigma(x_{n-1}))^d} \int_A r^{d-1} \exp\left[-\frac{r^2}{2(\sigma(x_{n-1}))^2}\right] dr.
\end{aligned}$$

(2) Corollary A.1 shows that $\frac{X}{|X|}$ is uniformly distributed on S, if the random vector X has a density $g(x)$ and $g(x)$ is only a function of $\mid x \mid$. Therefore in particular a normalized, standardized normally distributed random vector is equally distributed on S. A realization x of the distribution γ is thus obtained if realizations $x_1, \ldots, x_d$ of d independent and identically standardized normally distributed random vectors $X_1, \ldots, X_d$ are generated and if $x = \frac{(x_1, \ldots, x_d)^T}{\sqrt{x_1^2 + \ldots + x_d^2}}$.

The following lemma which is of crucial importance to future considerations of the rate of convergence provides an important common integration property of R-D-M and methods with absolutely continuous mutation sequence.

Lemma 10.1 *Let the mutation sequence* $(p_n : n \in \mathbb{N})$ *be of the random direction type with the step length distributions* λ_n *or absolutely continuous with the associated density sequence* $(g_n : n \in \mathbb{N})$.

In the first case let it be defined that

$$\begin{cases} \rho_n(x_0,\ldots,x_{n-1},A) & := \gamma(A) \\ \mu_n(x_0,\ldots,x_{n-1},s,B) & := \lambda_n(x_0,\ldots,x_{n-1},B) \end{cases} \tag{10.2}$$

$$\left(x_0,\ldots,x_{n-1}\in\mathbb{R}^d,\quad s\in S,\quad A\in\mathcal{B}^d\mid S,\quad B\in\mathcal{B}^1\mid\mathbb{R}_+\right).$$

In the second case let it be defined that

$$\begin{cases} \rho_n(x_0,\ldots,x_{n-1},A) & := \frac{2\pi^{\frac{d}{2}}}{\Gamma(\frac{d}{2})}\int_A\int_{\mathbb{R}_+} r^{d-1}g_n(x_0,\ldots,x_{n-1},x_{n-1}+rs)\cdot \\ & \quad \cdot\lambda^1(dr)\gamma(ds) \\ \text{and} & \\ \mu_n(x_0,\ldots,x_{n-1},s,B) & := \frac{\int_B r^{d-1}g_n(x_0,\ldots,x_{n-1},x_{n-1}+rs)\lambda^1(dr)}{\int_{\mathbb{R}_+} r^{d-1}g_n(x_0,\ldots,x_{n-1},x_{n-1}+rs)\lambda^1(dr)} \\ \text{if the denominator is} & \text{not zero or infinite.} \\ \text{Otherwise we have} & \\ \mu_n(x_0,\ldots,x_{n-1},s,B) & := \varepsilon_0(B). \end{cases} \tag{10.3}$$

Then in both cases the following integration property is valid:

$$\int_{\mathbb{R}^d} h(y_n-x_{n-1})p_n(x_0,\ldots,x_{n-1},dy_n) =$$
$$\int_S\int_{\mathbb{R}_+} h(rs)\mu_n(x_0,\ldots,x_{n-1},s,dr)\rho_n(x_0,\ldots,x_{n-1},ds) \tag{10.4}$$

for every measurable, nonnegative function $h:\mathbb{R}^d\to\mathbb{R}$

and all $n\in\mathbb{N}$, $x_0,\ldots,x_{n-1}\in\mathbb{R}^d$.

Proof

(i) Let $(p_n : n \in \mathbb{N})$ be of the random direction type with the step length distributions λ_n and let ρ_n and μ_n be define according to (10.2).
Then we have for the indicator functions $h(x) = 1_B(x)$ $(B \in \mathcal{B}^d)$:

$$\int_{\mathbb{R}^d} h(y_n-x_{n-1})p_n(x_0,\ldots,x_{n-1},dy_n) =$$
$$\int_{\mathbb{R}^d} 1_{x_{n-1}+B}(y_n)p_n(x_0,\ldots,x_{n-1},dy_n) =$$
$$\int_S\int_{\mathbb{R}_+} 1_B(rs)(\gamma\otimes\lambda_n(x_0,\ldots,x_{n-1},\cdot))(d(s,r)) =$$
$$\int_S\int_{\mathbb{R}_+} h(rs)\mu_n(x_0,\ldots,x_{n-1},s,dr)\rho_n(x_0,\ldots,x_{n-1},ds).$$

(10.4) is therefore satisfied for indicator functions and thus also for staircase functions. In the first case therefore the assertion follows from the monotone convergence theorem.

(ii) Let $(p_n : n \in \mathbb{N})$ be absolutely continuous with the density sequence $(g_n : n \in \mathbb{N})$.

Let ρ_n and μ_n be defined in accordance with (10.3).

Since g_n is a probability density, it follows from Lemma A.2 that

$$\int_S \int_{\mathbb{R}_+} r^{d-1} g_n(x_0, \ldots, x_{n-1}, x_{n-1} + rs) dr \lambda(ds) = \frac{\Gamma(\frac{d}{2})}{2\pi^{\frac{d}{2}}}.$$

Therefore

$$\int_{\mathbb{R}_+} r^{d-1} g_n(x_0, \ldots, x_{n-1}, x_{n-1} + rs) dr < \infty \ \lambda - a.s. \quad s \in S.$$

If $A := \left\{ s \in S : \int_{\mathbb{R}_+} r^{d-1} g_n(x_0, \ldots, x_{n-1}, x_{n-1} + rs) dr \in (0, \infty) \right\}$.
Then:

$$\int_S \int_{\mathbb{R}_+} h(rs) \mu_n(x_0, \ldots, x_{n-1}, s, dr) \rho_n(x_0, \ldots, x_{n-1}, ds)$$

$$= \frac{2\pi^{\frac{d}{2}}}{\Gamma(\frac{d}{2})} \int_A \int_{\mathbb{R}_+} h(rs) \mu_n(x_0, \ldots, x_{n-1}, s, dr) \cdot$$

$$\cdot \left[\int_{\mathbb{R}_+} t^{d-1} g_n(x_0, \ldots, x_{n-1}, x_{n-1} + ts) dt \right] \gamma(ds)$$

$$= \frac{2\pi^{\frac{d}{2}}}{\Gamma(\frac{d}{2})} \int_A \int_{\mathbb{R}_+} h(rs) \frac{r^{d-1} g_n(x_0, \ldots, x_{n-1}, x_{n-1} + rs)}{\int_{\mathbb{R}_+} t^{d-1} g_n(x_0, \ldots, x_{n-1}, x_{n-1} + ts) dt} dr \cdot$$

$$\cdot \left[\int_{\mathbb{R}_+} t^{d-1} g_n(x_0, \ldots, x_{n-1}, x_{n-1} + ts) dt \right] \gamma(ds)$$

$$= \frac{2\pi^{\frac{d}{2}}}{\Gamma(\frac{d}{2})} \int_A \int_{\mathbb{R}_+} h(rs) r^{d-1} g_n(x_0, \ldots, x_{n-1}, x_{n-1} + rs) dr \gamma(ds)$$

$$= \frac{2\pi^{\frac{d}{2}}}{\Gamma(\frac{d}{2})} \int_S \int_{\mathbb{R}_+} h(rs) r^{d-1} g_n(x_0, \ldots, x_{n-1}, x_{n-1} + rs) dr \gamma(ds)$$

$$= \int_{\mathbb{R}^d} h(y_n) g_n(x_0, \ldots, x_{n-1}, x_{n-1} + y_n) \lambda^d(dy_n)$$

(according to Lemma A.2)

$$= \int_{\mathbb{R}^d} h(y_n - x_{n-1}) g_n(x_0, \ldots, x_{n-1}, y_n) \lambda^d(dy_n)$$

$$= \int_{\mathbb{R}^d} h(y_n - x_{n-1}) p_n(x_0, \ldots, x_{n-1}, y_n).$$

□

References

1. Beltrami, E.J., Indusi, J.P.: An adaptive random search algorithm for constrained minimization. IEEE Trans. Comput. **C-21**(9), 1004–1008 (1972)
2. Gaviano, M.: Some General Results on the Convergence of Random Search Algorithms in Minimization Problems, pp. 149–157. North Holland, Amsterdam (1975)
3. Guseva, O.: Convergence of a random search algorithm. Kibernetika **6**, 143–145 (1971)
4. Lawrence, J., Steiglitz, K.: Randomized pattern search. IEEE Trans. Comput. **100**(4), 382–385 (1972)
5. Lawrence, J., Emad, F.: An analytic comparison of random searching and gradient searching for the extremum of a known objective function. IEEE Trans. Autom. Control **18**(6), 669–671 (1973)
6. Rastrigin, L.: The convergence of the random search method in the extremal control of a many parameter system. Autom. Remote. Control. **24**, 1337–1342 (1963)

Chapter 11
Accessibility Theorems

The central topic in the next chapters is to find sufficient conditions concerning the feasible domain D, the objective function f and the sequence $(p_n : n \in \mathbb{N}_0)$ of mutation probability distributions guaranteeing the convergence of the corresponding random search procedure $(X_n : n \in \mathbb{N}_0)$, hence,

$$f(X_n) \to \inf_{x \in D} f(x) \quad P-\text{a.s.}, \; n \to \infty, \quad \text{for all starting points } x_0 \in D.$$

The decisive tool to solve this problem is given in the present chapter: Here, sufficient conditions are derived such that the procedure $(X_n : n \in \mathbb{N}_0)$ attains sometimes a set of ϵ-minimal points $\{x \in D : f(x) < \epsilon + \inf_{x \in D} f(x)\}$ for any $\epsilon > 0$, see, e.g., [1, 3, 4].

Based on these so-called *accessibility theorems*, subsequently convergence results are derived for special random search methods. Moreover, we consider whether for stationary random search methods positive probabilities of search successes imply the convergence of the search procedure with probability 1.

In the further course of the considerations the additional condition

$$f^* := \inf_{x \in D} \quad f(x) > -\infty$$

is established for the set D and the function f. Furthermore,

$$D^* := \{x \in D : f(x) = f^*\} \text{ and } D_a := \{x \in D : f(x) \leq a\} \qquad (a \in \mathbb{R}).$$

If for all $a \in \mathbb{R}$ the set D_a is nonempty and compact, and if f is continuous on D_a, then, as it is generally known, it follows that D^* is nonempty. Since f was assumed to be nonconstant, $D^* \subsetneqq D$ applies at any time.

K. Marti, *Optimization Under Stochastic Uncertainty*, International Series in Operations Research & Management Science 296,
https://doi.org/10.1007/978-3-030-55662-4_11

Remark 11.1 Let $(X_n : n \in \mathbb{N}_0)$ be an R-S-M with an arbitrary mutation sequence $(p_n : n \in \mathbb{N})$ and an arbitrary starting point x_0 then the following statements are equivalent:

$$(1) \quad \lim_{n\to\infty} P(X_n \in D_a) = 1$$

$$(2) \quad \lim_{n\to\infty} P\left(\bigcap_{m\geq n}\{X_m \in D_a\}\right) = 1$$

$$(3) \quad P\left(\bigcup_{n\in\mathbb{N}}\bigcap_{m\geq n}\{X_m \in D_a\}\right) = 1$$

$$(4) \quad P\left(\bigcup_{n\in\mathbb{N}}\{X_n \in D_a\}\right) = 1.$$

Proof

$$\text{From} \qquad f(X_0) \geq f(X_1) \geq f(X_2) \geq \ldots \quad P-\text{a.s.}$$

$$\text{follows that} \qquad \left\{X_n \in D_a\} = \bigcap_{m\geq n}\{X_m \in D_a\right\} \quad P-\text{a.s.}$$

From this it follows that (1) and (2) are equivalent and that (3) and (4) are equivalent.

With the definition $A_n := \bigcap_{m\geq n}\{X_m \in D_a\}$, the isotony of the sequence $(A_n : n \in \mathbb{N})$ implies the equation

$$P\left(\bigcup_{n=1}^{\infty} A_n\right) = \lim_{n\to\infty} P(A_n).$$

This shows that (2) and (3) are equivalent. □

Theorem 11.1 *Let $(X_n : n \in \mathbb{N}_0)$ be an R-S-M with the mutation sequence $(p_n : n \in \mathbb{N})$. Then we have for every starting point $x_0 \in D$, $a \geq f^*$ and all $n \in \mathbb{N}$*

$$P\left(\bigcup_{i=1}^{n}\{X_i \in D_a\}\right) \geq 1 - \prod_{i=1}^{n}(1-b_i), \text{ where} \tag{11.1}$$
$$b_i := \inf\{p_i(x_0,\ldots,x_{i-1},D_a) : x_1,\ldots,x_{i-1} \in D_{f(x_0)} \setminus D_a\}.$$

If additionally

$$\sum_{i=1}^{\infty} b_i = \infty, \tag{11.2}$$

then

$$\lim_{n\to\infty} P(X_n \in D_a) = 1.$$

Proof

$$1 - P\left(\bigcup_{i=1}^{n}\{X_i \in D_a\}\right) = \quad P\left(\bigcap_{i=1}^{n}\{X_i \in D_{f(x_0)} \setminus D_a\}\right)$$

$$= E(1_{D_{f(x_0)}\setminus D_a}(X_1)\dots 1_{D_{f(x_0)}\setminus D_a}(X_n))$$

$$= \int_{D_{f(x_0)}\setminus D_a} \dots \int_{D_{f(x_0)}\setminus D_a} q_n(x_0, x_1, \dots, x_{n-1}, dx_n)\dots q_1(x_0, dx_1)$$

according to Lemma 9.1.

Because of $q_n(x_0, x_1, \dots, x_{n-1}, D_a) = p_n(x_0, \dots, x_{n-1}, D_a) \geq b_n$ and $q_n(x_0, \dots, x_{n-1}, D_{f(x_0)}) = 1$ if $x_1, \dots, x_{n-1} \in D_{f(x_0)} \setminus D_a$ it follows that

$$\int_{D_{f(x_0)}\setminus D_a} q_n(x_0, x_1, \dots, x_{n-1}, dx_n) \leq 1 - b_n.$$

Iteration yields

$$1 - P\left(\bigcup_{i=1}^{n}\{X_i \in D_a\}\right) \leq (1 - b_n)\cdots(1 - b_1),$$

and this gives (11.1).

The second assertion follows directly from (11.1) and (11.2). □

Remark 11.2 It is trivial that Condition (11.2) is satisfied if

$$\limsup_{n\to\infty} b_n > 0. \tag{11.3}$$

The condition $b_i \geq c > 0$ means that—with the history "$X_1 = x_1, \dots, X_{n-1} = x_{n-1}$" ($x_1, \dots, x_{n-1}\quad D_{f(x_0)} \setminus D_a$)—the probability that the n-th mutation point y_n lies in D_a must be at least equal to c. Naturally, many R-S-Ms do not satisfy this condition; in particular those methods, whose mutation transition probabilities p_n are concentrated only in relatively small areas, e.g. in neighborhoods of the last selection point x_{n-1}, where as a rule $b_i = 0$ (for each $i = 1, 2, \dots$). This means that there is no instant i at which the accessibility of the area D_a in one step from $\mathbb{R}^d \setminus D_a$ is guaranteed with a positive probability.

If convergence considerations are intended for methods of this type as well—which are frequently called local methods—Theorem 11.1 is usually not of much use. In many cases a way out of this situation is offered by the following theorem.

Beforehand it should be noted that Theorem 11.1 can also be proven using 0-1 laws. However, the consequence of this would be that we would have to do without statement (11.1), which will be used substantially in Chap. 14.

Theorem 11.2 *Let $(X_n : n \in \mathbb{N}_0)$ be an R-S-M with the mutation sequence $(p_n : n \in \mathbb{N})$. If for $x_0 \in D$, $a \geq f^*$ and $i \in \mathbb{N}$*

$$c_i := \inf\{E(f(X_{i-1}) - f(X_i) \mid X_{i-1} = x_{i-1}, \dots, X_1 = x_1, X_0 = x_0) : x_1, \dots, x_{i-1} \in D_{f(x_0)} \setminus D_a\}.$$

Then we have for every starting point $x_0 \in D$, $a \geq f^*$ and all $n \in \mathbb{N}$ with $f^* - f(x_0) + \sum_{i=1}^n c_i > 0$

$$P\left(\bigcap_{i=1}^{n}\{X_i \in \mathbb{R}^d \setminus D_a\}\right) \leq \frac{\sum_{i=1}^n Var(f(X_{i-1}) - f(X_i))}{(f^* - f(x_0) + \sum_{i=1}^n c_i)^2} \leq \leq \frac{E(f(x_0) - f(X_n))^2}{(f^* - f(x_0) + \sum_{i=1}^n c_i)^2}. \tag{11.4}$$

If additionally the condition

$$\sum_{i=1}^{\infty} c_i = \infty \tag{11.5}$$

is satisfied, it follows that

$$\lim_{n\to\infty} P(X_n \in D_a) = 1.$$

Proof

(i) $P(X_0 = x_0) = 1$ for every starting point $x_0 \in D$.
Therefore $c_1 = E(f(x_0) - f(X_1))$, and

$$c_n := \inf\{E(f(x_{n-1}) - f(X_n) \mid X_{n-1} = x_{n-1}, \dots, X_1 = x_1) : a < f(x_i) \leq f(x_0), i = 1, \dots, n-1\}$$

for $n \geq 2$.
For $n \in \mathbb{N}_0$ let $\mathcal{A}_n$ be the sigma-algebra generated by $X_0, \dots, X_n$.
Then

$$\begin{aligned}c_n &= \inf\{E(f(X_{n-1}) - f(X_n) \mid \mathcal{A}_{n-1})(\omega) :\\ &\quad : a < f(X_i(\omega)) \le f(x_0), i = 1, \ldots, n-1\}\\ &= \inf\left\{E(f(X_{n-1}) - f(X_n) \mid \mathcal{A}_{n-1})(\omega) :\right.\\ &\quad \left. : \omega \in \bigcap_{i=1}^{n-1}\{X_i \in (D_{f(x_0)} \backslash D_a)\}\right\}\end{aligned}$$

for all $n \in \mathbb{N}$.

(ii) Let $Z_n := f(X_{n-1}) - f(X_n)$, $\hat{Z}_n := E(Z_n \mid \mathcal{A}_{n-1})$ and $\xi_n := Z_n - \hat{Z}_n$ ($n \in \mathbb{N}$).

From $E(\xi_{n+1} \mid \mathcal{A}_n) = 0$ and the $\mathcal{A}_n$-measurability of ξ_n it follows for $k \in \mathbb{N}$ that

$$E(\xi_n \xi_{n+k}) = E(\xi_{n+k} E(\xi_n \mid \mathcal{A}_{n+k-1})) = 0.$$

The variables $\xi_1, \xi_2, \ldots$ are thus uncorrelated.

Therefore

$$Var\left(\sum_{i=1}^{n} \xi_i\right) = \sum_{i=1}^{n} \xi_i.$$

(iii) Let $x_0 \in D$, $a \ge f^*$ and $n \in \mathbb{N}$ with $f^* - f(x_0) + \sum_{i=1}^{n} c_i > 0$.

Let $C := \bigcap_{i=1}^{n}\{X_i \in (D_{f(x_0)} \backslash D_a)\}$.

Then we have for $m \le n$

$$C \subset \bigcap_{i=1}^{m}\{X_i \in (D_{f(x_0)} \backslash D_a)\}$$

and therefore

$$c_m := \inf\left\{\hat{Z}_m(\omega) : \omega \in \bigcap_{i=1}^{m-1}\{X_i \in (D_{f(x_0)} \backslash D_a)\}\right\} \le \inf\left\{\hat{Z}_m(\omega) : \omega \in C\right\}.$$

From $P\left(\bigcap_{i=1}^{n}\{X_i \in D_{f(x_0)}\}\right) = 1$ (cf. Lemma 9.1) it follows that

$$
\begin{aligned}
P\left(\bigcap_{i=1}^{n}\left\{X_i \in \mathbb{R}^d \setminus D_a\right\}\right) &= P(C) = P\left(C \cap \{f(X_n) > a\}\right) \\
&\le P(C \cap \{f(X_n) \ge f^*\}) = P\left(C \cap \left\{\sum_{i=1}^{n} Z_i \le f(x_0) - f^*\right\}\right) \\
&= P\left(C \cap \left\{\sum_{i=1}^{n} \xi_i \le f(x_0) - f^* - \sum_{i=1}^{n} \hat{Z}_i\right\}\right) \\
&\le P\left(C \cap \left\{\sum_{i=1}^{n} \xi_i \le f(x_0) - f^* - \sum_{i=1}^{n} \inf\left\{\hat{Z}_i(\omega) : \omega \in C\right\}\right\}\right) \\
&\le P\left(C \cap \left\{\sum_{i=1}^{n} \xi_i \le f(x_0) - f^* - \sum_{i=1}^{n} c_i\right\}\right) \\
&\le P\left(\left|\sum_{i=1}^{n} \xi_i\right| \ge f^* - f(x_0) + \sum_{i=1}^{n} c_i\right)
\end{aligned}
$$

because $f(x_0) - f^* - \sum_{i=1}^{n} c_i < 0.$

Now the first inequality (11.4), follows from the Chebyshev inequality if it is taken into consideration that

$$
\begin{aligned}
E\left(\xi_n^2 \mid \mathcal{A}_{n-1}\right) &= E\left(Z_n^2 \mid \mathcal{A}_{n-1}\right) - 2\hat{Z}_n E(Z_n \mid \mathcal{A}_{n-1}) + \hat{Z}_n^2 \\
&= E\left(Z_n^2 \mid \mathcal{A}_{n-1}\right) - \hat{Z}_n^2
\end{aligned}
$$

furthermore $E(\xi_n) = 0$ and therefore

$$
Var\,(\xi_n) = E\left(\xi_n^2\right) = E\left(Z_n^2\right) - E\left(\hat{Z}_n^2\right)
$$

so that

$$
\begin{aligned}
Var(\xi_n) \le& E\left(Z_n^2\right) - \left(E\hat{Z}_n\right)^2 \\
=& E\left(Z_n^2\right) - (EZ_n)^2 = Var \quad Z_n
\end{aligned}
$$

is obtained from the Jensen inequality.

The second inequality of (11.4) follows from

$$\sum_{i=1}^{n} Var Z_i \le \sum_{i=1}^{n} E(Z_i^2) = E\left(\sum_{i=1}^{n} Z_i^2\right)$$
$$\le E\left(\sum_{i=1}^{n} Z_i\right)^2 = E(f(x_0) - f(X_n))^2$$

with the last inequality following because of $Z_i \ge 0 \quad P-$a.s..

(iv) The second assertion of the theorem follows from (11.4) and (11.5) taking into account the inequality

$$E(f(x_0) - f(X_n))^2 \le (f(x_0) - f^*)^2.$$

□

Remark 11.3 Condition1 (11.5) of Theorem 11.2 is satisfied in particular if sequences of positive numbers $(\alpha_n : n \in \mathbb{N})$ and $(\beta_n : n \in \mathbb{N})$ exist with

$$\begin{cases} \sum_{n=1}^{\infty} \alpha_n \beta_n = \infty \qquad \text{and} \\ \inf\Big\{ p_n(x_0, x_1, \ldots, x_{n-1}, \{x \in D : f(x) \le f(x_{n-1}) - \alpha_n\}) : \\ \qquad a < f(x_i) \le f(x_0), i = 1, 2, \ldots, n-1 \Big\} \ge \beta_n. \end{cases} \tag{11.6}$$

For with the Chebyshev inequality it follows that

$$\begin{aligned} c_n &= \inf\Big\{ E(f(x_{n-1}) - f(X_n) \mid X_{n-1} = x_{n-1}, \ldots, X_1 = x_1, X_0 = x_0) : \\ &\qquad a < f(x_i) \le f(x_0), \quad i = 1, 2, \ldots, n-1 \Big\} \\ &\ge \alpha_n \inf\Big\{ P(f(x_{n-1}) - f(X_n) \ge \alpha_n \mid X_{n-1} = x_{n-1}, \ldots, X_0 = x_0) : \\ &\qquad a < f(x_i) \le f(x_0), \quad i = 1, 2, \ldots, n-1 \Big\} \\ &= \alpha_n \inf\Big\{ q_n(x_0, x_1, \ldots, x_{n-1}, \{x \in D : f(x) \le f(x_{n-1}) - \alpha_n\}) : \\ &\qquad a < f(x_i) \le f(x_0), \quad i = 1, 2, \ldots, n-1 \Big\} \\ &= \alpha_n \inf\Big\{ p_n(x_0, x_1, \ldots, x_{n-1}, \{x \in D : f(x) \le f(x_{n-1}) - \alpha_n\}) : \\ &\qquad a < f(x_i) \le f(x_0), \quad i = 1, 2, \ldots, n-1 \Big\} \\ &\ge \alpha_n \beta_n. \end{aligned}$$

The next lemma establishes the relationship between the convergence

$$f(X_n) \to f^* \qquad P - \text{a.s.}$$

and the accessibility of the areas D_a for all $a > f^*$, which is crucial for the next chapters.

Lemma 11.1 *Let $(X_n : n \in \mathbb{N}_0)$ be an R-S-M with an arbitrary mutation sequence and an arbitrary starting point $x_0 \in D$.*

Then the following statements are equivalent:

(1)

$$f(X_n) \to f^* \qquad P - a.s.$$

(2)

$$\lim_{n\to\infty} P(\{X_n \in D_a\}) = 1 \qquad \textit{for any } a > f^*.$$

Proof It is generally known (cf. [2], p. 171) that

$$f(X_n) - f^* \to 0 \qquad P - \text{a.s.} \Longleftrightarrow \sup_{m\geq n} \mid f(X_m) - f^* \mid \underset{n\to\infty}{\longrightarrow} 0 \qquad \text{P-stoch.}$$

From this it follows that

$$\begin{aligned} &f(X_n) - f^* \to 0 \qquad P - \text{a.s.} \Longleftrightarrow \\ &\lim_{n\to\infty} P\left(\sup_{m\geq n} \mid f(X_m) - f^* \mid \leq a\right) = 1 \text{ for all } a > 0 \Longleftrightarrow \\ &\lim_{n\to\infty} P\left(\bigcap_{m\geq n} \{X_m \in D_a\}\right) = 1 \text{ for all } a > f^*. \end{aligned}$$

Now the assertion follows from Remark 11.1. □

References

1. Driml, M., Hanš, O.: On a randomized optimization procedure. In: Transactions of the Fourth Prague Conference on Information Theory, Statistical Decision Functions and Random Processes (held at Prague 1965), pp. 273–276 (1967)
2. Gänssler, P., Stute, W.: Wahrscheinlichkeitstheorie. Springer, Berlin (1977)
3. Marti, K.: On accelerations of the convergence in random search methods. Methods Oper. Res. **37**, 391–406 (1980)
4. Oppel, U.: Random search and evolution. In: Trans. Symp. Applied Math. Thessaloniki (1976)

Chapter 12
Convergence Theorems

12.1 Convergence of Random Search Methods with an Absolutely Continuous Mutation Sequence

Theorem 12.1 *Let $(X_n : n \in \mathbb{N}_0)$ be an R-S-M with an absolutely continuous mutation sequence $(p_n : n \in \mathbb{N})$. Let the associated density sequence $(g_n : n \in \mathbb{N})$ be uniformly bounded towards zero, i.e.:*

For each $x_0 \in D \setminus D^*$ and $a \in (f^*, f(x_0))$ there exists a $c > 0$ with

$$g_n(x_0, \ldots, x_{n-1}, y_n) \geq c \text{ for all } n \in \mathbb{N}, \tag{12.1}$$
$$x_1, \ldots, x_{n-1} \in D_{f(x_0)} \setminus D_a, \quad y_n \in D_a.$$

If furthermore

$$\lambda^d(D_a) > 0 \quad \text{for all } a > f^*, \tag{12.2}$$

then we have for every starting point $x_0 \in D$

$$f(X_n) \underset{n\to\infty}{\longrightarrow} f^* \qquad P-a.s.$$

Proof Let $a > f^*$ be arbitrary.

To prove the assertion, it is sufficient to demonstrate according to Lemma 11.1 that the condition $\lim_{n\to\infty} P(X_n \in D_a) = 1$ is satisfied.

However, this follows from Theorem 11.1, because for all starting points $x_0 \in D \setminus D^*$ and all $x_1, \ldots, x_{n-1} \in D_{f(x_0)} \setminus D_a$ we have

K. Marti, *Optimization Under Stochastic Uncertainty*, International Series in Operations Research & Management Science 296,
https://doi.org/10.1007/978-3-030-55662-4_12

$$p_n(x_0, x_1, \ldots, x_{n-1}, D_a) = \int_{D_a} g_n(x_0, \ldots, x_{n-1}, y_n)\lambda^d(dy_n)$$
$$\geq c\lambda^d(D_a)$$

and this, which is trivial, satisfies condition (11.2).

For starting points $x_0 \in D^*$ the assertion is trivial. □

Remark 12.1 If f is continuous and D_a is compact for all $a \geq f^*$, then for all $x_0 \in D$ and sequences $(x_n : n \in \mathbb{N})$ from $D_{f(x_0)}$ with

$$\lim_{n\to\infty} f(x_n) = f^* \qquad \text{also} \qquad \lim_{n\to\infty} \text{dist}(x_n, D^*) = 0.$$

Proof Let $(x_n : n \in \mathbb{N})$ be a sequence from $D_{f(x_0)}$ with the above property.

If it is assumed that $\limsup_{n\to\infty} \text{dist}(x_n, D^*) > 0$, then there is an $\varepsilon > 0$ and one subsequence of the natural numbers $(n_k : k \in \mathbb{N})$ so that $\text{dist}(x_{n_k}, D^*) \geq \varepsilon$ for each $k \in \mathbb{N}$.

Due to the compactness of the set $D_{f(x_0)}$ a subsequence $(x_{n_{k'}}) \subset (x_{n_k})$ converges towards an $x \in D$.

From the continuity of f it follows that $\lim_{k\to\infty} f(x_{n_{k'}}) = f(x)$.

From the continuity of the representation $z \to \text{dist}(z, D^*)$ (cf. [4, p. 78]) it follows that $\text{dist}(x, D^*) \geq \varepsilon$, i.e. $f(x) > f^*$ in contradiction to $f(x) = \lim_{k\to\infty} f(x_{n_{k'}}) = \lim_{n\to\infty} f(x_n) = f^*$. □

Remark 12.2 If $D^* \cap \overline{\overset{\circ}{D}} \neq \emptyset$ and if f is continuous on D^*, then there is a $y \in \overset{\circ}{D}$ and an $r > 0$ with $K(y, r) \subset D_a$ for all $a > f^*$.

Therefore in particular condition (12.2) is satisfied.

Proof Let $a > f^*$ be arbitrary and $x^* \in D^* \cap \overline{\overset{\circ}{D}}$.

Then for each $\varepsilon > 0$ there is a $y \in \overset{\circ}{D}$ with $| y - x^* | < \varepsilon$.

Since f is continuous at x^*, $\varepsilon > 0$ can be chosen in a way that $z \in D_a$, if $| z - x^* | < \varepsilon$.

Let now be $y \in \overset{\circ}{D}$ with $| y - x^* | < \frac{\varepsilon}{2}$ and $\sigma \in (0, \frac{\varepsilon}{2})$ with $K(y, \sigma) \subset \overset{\circ}{D}$. From the triangular inequality it follows that $K(y, \sigma) \subset K(x^*, \varepsilon)$ and therefore $K(y, \sigma) \subset D_a$. □

Note 12.1 The condition $D^* \cap \overline{\overset{\circ}{D}} \neq \emptyset$ states that the points $x^* \in D^*$ must not be isolated points with regard to $\overset{\circ}{D}$.

For, if $D^* \cap \overline{\overset{\circ}{D}} = \emptyset$, then any $x^* \in D^*$ would have a neighborhood $K(x^*, r)$ $(r > 0)$ with $K(x^*, r) \cap \overset{\circ}{D} = \emptyset$.

Remark 12.3 Condition (12.1) is satisfied if any $D_a \quad (a > f^*)$ is compact, f is continuous and, furthermore, in the stationary case, i.e. if $g_n(x_0, \ldots, x_{n-1}, y_n) = g(x_{n-1}, y_n)$, and g is continuous and positive on $(D \setminus D^*) \times D$.

Proof Let $x_0 \in D \setminus D^*$ and $a \in (f^*, f(x_0))$.

The compactness of the set $\{x \in D : f(x_0) \geq f(x) \geq a\}$ follows from the fact that f is continuous and $D_{f(x_0)}$ is compact.

Since $D_{f(x_0)} \setminus D_a$ is included in this set, it follows that there is a $c > 0$ with $g(x, y) \geq c$ for all $x \in D_{f(x_0)} \setminus D_a$ and $y \in D_a$. □

Example 12.1

$$
\begin{aligned}
g_n(x_0, \ldots, x_{n-1}, y_n) =& g(x_{n-1}, y_n) = \\
=& (2\pi(\sigma(x_{n-1}))^2)^{-\frac{d}{2}} \exp\left[-\frac{| y_n - x_{n-1} |^2}{2(\sigma(x_{n-1}))^2}\right]
\end{aligned}
$$

with $\sigma : D \to \mathbb{R}_+$ measurable (cf. Example 10.1).

If f is continuously and D_a is compact for all $a > f^*$, then

$$\sigma(x) := \sigma > 0$$

Condition (12.1) is satisfied according to Note 12.3.

If f is continuously differentiable and $\nabla f(x) \neq 0$ is satisfied for all $x \in D \setminus D^*$ (which is the case, e.g., for a convex function), the control

$$\sigma(x) :=| \nabla f(x) |$$

also satisfies condition (12.1) according to Note 12.3.

This follows from the fact that σ is continuous and positive on $D \setminus D^*$. If therefore additionally $D^* \cap \overset{\circ}{\overline{D}} \neq \emptyset$, then the conditions (12.1), respectively, (12.2) are met in both cases according to the Remark 12.3 and 12.2, and it follows from Theorem 12.1 and Remark 12.1 for all starting points $x_0 \in D$ that:

$$f(X_n) - f^* \to 0 \qquad P-\text{a.s.} \qquad \text{and} \qquad \text{dist}(X_n, D^*) \to 0 \qquad P-\text{a.s.}$$

The next example shows that the convergence of $f(X_n)$ towards f^* does not generally imply the convergence of $\text{dist}(X_n, D^*)$ towards zero.

Example 12.2 Let $D = [0, 1], \quad g_n(x_0, \ldots, x_{n-1}, y_n) = 1_{[0,1]}(y_n)$ and

$$f(x) := \begin{cases} 1 & \text{if} \quad x = 0 \\ x & \text{if} \quad 0 < x < 1 \\ 0 & \text{if} \quad x = 1. \end{cases}$$

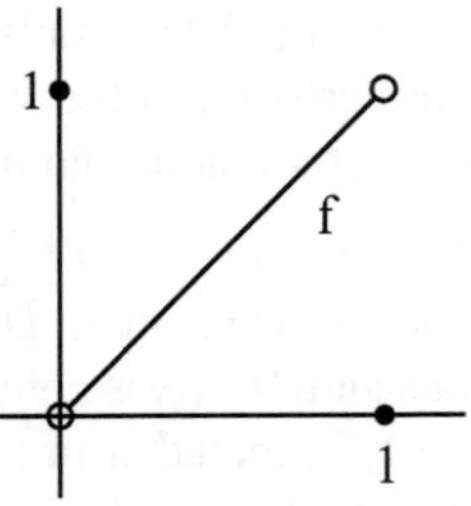

In this case therefore $p(x, \cdot)$ is the uniform distribution on D for any $x_0 \in D$. From Theorem 12.1 it follows that for all starting points $x_0 \in D$

$$f(X_n) \to f^* \qquad P - \text{a.s.} \qquad \text{, i.e.} \qquad X_n \to 0 \qquad P - \text{a.s.}$$

and therefore

$$\text{dist}(X_n, D^*) \to 1 \qquad P - \text{a.s.}$$

12.2 Convergence of Random Direction Methods

Theorem 12.2 *Let $(X_n : n \in \mathbb{N}_0)$ be an R-D-M. Let the step length distributions λ_n be absolutely continuous with the Lebesgue densities h_n.*

For all $x_0 \in D, a \in (f^*, f(x_0))$ and $K > 0$:

$$\sum_{n=1}^{\infty} \inf\big\{h_n(x_0, x_1, \ldots, x_{n-1}, r) : \\ x_1, \ldots, x_{n-1} \in D_{f(x_0)} \setminus D_a, r \in (0, K]\big\} = \infty. \tag{12.3}$$

Furthermore, if (12.2) is satisfied and any D_a ($a \geq f^*$) is bounded, we have for all starting points $x_0 \in D$:

$$f(X_n) \to f^* \qquad P - a.s.$$

Proof According to Theorem 10.1, $(X_n : n \in \mathbb{N}_0)$ is an R-S-M with the absolutely continuous mutation sequence $(p_n : n \in \mathbb{N})$ and associated density sequence $(g_n : n \in \mathbb{N})$ where

$$g_n(x_0, x_1, \ldots, x_{n-1}, y_n) := \frac{\Gamma(\frac{d}{2})}{2\pi^{\frac{d}{2}}} \cdot \frac{h_n(x_0, x_1, \ldots, x_{n-1}, \mid y_n - x_{n-1} \mid)}{\mid y_n - x_{n-1} \mid^{d-1}} \cdot \\ \cdot 1_{\mathbb{R}^d \setminus \{x_{n-1}\}}(y_n)$$

can be selected as a possible value for g_n.

Let now $x_0 \in D \setminus D^*$ and $a \in (f^*, f(x_0))$ be arbitrary.
Furthermore, let $\tilde{a} \in (f^*, a)$ and $K := \sup\left\{ \mid x - y \mid : x, y \in D_{f(x_0)}\right\}$.
According to the condition, $D_{f(x_0)}$ is bounded, i.e. $K < \infty$ and therefore because of (12.2)

$$c := \frac{\Gamma(\frac{d}{2})}{2\pi^{\frac{d}{2}}}(a - \tilde{a})\frac{\lambda^d(D_{\tilde{a}})}{K^{d-1}} > 0.$$

Let now $x_1, \ldots, x_{n-1} \in D_{f(x_0)} \setminus D_a$ be arbitrary. Then:

$$\begin{aligned}
&E\Big(f(x_{n-1}) - f(X_n) \mid X_{n-1} = x_{n-1}, \ldots, X_1 = x_1, X_0 = x_0\Big)\\
&= \int \Big(f(x_{n-1}) - f(x_n)\Big) q_n(x_0, x_1, \ldots, x_{n-1}, dx_n)\\
&= \int_{D_{f(x_{n-1})}} \Big(f(x_{n-1}) - f(y_n)\Big) p_n(x_0, x_1, \ldots, x_{n-1}, dy_n)\\
&\geq \int_{D_{\tilde{a}}} \Big(f(x_{n-1}) - f(y_n)\Big) p_n(x_0, x_1, \ldots, x_{n-1}, dy_n)\\
&\geq c \cdot \inf\Big\{h_n(x_0, x_1, \ldots, x_{n-1}, \mid y_n - x_{n-1} \mid) : y_n \in D_{\tilde{a}}\Big\}\\
&\geq c \cdot \inf\Big\{h_n(x_0, x_1, \ldots, x_{n-1}, r) : r \in (0, K)\Big\}.
\end{aligned}$$

Therefore:

$$\begin{aligned}
&\inf\Big\{E\big(f(x_{n-1}) - f(X_n) \mid X_{n-1} = x_{n-1}, \ldots, X_1 = x_1, X_0 = x_0\big) :\\
&\qquad x_1, \ldots, x_{n-1} \in D_{f(x_0)} \setminus D_a\Big\}\\
&\qquad \geq c \cdot \inf\Big\{h_n(x_0, x_1, \ldots, x_{n-1}, r) : x_1, \ldots, x_{n-1} \in D_{f(x_0)} \setminus D_a, r \in (0, K)\Big\}.
\end{aligned}$$

From (12.3) it follows that condition (11.5) is valid.
Then the assertion follows from Theorem 11.2 and Lemma 11.1 for all $x_0 \in D \setminus D^*$.
For $x_0 \in D^*$ the assertion is trivial. □

Example 12.3

(1) Let $(a_n : n \in \mathbb{N})$ be a bounded sequence of positive numbers with $\sum\limits_{n=1}^{\infty} a_n = \infty$.
Let $\lambda_n(x_0, x_1, \ldots, x_{n-1}, \cdot) = \lambda_n(\cdot)$ be the exponential distribution with the parameter a_n, i.e. $h_n(x_0, \ldots, x_{n-1}, r) = a_n \exp(-a_n r)$.
With $a := \sup\{a_n : n \in \mathbb{N}\}$ it follows for all $K > 0$ that

$$\sum_{n=1}^{\infty} \inf \left\{h_n(r) : r \in (0, K]\right\} \geq \sum_{n=1}^{\infty} a_n \exp(-a_n K) \geq \exp(-aK) \sum_{n=1}^{\infty} a_n.$$

Thus (12.3) is satisfied.

(2) Let $h_n(x_0, x_1, \ldots, x_{n-1}, r) := \sqrt{\frac{2}{\pi}} \frac{1}{\sigma(x_{n-1})} \exp\left[-\frac{r^2}{2\left(\sigma(x_{n-1})\right)^2}\right]$ with $\sigma : D \to (0, \infty)$.
   Let σ be continuous on $D \setminus D^*$.
   Because of

$$\inf\left\{\frac{1}{\sigma(x)} \exp\left[-\frac{r^2}{2\sigma^2(x)}\right] : x \in D_{f(x_0)} \setminus D_a, r \in (0, K)\right\}$$
$$= \inf\left\{\frac{1}{\sigma(x)} \exp\left[-\frac{K^2}{2\sigma^2(x)}\right] : x \in D_{f(x_0)} \setminus D_a\right\}$$

it follows from Note 12.3 that (12.3) f is valid, if f is continuous and any $D_a \quad (a > f^*)$ is compact.

Theorem 12.3 *Let $(X_n : n \in \mathbb{N}_0)$ be an R-D-M with the step length distributions λ_n. Let D be convex, $D^* \neq \emptyset$, $\overset{\circ}{D} \neq \emptyset$ and $f : D \to \mathbb{R}$ continuous and convex. Furthermore, let D_a be bounded for all $a \geq f^*$. It is also assumed that we have for all $x_0 \in D$, $a \in (f^*, f(x_0))$ and $c > 0$*

$$\sum_{n=1}^{\infty} \inf\left\{ \int_{(0,c)} r\lambda_n(x_0, \ldots, x_{n-1}, dr) : \right.$$
$$\left. x_1, \ldots, x_{n-1} \in D_{f(x_0)} \setminus D_a \right\} = \infty. \tag{12.4}$$

Then for all starting points $x_0 \in D$

$$f(X_n) \to f^* \qquad P - a.s.$$

Proof It is assumed that $x_0 \in D \setminus D^*$ without loss of generality (for $x_0 \in D^*$ the assertion is trivial).

Let $a \in (f^*, f(x_0))$ and $x_1, \ldots, x_{n-1} \in D_{f(x_0)} \setminus D_a$ be arbitrary. Furthermore, let $\tilde{a} \in (f^*, a)$.

From the convexity of D it follows that $\overline{\overset{\circ}{D}} = \overline{D}$ (cf. [2, p. 16]) and therefore $D^* \subset \overline{\overset{\circ}{D}}$.

Because of Remark 12.2 there exist a $z \in \overset{\circ}{D}$ and an $\varepsilon > 0$ with

$$\overline{K(z, 2\varepsilon)} \subset D_{\tilde{a}}.$$

From the boundedness of $D_{f(x_0)}$ it follows that

$$K := \sup \left\{ | x - y | : x, y \in D_{f(x_0)} \right\} < \infty.$$

For $x \in D_{f(x_0)} \setminus D_a$ let

$$\beta(x) := \arcsin \frac{\varepsilon}{| x - z |}.$$

Because of $x_{n-1} \in D_{f(x_0)} \setminus D_a$ and $\overline{K(z, 2\varepsilon)} \subset D_{\tilde{a}} \subset D_a$, it follows that

$$| z - x_{n-1} | > 2\varepsilon \text{ and therefore } | z - x_{n-1} | - \varepsilon > \varepsilon > 0.$$

From Lemma B.10 it follows that

$$\begin{aligned} & x_{n-1} + C\,(\beta(x_{n-1}), z - x_{n-1}, \varepsilon) \\ & \subset x_{n-1} + C\,(\beta(x_{n-1}), z - x_{n-1}, | z - x_{n-1} | - \varepsilon) \\ & \subset \operatorname{conv}\left(\{x_{n-1}\} \cup \overline{K(z, \varepsilon)} \right). \end{aligned}$$

From the convexity of D and from $\{x_{n-1}\} \cup \overline{K(z, \varepsilon)} \subset D$ it follows that:

$$x_{n-1} + C\,(\beta(x_{n-1}), z - x_{n-1}, \varepsilon) \subset D.$$

Let now $y \in x_{n-1} + C\,(\beta(x_{n-1}), z - x_{n-1}, \varepsilon)$ be arbitrary.
Then according to Lemma B.10, there is a $t > 0$ with

$$| x_{n-1} + t(y - x_{n-1}) - z | \leq \varepsilon.$$

Let $z' := x_{n-1} + t(y - x_{n-1})$.
Because of

$$| z' - x_{n-1} | \geq | z - x_{n-1} | - | z' - z |$$

and

$$| y - x_{n-1} | \leq \varepsilon \leq | z - x_{n-1} | - \varepsilon$$

it follows that

$$| z' - x_{n-1} | \geq | y - x_{n-1} | + \varepsilon - | z' - z | \geq | y - x_{n-1} |$$

and therefore $t = \frac{|z' - x_{n-1}|}{|y - x_{n-1}|} \geq 1$.

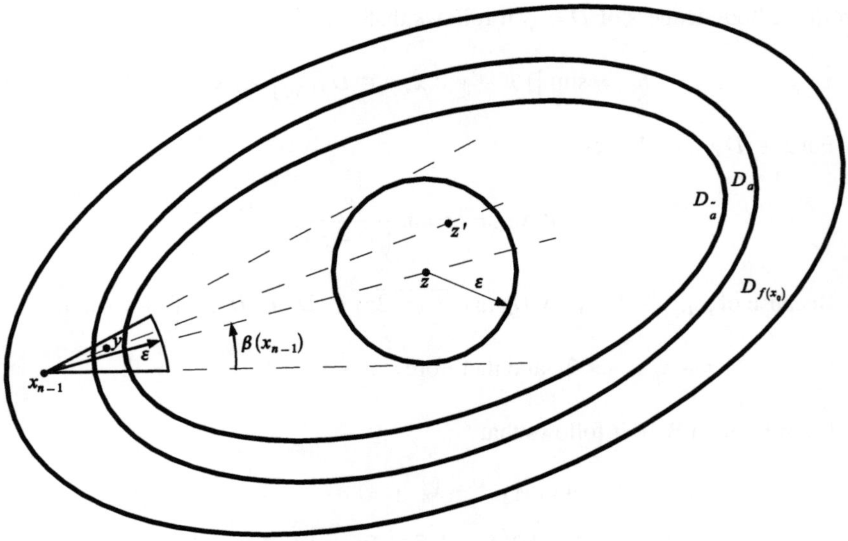

From the central inequality for convex functions (cf. Lemma B.1) it follows now that

$$\frac{f(y) - f(x_{n-1})}{| y - x_{n-1} |} \le \frac{f(z') - f(x_{n-1})}{| z' - x_{n-1} |}$$

therefore

$$\begin{aligned} f(x_{n-1}) - f(y) &\ge | y - x_{n-1} | \frac{f(x_{n-1}) - f(z')}{| x_{n-1} - z' |} \\ &\ge | y - x_{n-1} | \frac{a - \tilde{a}}{K}. \end{aligned}$$

If $\beta := \arcsin \frac{\varepsilon}{K}$, then

$$\beta \le \beta(x) \text{ for all } x \in D_{f(x_0)} \setminus D_a.$$

Therefore

$$\begin{aligned} & x_{n-1} + C(\beta, z - x_{n-1}, \varepsilon) \\ & \quad \subset \left\{ y \in D : f(x_{n-1}) - f(y) \ge \frac{a - \tilde{a}}{K} \, | y - x_{n-1} | \right\} \subset D_{f(x_{n-1})}. \end{aligned}$$

From this it follows that

$$\begin{aligned}
&E\big(f(x_{n-1}) - f(X_n) \mid X_{n-1} = x_{n-1}, \ldots, X_1 = x_1, X_0 = x_0\big) \\
&= \int_{D_{f(x_{n-1})}} \big(f(x_{n-1}) - f(y)\big) p_n(x_0, \ldots, x_{n-1}, dy) \\
&\geq \frac{a - \tilde{a}}{K} \int_{x_{n-1}+C(\beta, z-x_{n-1}, \varepsilon)} \mid y - x_{n-1} \mid p_n(x_0, \ldots, x_{n-1}, dy) \\
&= \frac{a - \tilde{a}}{K} \int_S \int_{\mathbb{R}_+} 1_{C(\beta, z-x_{n-1}, \varepsilon)}(rs) \mid rs \mid \lambda_n(x_0, \ldots, x_{n-1}, dr) \gamma(ds) \\
&= \frac{a - \tilde{a}}{K} \gamma\big(S \cap C(\beta, z - x_{n-1})\big) \int_{(0,\varepsilon]} r \lambda_n(x_0, \ldots, x_{n-1}, dr)
\end{aligned}$$

with the last but one equation following from the integration property (10.4).

According to Lemma A.3, $\lambda(S \cap C(\beta, z - x_{n-1}))$ does not depend on $z - x_{n-1}$. Therefore a positive constant M exists with

$$\begin{aligned}
&E\big(f(x_{n-1}) - f(X_n) \mid X_{n-1} = x_{n-1}, \ldots, X_1 = x_1, X_0 = x_0\big) \\
&\geq M \int_{(0,\varepsilon]} r \lambda_n(x_0, \ldots, x_{n-1}, dr)
\end{aligned}$$

for all $x_1, \ldots, x_{n-1} \in D_{f(x_0)} \setminus D_a$.

Because of (12.4) the condition (11.5) is satisfied.

Now the assertion follows from Theorem 11.2 and Lemma 11.1. □

Example 12.4

(1) Let $(a_n : n \in \mathbb{N})$ be a null sequence of positive numbers.
Let $\lambda_n(x_0, \ldots, x_{n-1}, \cdot) = \varepsilon_{a_n}$.
In this case step length is therefore selected deterministically, namely a_n in the nth step (cf. [1], where $a_n = n^{-s}$).
Condition (12.4) is therefore satisfied if and only if

$$\sum_{n=1}^{\infty} a_n = \infty.$$

(2) Let $(a_n : n \in \mathbb{N})$ be a bounded sequence of positive numbers with

$$\sum_{n=1}^{\infty} a_n = \infty.$$

Let $\lambda_n(x_0, \ldots, x_{n-1}, \cdot)$ be the uniform distribution on the Interval $(0, a_n)$. Then the condition (12.4) is satisfied.

This follows from

$$\int_{(0,c)} r\lambda_n(x_0, \ldots, x_{n-1}, dr) \geq \frac{1}{2} \min\left\{\frac{c^2}{a_n}, a_n\right\}$$
$$\geq \frac{1}{2} \min\left\{\frac{c^2}{a}, a_n\right\} \qquad \text{where} \qquad a := \sup\{a_n : n \in \mathbb{N}\}.$$

(3) Let $\lambda_n(x_0, \ldots, x_{n-1}, \cdot)$ have the density

$$h_n(r) = \sqrt{\frac{2}{\pi}} \frac{1}{\sigma_n} \exp\left(-\frac{r^2}{2{\sigma_n}^2}\right) \qquad (\sigma_n > 0).$$

Because of $\int_{(0,c)} rh_n(r)dr = \sqrt{\frac{2}{\pi}}\left(1 - \exp\left(-\frac{c^2}{2{\sigma_n}^2}\right)\right)\sigma_n$ (12.4) is satisfied if an only if the sequence

$$\sum_{n-1}^{\infty} \sigma_n \left(1 - \exp\left(-\frac{c^2}{2{\sigma_n}^2}\right)\right)$$

is divergent.

This is, for example, the case if

$$\sigma_n \to 0 \qquad \text{and} \qquad \sum_{n-1}^{\infty} \sigma_n = \infty.$$

(4) Let $h : \mathbb{R}_+ \to \mathbb{R}_+$ be a Lebesgue density.
Let there be a $\varepsilon > 0$ such that h is positive on $(0, \varepsilon)$.
Let $\sigma : D \to (0, \infty)$ on $D \setminus D^*$ be continuous.
Let $\lambda_n(x_0, \ldots, x_{n-1}, \cdot) = \lambda(x_{n-1}, \cdot)$ have the density

$$g(x_{n-1}, r) := \frac{1}{\sigma(x_{n-1})} h\left(\frac{r}{\sigma(x_{n-1})}\right).$$

Then we have for all $c > 0$

$$\int_{(0,c)} r\lambda_n(x_0, \ldots, x_{n-1}, dr) = \sigma(x_{n-1}) \int_{\left(0, \frac{c}{\sigma(x_{n-1})}\right)} rh(r)dr.$$

If f is continuous and D_a is compact for all $a > f^*$, then for all $a \in (f^*, f(x_0))$ there are constants $\sigma_1 > 0$ and $\sigma_2 < \infty$ with

$$\sigma_1 \leq \sigma(x) \leq \sigma_2 \qquad \text{for all} \qquad x \in D_{f(x_0)} \setminus D_a.$$

With $m := \sigma_1 \int_{\left(0,\frac{c}{\sigma_2}\right)} rh(r)dr$ it follows that

$$\inf\left\{\int_{(0,c)} r\lambda_n(x_0,\ldots,x_{n-1},dr) : x_1,\ldots,x_{n-1} \in D_{f(x_0)} \setminus D_a\right\} \geq m.$$

Since h is positive on $(0,\varepsilon)$, we have $m > 0$.

Therefore condition (12.4) is satisfied.

Additional examples and convergence statements with regard to R-D-M can be found in [3].

References

1. Guseva, O.: Convergence of a random search algorithm. Kibernetika **7**, 1104–1106 (1971)
2. Marti, J.: Konvexe Analysis. Birkhäuser, Basel (1977)
3. Rastrigin, L.: The convergence of the random search method in the extremal control of a many parameter system. Autom. Remote Control **24**, 1337–1342 (1963)
4. Schwefel, K.P.: Numerische Optimierung von Computer-Modellen mittels der Evolutions-Strategie. Birkhäuser, Basel-Stuttgart (1977)

Chapter 13
Convergence of Stationary Random Search Methods for Positive Success Probability

Let $(X_n : n \in \mathbb{N}_0)$ be in this section—in the sense of Definition 9.3—a stationary R-S-M with the mutation transition probability p. Let us now again pose the question on which conditions $(X_n : n \in \mathbb{N}_0)$ converges, i.e. when does

$$f(X_n) \to f^* \qquad P - \text{a.s.}$$

apply to any starting point $x_0 \in D$?

For this it is obviously necessary that the following condition is satisfied:

$$p\Big(x, \Big\{y \in D : f(y) < f(x)\Big\}\Big) > 0 \qquad \text{for all } x \in D \setminus D^*. \tag{13.1}$$

For if an $x \in D \setminus D^*$ would exist with $p\big(x, \{f < f(x)\}\big) = 0$, then

$$\begin{aligned} P\big(X_{n+1} = x \mid X_n = x\big) &= q\big(x, \{x\}\big) \\ &= p\big(x, \{x\} \cap \{f < f(x)\}\big) + p\big(x, \mathbb{R}^d \setminus \{f < f(x)\}\big) \\ &= 1 - p\big(x, \{f < f(x)\}\big) = 1 \end{aligned}$$

and therefore

$$P(X_n = x \mid X_0 = x) = 1 \qquad \text{for all } n \in \mathbb{N}.$$

Then because of $P(X_0 = x)$ the equality

$$P(X_n = x) = 1$$

K. Marti, *Optimization Under Stochastic Uncertainty*, International Series in Operations Research & Management Science 296,
https://doi.org/10.1007/978-3-030-55662-4_13

applies to the R-S-M that starts to $x_0 = x$ for all $n \in \mathbb{N}_0$.

Therefore

$$P\bigl(\lim_{n\to\infty} f(X_n) = f(x)\bigr) = 1$$

in contradiction to $f(x) > f^*$ and $P\bigl(\lim_{n\to\infty} f(X_n) = f^*\bigr) = 1$.

It remains to be determined when (13.1), i.e. the requirement for a positive success probability, is sufficient for the convergence. For this purpose initially the following lemma will be used.

Lemma 13.1

$$\inf\left\{\int_{D_{f(y)}} \bigl(f(y) - f(z)\bigr)p(y, dz) : y \in D_{f(x_0)} \setminus D_a\right\} > 0. \tag{13.2}$$

For all $x_0 \in D \setminus D^*$ and all $a \in (f^*, f(x_0))$.

Then we have $f(X_n) \to f^* \quad P - a.s.$ for all starting points $x_0 \in D$.

Proof Because of

$$\begin{aligned}
&\int_{D_{f(y)}} \bigl(f(y) - f(z)\bigr)p(y, dz)\\
&= \int_{\{u\in D: f(u)<f(y)\}} \bigl(f(y) - f(z)\bigr)p(y, dz)\\
&= E\bigl(f(X_{n-1}) - f(X_n) \mid X_{n-1} = y\bigr) \quad \text{for all} \quad n \in \mathbb{N}\\
&\text{(cf. Lemma 9.1)}
\end{aligned}$$

it follows from (13.2) that the condition (11.5) of Theorem 11.2 is satisfied. Then the assertion follows from Lemma 11.1. □

Theorem 13.1 *Let condition (13.1) be satisfied and D be finite.*

Then for all starting points $x_0 \in D$

$$f(X_n) \to f^* \qquad P - a.s.$$

Proof Since D is finite, there is a $b > 0$ with

$$\bigl\{y \in D : f(y) < f(x)\bigr\} = \bigl\{y \in D : f(y) \le f(x) - b\bigr\} \quad \text{for all } x \in D.$$

There is furthermore a $c > 0$ with $p(x, D_{f(x)}) \ge c$ for all $x \in D \setminus D^*$.

Therefore $\int_{D_{f(y)}} (f(y) - f(z))p(y, dz) \ge bc$ for all $y \in D \setminus D^*$.

Now the assertion follows from Lemma 10.1. □

Definition 13.1 Let E be a topological space and let $\mathcal{B}_E$ be the Borel-σ-algebra on E.

(1) A function $g : E \to \mathbb{R}$ is called *lower semi-continuous*, if

$$\liminf_{n\to\infty} g(x_n) \geq g(x)$$

for all $x \in E$ and all sequences $(x_n : n \in \mathbb{N})$ from E with $\lim_{n\to\infty} x_n = x$.

(2) A transition probability K from $(E, \mathcal{B}_E)$ to $(E, \mathcal{B}_E)$ is called *continuous*, if the representation

$$x \to \int h(y)K(x, dy)$$

is continuous for all continuous and bounded functions $h : E \to \mathbb{R}$.

Lemma 13.2 *Let E be a Polish space (i.e. completely metrisable and separable) and let $h : E \times E \to \mathbb{R}$ be a lower semi-continuous and bounded from below.*
Let K be a continuous transition probability from E to E.
Then the representation

$$x \to \int h(x, y)K(x, dy)$$

is lower semi-continuous.

Proof cf. [1, p. 18]. □

Theorem 13.2 *Let the condition (13.1) be satisfied. Also let f and p be continuous and D_a compact for all $a \geq f^*$.*
Then for all starting points $x_0 \in D$

$$f(X_n) \to f^* \qquad P-a.s.$$

Proof Let $x_0 \in D \setminus D^*$ and let $a \in (f^*, f(x_0))$ be arbitrary.
Let $h : D_{f(x_0)} \times D_{f(x_0)} \to \mathbb{R}_+$

$$h(y, z) := 1_{\{(y,z):f(z)<f(y)\}}(y, z) \cdot (f(y) - f(z)).$$

Since f is continuous, the set

$$\{(y, z) \in D_{f(x_0)} \times D_{f(x_0)} : f(z) < f(y)\}$$

is open in $D_{f(x_0)} \times D_{f(x_0)}$ and therefore the restriction of the indicator function $1_{\{(y,z):f(z)<f(y)\}}$ on the set $D_{f(x_0)} \times D_{f(x_0)}$ is lower semi-continuous.

Since $D_{f(x_0)}$—as a closed subset of the $\mathbb{R}^d$ (provided with the relative topology)—is a Polish topological space, Lemma 13.2 implies that the map $g : D_{f(x_0)} \to \mathbb{R}_+$, $g(y) := \int h(y, z)p(y, dz)$ is lower semi-continuous and therefore attains its infimum on any compact subset of $D_{f(x_0)}$.

Since $D_{f(x_0)}$ is compact (this means in particular that it is closed), due to the continuity of f also the set $\{y \in D_{f(x_0)} : f(y) \geq a\}$ and thus

$$D_{f(x_0)} \cap \{y \in D : f(y) \geq a\} = D_{f(x_0)} \cap \{y \in D_{f(x_0)} : f(y) \geq a\}$$

as the product of a compact and a closed set is again compact. Furthermore

$$D_{f(x_0)} \setminus D_a \subset D_{f(x_0)} \cap \{y \in D : f(y) \geq a\}$$

and therefore

$$\begin{aligned} &\inf \left\{g(y) : y \in D_{f(x_0)} \setminus D_a\right\} \\ &\quad \geq \inf \left\{g(y) : y \in D_{f(x_0)} \cap \{z \in D : f(z) \geq a\}\right\}. \end{aligned}$$

From condition (13.1) it follows that the representation

$$y \rightarrow \int_{D_{f(y)}} (f(y) - f(z))p(y, dz)$$

is positive on $D \setminus D^*$.

Because of $g(y) = \int_{D_{f(y)}} (f(y) - f(z))p(y, dz)$, condition (13.2) is satisfied, and the assertion follows from Lemma 13.1. □

The following examples show that on the whole the convergence of the R-S-M $(X_n : n \in \mathbb{N}_0)$ will be lost in spite of a positive success probability if the additional conditions formulated in Theorem 13.2 are dropped.

Example 13.1

(1) Let $D := [0, 1]$ and $f(x) := 1 - x$.
For $n \in \mathbb{N}_0$ let $a_n := \frac{2^n - 1}{2^{n+1}}$.
Let $\sigma : [0, 1] \rightarrow [0, 1]$

$$\sigma(a_n) := a_{n+1} \text{ and } \sigma(x) := 1 \text{ if } x \notin \{a_n : n \in \mathbb{N}_0\}.$$

Furthermore, let $p(x, \cdot) := \varepsilon_{\sigma(x)}$.
Because of $f(a_{n+1}) < f(a_n)$ and $f(1) = f^* = 0$, it follows that

$$p(x, \{y \in D : f(y) < f(x)\}) = 1, \quad \text{if } x \notin D^* = \{1\}.$$

Except for the continuity of p all conditions of Theorem 13.2 are satisfied.
Because of

$$\begin{aligned} P(X_{n+1} = a_{n+1} \mid X_n = a_n) &= q(a_n, \{a_{n+1}\}) \\ &= \varepsilon_{\sigma(a_n)}(\{a_{n+1}\}) = \varepsilon_{a_{n+1}}(\{a_{n+1}\}) = 1, \end{aligned}$$

we have

$$P(X_n = a_n) = 1 \quad \text{for all } n \in \mathbb{N}_0$$

for the starting point $x_0 = a_0 = 0$.

From $\lim\limits_{n\to\infty} a_n = \frac{1}{2}$, it follows that $f(X_n) \to \frac{1}{2} \quad P-\text{a.s.}$

Thus because of $f^* = 0$ there is no convergence for the starting point $x_0 = 0$.

(2) Let $D := \mathcal{Z}$ and

$$f(x) := \begin{cases} x & \text{if} \quad |x| < 1 \\ \frac{1}{x} & \text{if} \quad |x| \geq 1. \end{cases}$$

Let $p(x, B) := \frac{1}{2}\varepsilon_{n-1}(B) + \varepsilon_{n+1}(B)$.

It is then trivial that p is continuous.

Because of $p(x, \{y \in D : f(y) < f(x)\}) \geq \frac{1}{2}$ for all $x \in D \setminus D^*$ condition (13.1) is satisfied.

Except for the compactness of the set D_a all conditions of Theorem 13.2 are satisfied.

For all starting points $x_0 \geq 2$ we have

$$P(X_n \geq 2) = 1 \quad \text{for all } n \in \mathbb{N}_0$$

and therefore

$$P\left(\lim_{n\to\infty} f(X_n) = f^*\right) = 0.$$

(3) Let $D = [0, 1]$ and

$$f(x) := \begin{cases} x & \text{if} \quad x > 0 \\ -1 & \text{if} \quad x = 0. \end{cases}$$

Let $p(x, \cdot) := \lambda_{[0,1]}$.

Except for the continuity of f, all conditions of Theorem 13.2 are satisfied. For $x_0 \neq 0$ we have

$$P\left(\lim_{n\to\infty} f(X_n) = f^*\right) = 0.$$

Reference

1. Gihman, I., Skorohod, A.: Controlled Stochastic Processes. Springer, Berlin (1979)

Chapter 14
Random Search Methods of Convergence Order $O(n^{-\alpha})$

At the beginning of this section we would like to provide a small example that illustrates some of the most important ideas in simplified form.

Let $D = \mathbb{R}$, $f(x) = | x |^s$ ($s > 0$ arbitrary) and

$$p(x, B) := \int_B \frac{1}{2r(x)} 1_{[x-r(x),x+r(x)]}(y)dy,$$

where r denotes a measurable positive function. Hence, $p(x, \cdot)$ denotes the uniform distribution on the interval

$$[x - r(x), x + r(x)].$$

As already mentioned, we will deal in Chap. 14 with uncontrolled methods, and we will therefore assume that the span $r(x)$ does not depend on x, hence,

$$r(x) = r > 0 \qquad \text{for all } x \in \mathbb{R}.$$

In the next chapter we will then consider the controlled case.

The first and decisive step when determining the rate of convergence of the sequence $E(f(X_n) - f^*)$ is to calculate the probability distribution function F_n of the variable $f(X_n) - f^*$.

For this purpose we will assume that the starting point x_0 satisfies the condition $| x_0 | \leq r$. This makes the calculation much easier.

If $z = x^s$, then for $x \in (0, | x_0 |)$ (without loss of generality it is assumed that $x_0 \neq x^* = 0$) we have

K. Marti, *Optimization Under Stochastic Uncertainty*, International Series in Operations Research & Management Science 296,
https://doi.org/10.1007/978-3-030-55662-4_14

$$\begin{aligned}1 - F_n(z) &= P\left(f(X_n) - f^* > z\right) = P\left(f(X_n) > z\right) \\ &= P\left(\bigcap_{i=1}^{n}\{f(X_i) > z\}\right) = P\left(\bigcap_{i=1}^{n}\{X_i \notin [-x, x]\}\right) \\ &= \int_{[-|x_0|,|x_0|]\setminus[-x,x]} \cdots \int_{[-|x_0|,|x_0|]\setminus[-x,x]} q(x_{n-1}, dx_n)\dots q(x_0, dx_1).\end{aligned}$$

Furthermore, for $x_{n-1} \in [-\mid x_0 \mid, \mid x_0 \mid]\setminus[-x, x]$ it holds

$$\begin{aligned}q(x_{n-1}, \mathbb{R}\setminus[-x, x]) = 1 - q(x_{n-1}, [-x, x]) &= 1 - p(x_{n-1}, [-x, x]) \\ &= 1 - \frac{1}{2r}\int_{-x}^{x} 1_{[x_{n-1}-r, x_{n-1}+r]}(y)dy.\end{aligned}$$

Since $\mid x_0 \mid \leq r$ it holds

$$x_{n-1} + r \geq -\mid x_{n-1} \mid + r \geq -\mid x_0 \mid + r \geq \mid x_0 \mid > x;$$

moreover, $x_{n-1} - r < -x$.

Thus,

$$q(x_{n-1}, \mathbb{R}\setminus[-x, x]) = 1 - \frac{1}{2r}\int_{-x}^{x} dy = 1 - \frac{x}{r}.$$

Iteration yields

$$1 - F_n(z) = \left(1 - \frac{x}{r}\right)^n, \text{ and}$$

$$1 - F_n(x) = \left(1 - \frac{x^{\frac{1}{s}}}{r}\right)^n \quad \text{for } x \in (0, \mid x_0 \mid^s).$$

Obviously,

$$\begin{aligned}1 - F_n(x) &= 1 \qquad \text{for } x \leq 0 \qquad \text{and} \\ 1 - F_n(x) &= 0 \qquad \text{for } x \geq \mid x_0 \mid^s.\end{aligned}$$

From this it is now possible to calculate $E(f(X_n) - f^*)$.

Then, cf. [1, p. 47]:

$$E\left(f(X_n) - f^*\right) = \int_0^{|x_0|^s} (1 - F_n(x))\, dx = \int_0^{|x_0|^s} \left(1 - \frac{x^{\frac{1}{s}}}{r}\right)^n dx.$$

This integral is evaluated using Lemma B.16.

For special values of s an accurate calculation of the integral is easily possible. For example, for $s = 1$ we have

$$E\left(f(X_n) - f^*\right) = \frac{r}{n+1}\left(1 - \left(1 - \frac{|\, x_0 \,|}{r}\right)^{n+1}\right).$$

For $s = 2$ we get

$$E\left(f(X_n) - f^*\right) = \frac{2r^2}{(n+1)(n+2)}\left(1 - \left(1 - \frac{|\, x_0 \,|}{r}\right)^{n+2} - (n+2)\frac{|\, x_0 \,|}{r}\left(1 - \frac{|\, x_0 \,|}{r}\right)^{n+1}\right).$$

Therefore in both cases:

$$E\left(f(X_n) - f^*\right) = O(n^{-s}).$$

Since for all $x \geq 0$

$$\left\{|\, X_n - x^* \,| > x\right\} = \left\{f(X_n) > x^s\right\},$$

for $x \in (0, |\, x_0 \,|)$ we have

$$P\left(|\, X_n - x^* \,| > x\right) = \left(1 - \frac{x}{r}\right)^n.$$

Using the same argument as above, if follows now that

$$E \mid X_n - x^* \mid = O(\frac{1}{n}) \qquad \text{independent on s.}$$

For the rest of this chapter let $m := \lambda^d(K(0, 1))$ denote the volume of the d-dimensional unit sphere.

(We have $m = \frac{\pi^{\frac{d}{2}}}{\Gamma(1+\frac{d}{2})}$, where Γ denotes the Eulerian Gamma function.)

To avoid unnecessary paper work, we will use for the set $\{x \in D : f(x) < c\}$ the short notation

$$\{f < c\}.$$

The following Theorem 14.1 permits an estimate of the rate of convergence from below, i.e. it contains a result about the minimum rate of convergence to be expected.

Exactly the opposite can be found in Theorem 14.2, where estimates of the convergence rate from above are given.

Theorem 14.1 *Let the following conditions (14.1) and (14.2) be fulfilled:*

$$\begin{cases} \text{For all } x_0 \in D \setminus D^* \text{ there exists an } a_0 > 0 \text{ with} \\ p_n(x_0, x_1, \dots, x_{n-1}, B) \geq a_0 \lambda^d(B) \\ \text{for all } n \in \mathbb{N}, \quad x_1, \dots, x_{n-1} \in D_{f(x_0)} \setminus D^* \\ \text{ and } B \subset D_{f(x_0)} \text{ measurable;} \end{cases} \tag{14.1}$$

$$\begin{cases} \text{There exist } x^* \in D \text{ and positive constants } b_1, c_1, \rho_1 \\ \text{with } K(x^*, b_1 \varepsilon^{\rho_1}) \subset \{f \leq f^* + \varepsilon\} \text{ for all } \varepsilon \in (0, c_1]. \end{cases} \tag{14.2}$$

Then

(i) For all $x_0 \in D$ *there exists a* $c(x_0) > 0$ *such that*

$$P\left(f(X_n) - f^* \leq \varepsilon\right) \geq 1 - \left(1 - c(x_0)\varepsilon^{d\rho_1}\right)^n$$

for all $n \in \mathbb{N}$ *and all* $\varepsilon \in \left[0, f(x_0) - f^*\right]$.

(ii) $n^\alpha \left(f(X_n) - f^*\right) \to 0 \qquad P-$*a.s. for all* $x_0 \in D$ *and any* $\alpha < (d\rho_1)^{-1}$.

(iii) For all $x_0 \in D$ *there exists* $\tau_1(x_0) < \infty$ *with*

$$\limsup_{n\to\infty} \left(n^{\frac{1}{d\rho_1}} E\left(f(X_n) - f^*\right)\right) \leq \tau_1(x_0).$$

If in addition to (14.1) and (14.2) the following condition

$$\begin{cases} \text{There exist } x^* \in D \text{ and positive constants } b_2, c_2, \rho_2 \\ \text{with } K(x^*, b_2 \varepsilon^{\rho_2}) \supset \{f \leq f^* + \varepsilon\} \text{ for all } \varepsilon \in (0, c_2] \end{cases} \tag{14.3}$$

holds and all $D_a \quad (a \geq f^*)$ *are bounded, then*

(iv) For all $x_0 \in D$ *there is a* $\tilde{c}(x_0) > 0$ *with*

$$P\left(\mid X_n - x^* \mid \leq \varepsilon\right) \geq 1 - \left(1 - \tilde{c}(x_0)\varepsilon^{\frac{d\rho_1}{\rho_2}}\right)^n$$

for all $n \in \mathbb{N}$ *and* $\varepsilon \in \left[0, \sup\{\mid x - x^* \mid : x \in D_{f(x_0)}\}\right]$.

(v) $n^\alpha \mid X_n - x^* \mid \to 0 \qquad P-$*a.s. for all* $x_0 \in D$ *and* $\alpha < \frac{\rho_2}{d\rho_1}$

(vi) For all $x_0 \in D$ *there is* $\tau_2(x_0) < \infty$ *with*

$$\limsup_{n\to\infty} \left(n^{\frac{\rho_2}{d\rho_1}} E \mid X_n - x^* \mid\right) \leq \tau_2(x_0).$$

If in addition to (14.1) and (14.2) condition

$$\mu := \lambda^d(D^*) > 0 \tag{14.4}$$

is fulfilled, then

(vii) $(1 - a_0\mu)^{-n} E\,(f(X_n) - f^*) \le f(x_0) - f^*$ *for all* $n \in \mathbb{N}$ *and* $x_0 \in D$ *as well as* $a_n\,(f(X_n) - f^*) \to 0$ $\quad P - a.s.$ *for each sequence* $(a_n : n \in \mathbb{N})$ *in* $\mathbb{R}$.

Remark 14.1

(1) The condition (14.2) implies $x^* \in D^*$.
(2) Let $D^* \neq \emptyset$. Then (14.3) yields that D^* is a singleton, hence, $D^* = \{x^*\}$.
(3) The proof of Theorem 14.1 shows that $\tau_1(x_0)$, $\tau_2(x_0)$, resp., can be chosen as follows:

$$\tau_1(x_0) = \left(a_0 m {b_1}^d\right)^{-\frac{1}{\rho_1 d}} \frac{f(x_0) - f^*}{\tilde{\alpha}} \Gamma\left(1 + \frac{1}{\rho_1 d}\right),$$

where

$$\tilde{\alpha} := \min\left\{\frac{c_1}{2}, \frac{f(x_0) - f^*}{2}, \left(a_0 m {b_1}^d\right)^{-\frac{1}{\rho_1 d}}\right\};$$

$$\tau_2 = \left(a_0 m {b_1}^d\right)^{-\frac{\rho_2}{\rho_1 d}} \left(\frac{f(x_0) - f^*}{\alpha}\right)^{\rho_2} \frac{b_2 r}{\hat{\beta}} \Gamma\left(1 + \frac{\rho_2}{d\rho_1}\right),$$

where

$$r := \sup\left\{| x - x^* | : x \in D_{f(x_0)}\right\},$$

$$\hat{\beta} := \min\left\{r, b_2(f(x_0) - f^*)^{\rho_2}, b_2 {c_2}^{\rho_2}, \left(a_0 m {b_1}^d\right)^{-\frac{\rho_2}{\rho_1 d}} b_2, \left(\frac{f(x_0) - f^*}{\alpha}\right)^{\rho_2}\right\}$$

$$\alpha := \frac{1}{2} \min\left\{c_1, f(x_0) - f^*\right\}.$$

Proof of Theorem 14.1

(a) If $x_0 \in D^*$, then $P(X_n = x_0) = 1$ for all $n \in \mathbb{N}$.
This implies the assertions (i), (ii), (iii), and (vii). Since the uniqueness of the minimum point x^* is given by the condition (14.3), also the remaining assertions follow.
Without loss of generality we may therefore assume that $f(x_0) > f^*$,
(b) For $x_0 \in D \setminus D^*$, let a_0 be chosen according to (14.1) and let $\varepsilon > 0$ be an arbitrary positive number.
If conditions (14.1) and (14.2) hold, then:

$$
\begin{aligned}
P\left(f(X_n) - f^* > \varepsilon\right) &= P\left(\bigcap_{i=1}^{n}\{f(X_i) - f^* > \varepsilon\}\right) \\
&= \int_{\{f>f^*+\varepsilon\}} \cdots \int_{\{f>f^*+\varepsilon\}} q_n(x_0, x_1, \ldots, x_{n-1}, dx_n) \ldots q_1(x_0, dx_1) \\
&= \int_{\{f>f^*+\varepsilon\}} \cdots \int_{\{f>f^*+\varepsilon\}} \left(1 - p_n\left(x_0, x_1, \ldots, x_{n-1}, \{f \le f^* + \varepsilon\}\right)\right) \cdot \\
&\quad q_{n-1}(x_0, \ldots, x_{n-2}, dx_{n-1}) \ldots q_1(x_0, dx_1) \\
&\le \left(1 - a_0\lambda^d\left(\{f \le f^* + \varepsilon\}\right)\right) \cdot \int_{\{f>f^*+\varepsilon\}} \cdots \\
&\quad \cdots \int_{\{f>f^*+\varepsilon\}} q_{n-1}(x_0, \ldots, x_{n-2}, dx_{n-1}) \ldots q_1(x_0, dx_1) \\
&\le \ldots \le \left(1 - a_0\lambda^d\left(\{f \le f^* + \varepsilon\}\right)\right)^{n-1} \left(1 - q_1\left(x_0, \{f \le f^* + \varepsilon\}\right)\right).
\end{aligned}
$$

If $\varepsilon < f(x_0) - f^*$ and therefore $f(x_0) > f^* + \varepsilon$, then

$$
q_1\left(x_0, \{f \le f^* + \varepsilon\}\right) = p_1\left(x_0, \{f \le f^* + \varepsilon\}\right) \ge a_0\lambda^d\{f \le f^* + \varepsilon\}.
$$

Thus,

$$
P\left(f(X_n) - f^* > \varepsilon\right) \le \left(1 - a_0\lambda^d\{f \le f^* + \varepsilon\}\right)^n
$$

If additionally $\varepsilon \le c_1$, then, according to (11.2),

$$
\begin{aligned}
P\left(f(X_n) - f^* > \varepsilon\right) &\le \left(1 - a_0\lambda^d\left(K\left(x^*, b_1\varepsilon^{\rho_1}\right)\right)\right)^n \\
&= \left(1 - a_0 m {b_1}^d \varepsilon^{\rho_1 d}\right)^n
\end{aligned}
$$

and therefore

$$
P\left(f(X_n) - f^* \le \varepsilon\right) \ge H_n(\varepsilon)
$$

for all $\varepsilon \in (0, \min\{c_1, f(x_0) - f^*\})$, with $H_n(\varepsilon) := 1 - \left(1 - a_0 m {b_1}^d \varepsilon^{\rho_1 d}\right)^n$.
Obviously,

$$
\begin{aligned}
&P\left(f(X_n) - f^* \le \varepsilon\right) = 1 \text{ for } \varepsilon \ge f(x_0) - f^* \qquad \text{and} \\
&P\left(f(X_n) - f^* \le \varepsilon\right) = 0 \text{ for } \varepsilon < 0.
\end{aligned}
$$

Defining $\alpha := \frac{1}{2}\min\{c_1, f(x_0) - f^*\}$ and

$$G_n(\varepsilon) := \begin{cases} 0 & \text{for } \varepsilon \leq 0 \\ H_n\left(\frac{\alpha\varepsilon}{f(x_0)-f^*}\right) & \text{for } 0 < \varepsilon \leq f(x_0) - f^* \\ 1 & \text{for } \varepsilon > f(x_0) - f^*, \end{cases}$$

G_n is a monotone increasing function, and for all $\varepsilon \in \mathbb{R}$ we have

$$G_n(\varepsilon) \leq F_n(\varepsilon) := P\left(f(X_n) - f^* \leq \varepsilon\right).$$

For $\varepsilon \leq 0$ and $\varepsilon \geq f(x_0) - f^*$, it is clear that this inequality holds. For $\varepsilon \in (0, \alpha)$, this inequality follows because of

$$\frac{\alpha}{f(x_0) - f^*} \leq 1.$$

For $\varepsilon \in [\alpha, f(x_0) - f^*)$ it follows from

$$G_n(\varepsilon) \leq G_n\left(f(x_0) - f^*\right) = H_n(\alpha) \leq F_n(\alpha) \leq F_n(\varepsilon).$$

Hence, assertion (i) is shown.

(c) In addition to (14.1) and (14.2), let also (14.3) be fulfilled.
Furthermore, assume that $D_{f(x_0)}$ is bounded.
If $0 \leq \varepsilon \leq b_2 {c_2}^{\rho_2}$, hence, $\left(\frac{\varepsilon}{b_2}\right)^{\frac{1}{\rho_2}} \leq c_2$, condition (14.3) yields

$$\left\{f \leq f^* + \left(\frac{\varepsilon}{b_2}\right)^{\frac{1}{\rho_2}}\right\} \subset D \cap K(x^*, \varepsilon)$$

and therefore

$$\left\{y \in D :\mid y - x^* \mid > \varepsilon\right\} \subset \left\{f > f^* + \left(\frac{\varepsilon}{b_2}\right)^{\frac{1}{\rho_2}}\right\}.$$

According to (b), for $\varepsilon \in [0, b_2 {c_2}^{\rho_2}]$ we get

$$\begin{aligned} P\left(\mid X_n - x^* \mid \leq \varepsilon\right) \geq & P\left(f(X_n) - f^* \leq \left(\frac{\varepsilon}{b_2}\right)^{\frac{1}{\rho_2}}\right) \\ \geq & G_n\left(\left(\frac{\varepsilon}{b_2}\right)^{\frac{1}{\rho_2}}\right). \end{aligned}$$

Furthermore, if $\left(\frac{\varepsilon}{b_2}\right)^{\frac{1}{\rho_2}} \le f(x_0) - f^*$, i.e. $\varepsilon \le b_2\,(f(x_0) - f^*)^{\rho_2}$, we get

$$\begin{aligned} P\left(\mid X_n - x^* \mid \le \varepsilon\right) \ge & G_n\left(\left(\frac{\varepsilon}{b_2}\right)^{\frac{1}{\rho_2}}\right) \\ = & H_n\left(\frac{\alpha}{f(x_0) - f^*}\left(\frac{\varepsilon}{b_2}\right)^{\frac{1}{\rho_2}}\right). \end{aligned}$$

From the boundedness of $D_{f(x_0)}$ it follows that

$$r := \sup\left\{\mid x - x^* \mid : x \in D_{f(x_0)}\right\} < \infty.$$

Defining $\beta := \min\left\{r, b_2\,(f(x_0) - f^*)^{\rho_2}, b_2 c_2{}^{\rho_2}\right\}$ and

$$\tilde{G}_n(\varepsilon) := \begin{cases} 0 & \text{for } \varepsilon \le 0 \\ H_n\left(\frac{\alpha}{f(x_0) - f^*}\left(\frac{\varepsilon\beta}{r b_2}\right)^{\frac{1}{\rho_2}}\right) & \text{for } 0 < \varepsilon \le r \\ 1 & \text{for } \varepsilon > r, \end{cases}$$

with the same argument as in (b), for all $\varepsilon \in \mathbb{R}$ we get the inequality

$$P\left(\mid X_n - x^* \mid \le \varepsilon\right) \ge \tilde{G}_n(\varepsilon).$$

This shows now assertion (iv).

(d) To prove assertion (ii), it is sufficient to show that for all $x_0 \in D$, $\alpha < \frac{1}{d\rho_2}$, and $\varepsilon > 0$ we have

$$\sum_{n=1}^{\infty} P\left(n^{\alpha}\left(f(X_n) - f^*\right) > \varepsilon\right) < \infty. \tag{14.5}$$

With the lemma of Borel–Cantelli we obtain then

$$P\left(n^{\alpha}\left(f(X_n) - f^*\right) > \varepsilon \text{ for infinite many } n \in \mathbb{N}\right) = 0.$$

Taking the complement, we have

$$P\left(n^{\alpha}\left(f(X_n) - f^*\right) \le \varepsilon \text{ for asymptotically all } n \in \mathbb{N}\right) = 1.$$

This yields

$$P\left(\limsup_{n\to\infty} n^{\alpha}\left(f(X_n) - f^*\right) \le \varepsilon\right) = 1.$$

Because this is satisfied for each $\varepsilon > 0$, we get (ii).
Now we show (14.5):
If $\varepsilon n^{-\alpha} \le f(x_0) - f^*$, then (i) yields

$$P\left(n^{\alpha}\left(f(X_n) - f^*\right) > \varepsilon\right) \le \left(1 - c(x_0)(\varepsilon n^{-\alpha})^{d\rho_1}\right)^n.$$

Because of $\alpha < (d\rho_1)^{-1}$ it is $\alpha d\rho_1 < 1$.
Now (14.5) follows from Lemma B.16.
Part (v) can be shown analogously by using (iv).

(e) Because of $F_n(x) = 0$ for $x < 0$, part (b) yields

$$\begin{aligned} E\left(f(X_n) - f^*\right) &= \int_0^{\infty} (1 - F_n(x))dx \le \int_0^{\infty} (1 - G_n(x))dx \\ &= \int_0^{f(x_0)-f^*} \left(1 - a_0 m b_1{}^d \left(\frac{\alpha x}{f(x_0) - f^*}\right)^{\rho_1 d}\right)^n dx. \end{aligned}$$

For $\hat{\alpha} := \min\left\{\alpha, \left(a_0 m b_1{}^d\right)^{-\frac{1}{\rho_1 d}}\right\}$ we have $\hat{\alpha} \le \alpha$ and therefore

$$\begin{aligned} E\left(f(X_n) - f^*\right) &\le \int_0^{f(x_0)-f^*} \left(1 - a_0 m b_1{}^d \left(\frac{\hat{\alpha} x}{f(x_0) - f^*}\right)^{\rho_1 d}\right)^n dx \\ &= \left(a_0 m b_1{}^d\right)^{-\frac{1}{\rho_1 d}} \frac{f(x_0) - f^*}{\hat{\alpha}} \int_0^{\hat{\alpha}\left(a_0 m b_1{}^d\right)^{\frac{1}{\rho_1 d}}} \left(1 - u^{\rho_1 d}\right)^n du \\ &\le \left(a_0 m b_1{}^d\right)^{-\frac{1}{\rho_1 d}} \frac{f(x_0) - f^*}{\hat{\alpha}} \int_0^1 \left(1 - u^{\rho_1 d}\right)^n du. \end{aligned}$$

Now it follows that

$$\begin{aligned} &\limsup_{n\to\infty} \left(n^{\frac{1}{\rho_1 d}} E\left(f(X_n) - f^*\right)\right) \\ &\le \left(a_0 m b_1{}^d\right)^{-\frac{1}{\rho_1 d}} \frac{f(x_0) - f^*}{\hat{\alpha}} \limsup_{n\to\infty} \left(n^{\frac{1}{\rho_1 d}} \int_0^1 \left(1 - u^{\rho_1 d}\right)^n du\right) \\ &= \left(a_0 m b_1{}^d\right)^{-\frac{1}{\rho_1 d}} \frac{f(x_0) - f^*}{\hat{\alpha}} \Gamma\left(1 + \frac{1}{\rho_1 d}\right). \end{aligned}$$

Here, the last equation follows again from Lemma B.16.
This proves now assertion (iii).
From (c) it follows that

$$
\begin{aligned}
E \mid X_n - x^* \mid &\leq \int_0^\infty \left(1 - \tilde{G}_n(x)\right) dx \\
&= \int_0^r \left(1 - a_0 m \left(\frac{\alpha}{f(x_0) - f^*} b_1^{\frac{1}{\rho_1}} b_2^{-\frac{1}{\rho_2}}\right)^{d\rho_1} \left(\frac{x\beta}{r}\right)^{\frac{d\rho_1}{\rho_2}}\right)^n dx \\
&\leq \left(a_0 m b_1{}^d\right)^{-\frac{\rho_2}{\rho_1 d}} \left(\frac{f(x_0) - f^*}{\alpha}\right)^{\rho_2} \frac{b_2 r}{\hat{\beta}} \int_0^1 \left(1 - u^{\frac{d\rho_1}{\rho_2}}\right)^n du,
\end{aligned}
$$

where $\hat{\beta} := \min \left\{\beta, \left(a_0 m b_1{}^d\right)^{-\frac{\rho_2}{\rho_1 d}} b_2 \left(\frac{f(x_0)-f^*}{\alpha}\right)^{\rho_2}\right\}$.

Corresponding to the above, part (vi) follows now again by using Lemma B.16.

(f) If in addition to (14.1) and (14.2) condition (14.4) holds, then analogously to part (a) we get

$$
\begin{aligned}
P\left(f(X_n) - f^* > 0\right) &= P\left(\bigcap_{i=0}^{n} \{f(X_i) - f^* > 0\}\right) \\
&\leq \left(1 - a_0 \lambda^d \left(\{f \leq f^*\}\right)\right)^n = (1 - a_0\mu)^n .
\end{aligned}
$$

From this it follows on the one hand that

$$
\begin{aligned}
E\left(f(X_n) - f^*\right) &= \int_{\{f(X_n)-f^*>0\}} \left(f(X_n) - f^*\right) dP \\
&\leq \left(f(x_0) - f^*\right) P\left(f(X_n) - f^* > 0\right) \\
&\leq \left(f(x_0) - f^*\right) (1 - a_0\mu)^n .
\end{aligned}
$$

Therefore, with $\beta = (1 - a_0\mu)^{-1}$, for all $n \in \mathbb{N}$ we have the inequality

$$
\beta^n E\left(f(X_n) - f^*\right) \leq f(x_0) - f^*.
$$

On the other hand, we have

$$
\begin{aligned}
&P\left(\bigcap_{i=0}^{\infty} \{f(X_i) > f^*\}\right) = 0, \quad \text{hence} \\
&P\left(\bigcup_{i=0}^{\infty} \{f(X_i) = f^*\}\right) = 1.
\end{aligned}
$$

Because of the monotonicity of the sequence $f(X_n)$ we even have that (cf. Remark 11.1)

$$P\left(\bigcup_{i=0}^{\infty}\bigcap_{j\geq i}\{f(X_j)=f^*\}\right)=1$$

and therefore

$$P\left(\bigcup_{i=0}^{\infty}\bigcap_{j\geq i}\{a_j(f(X_j)-f^*)=0\}\right)=1$$

for each sequence $(a_n : n \in \mathbb{N})$ of real numbers.

This proves assertion (vii).

□

Remark 14.2 Let $D^* \neq \emptyset$, and assume that f is convex on an open sphere $K \supsetneqq D^*$. Then condition (14.2) holds (with $\rho_1 = 1$).

Proof Let $x^* \in D^*$ and $r > 0$ be chosen such that $D^* \subsetneqq \overline{K(x^*, r)} \subset D$, and suppose that f is continuous and convex on $\overline{K(x^*, r)}$.

Let $c := \sup\left\{f(x) : x \in \overline{K(x^*, r)}\right\}$.

Because of $D^* \subsetneqq \overline{K(x^*, r)}$ it follows that $c - f^* > 0$.

Define $c_1 := c - f^*$ and $b_1 := \frac{r}{c-f^*}$.

Let $\varepsilon \in (0, c_1]$ be arbitrary.

Because of $b_1\varepsilon \leq b_1c_1 \leq r$ we get $K(x^*, b_1\varepsilon) \subset D$.

Let now $y \in K(x^*, b_1\varepsilon)\backslash\{x^*\}$ be arbitrary.

With $z := x^* + \frac{r}{|y-x^*|}(y-x^*)$ we find $| z-x^* | = r$, and, according to Lemma B.1 it follows that

$$\frac{f(y)-f^*}{| y-x^* |} \leq \frac{f(z)-f^*}{| z-x^* |}.$$

From this we obtain

$$f(y)-f^* \leq | y-x^* | \frac{f(z)-f^*}{r} < b_1\varepsilon\frac{c-f^*}{r} = \varepsilon$$

Hence, $K(x^*, b_1\varepsilon) \subset \{f \leq f^* + \varepsilon\}$ for all $\varepsilon \in (0, c_1]$ □

Remark 14.3 Assume that $D^* = \{x^*\} \subset \dot{D}$ with compact D and continuous f. Furthermore, assume that f is twice continuously differentiable on a neighborhood of x^*, and has a positive definite Hessian matrix at x^*.

Then the conditions (11.2) and (11.3) are fulfilled (with $\rho_1 = \rho_2 = \frac{1}{2}$). This follows directly from Lemma B.10, Lemma B.11, and Remark B.1.

Remark 14.4 Assume that the mutation sequence $(p_n : n \in \mathbb{N})$ is absolutely continuous with the sequence of densities $(g_n : n \in \mathbb{N})$. Furthermore, suppose that

$$\begin{cases} \text{For all } x_0 \in D\backslash D^* \text{there exists a } c_0 > 0 \text{ with} \\ g_n(x_0, x_1, \ldots, x_{n-1}, y_n) \geq c_0 \text{ for all } n \in \mathbb{N}, \\ \quad x_1, \ldots, x_{n-1} \in D_{f(x_0)}\backslash D^* \text{ and } y_n \in D_{f(x_0)}. \end{cases} \tag{14.6}$$

(Cf. Condition (12.1). Obviously, (12.1) follows from(14.6).)

Then Condition (14.1) holds.

Equation (14.6) is satisfied in particular in the stationary case, i.e. if $g_n(x_0, x_1, \ldots, x_{n-1}, y_n) = g(x_{n-1}, y_n)$, as well as all $D_a(a > f^*)$ are bounded, and g is positive and continuous on $\bar{D} \times \bar{D}$.

Proof Let $x_0 \in D\backslash D^*$, $n \in \mathbb{N}$, $x_1, \ldots, x_{n-1} \in D_{f(x_0)}$ and $B \subset D_{f(x_0)}$ be measurable.

If (14.6) is satisfied, then

$$p_n(x_0, x_1, \ldots, x_{n-1}, B) = \int_B g_n(x_0, x_1, \ldots, x_{n-1}, y_n)\lambda^d(dy_n) \geq c_0\lambda^d(B),$$

and also (14.1) holds.

Based on the above assumptions, in the stationary case the density g takes its minimum on $\overline{D_{f(x_0)}} \times \overline{D_{f(x_0)}}$.

However, this is positive according to the assumptions.

Hence, (14.6) is therefore satisfied. □

Theorem 14.2 *Assume that condition (14.3) of Theorem 14.1 be satisfied. Additionally, suppose:*

$$\begin{cases} \textit{There exist a } b \geq 0 \textit{ and an } R \in (0, \infty) \cup \{\infty\} \textit{ with} \\ p_n(x_0, x_1, \ldots, x_{n-1}, B \cap K(x_{n-1}, R)) \leq b\lambda^d(B \cap K(x_{n-1}, R)) \\ \textit{for all } n \in \mathbb{N}, x_0, \ldots, x_{n-1} \in D\backslash D^* \textit{ and measurable } B \subset D. \end{cases} \tag{14.7}$$

Then, there exists $f_0 > f^*$ *with*

(i)

$$P\left(\liminf_{n\to\infty} \left(n^\alpha \left(f(X_n) - f^*\right)\right) = \infty\right) = 1$$

$$\textit{for all } \alpha > \frac{2}{d\rho_2} \textit{ and } x_0 \in D_{f_0}\backslash D^*;$$

(ii)

$$P\left(\liminf_{n\to\infty} \left(n^\alpha \mid X_n - x^* \mid\right) = \infty\right) = 1$$

$$\textit{for all } \alpha > \frac{2\rho_1}{d\rho_2} \textit{ and } x_0 \in D_{f_0}\backslash D^*.$$

if additionally condition (14.2) of Theorem 14.1 is satisfied.

(iii)

$$\liminf_{n\to\infty}\left(n^{\frac{1}{d\rho_2}}E\left(f(X_n)-f^*\right)\right)\geq\left(f(x_0)-f^*\right)\Gamma\left(1+\frac{1}{d\rho_2}\right)$$

for all $x_0\in D_{f_0}\backslash D^*$.

In the stationary case there exists a function h with $h(x)>0$ *for* $x\in D\backslash D^*$ *and*

$$\liminf_{n\to\infty}\left(n^{\frac{1}{d\rho_2}}E\left(f(X_n)-f^*\right)\right)\geq h(x_0)$$

for all $x_0\in D\backslash D^*$.

Note 14.1 For $R=\infty$ and arbitrary $x\in\mathbb{R}^d$ we put $K(x,R):=\mathbb{R}^d$.

Proof of Theorem 14.2

(a) Choose sufficiently small $c>0$ such that

$$c\leq\min\left\{c_2,\left(\frac{R}{2b_2}\right)^{\frac{1}{\rho_2}}\right\}\text{ and }b\lambda^d\left(K\left(x^*,b_2c^{\rho_2}\right)\right)<1$$

(As already mentioned, we have $\lambda^d\left(K\left(x^*,b_2c^{\rho_2}\right)\right)=m{b_2}^d{c_2}^{d\rho_2}$,

where $m=\dfrac{\pi^{\frac{d}{2}}}{\Gamma\left(1+\frac{d}{2}\right)}$ is the volume of the unit sphere).

Let $f_0:=f^*+c$.

Because of $c\leq c_2$, according to (14.3) we have

$$D_{f_0}=\{f\leq f^*+c\}\subset K\left(x^*,b_2c^{\rho_2}\right)\subset K\left(x^*,\frac{R}{2}\right).$$

Therefore, for $x,y\in D_{f_0}$ we have $|x-y|<R$.

Hence, $D_{f_0}\subset K(x,R)$ for all $x\in D_{f_0}$.

(b) Let now $x_0\in D_{f_0}\backslash D^*$ and $\varepsilon\in(0,f(x_0)-f^*]$.

Moreover, $x_0,\ldots,x_{n-1}\in D_{f(x_0)}\backslash D_{f^*+\varepsilon}$. Then

$D_{f^*+\varepsilon}\subset D_{f(x_{i-1})}\subset D_{f(x_0)}\subset D_{f_0}\subset K(x_{i-1},R)$ for $i=1,\ldots,n$

Thus, for $i=1,\ldots,n$ we have

$$\begin{aligned}
&q_i(x_0,\ldots,x_{i-1},D_{f(x_0)}\backslash D_{f^*+\varepsilon})\\
&=1-\left(q_i(x_0,\ldots,x_{i-1},D_{f^*+\varepsilon})+q_i(x_0,\ldots,x_{i-1},\mathbb{R}^d\backslash D_{f(x_0)})\right)\\
&=1-q_i(x_0,\ldots,x_{i-1},D_{f^*+\varepsilon})=1-p_i(x_0,\ldots,x_{i-1},D_{f^*+\varepsilon})
\end{aligned}$$

$$
\begin{aligned}
&= 1 - p_i\left(x_0, \ldots, x_{i-1}, D_{f^*+\varepsilon} \cap K(x_{i-1}, R)\right) \\
&\geq 1 - b\lambda^d\left(D_{f^*+\varepsilon} \cap K(x_{i-1}, R)\right) \\
&\geq 1 - b\lambda^d\left(K(x^*, b_2\varepsilon^{\rho_2})\right) \geq 0.
\end{aligned}
$$

Here, the last but one inequality follows from (14.3), since

$$\varepsilon \leq f(x_0) - f^* \leq f_0 - f^* = c \leq c_2.$$

Applying this inequality to $i = n, n-1, \ldots, 1$, we obtain

$$
\begin{aligned}
&P\left(f(X_n) - f^* \leq \varepsilon\right) \\
&= 1 - P\left(\bigcap_{i=1}^{n}\{X_i \in (D_{f(x_0)} \backslash D_{f^*+\varepsilon})\}\right) \\
&= 1 - \int_{D_{f(x_0)} \backslash D_{f^*+\varepsilon}} \cdots \\
&\qquad \int_{D_{f(x_0)} \backslash D_{f^*+\varepsilon}} q_n(x_0, \ldots, x_{n-1}, dx_n) \ldots q_1(x_0, dx_1) \\
&\leq 1 - \left(1 - b\lambda^d\left(K(x^*, b_2\varepsilon^{\rho_2})\right)\right)^n = 1 - \left(1 - bm{b_2}^d\varepsilon^{d\rho_2}\right)^n.
\end{aligned}
$$

(c) If additionally condition (14.2) holds, then for $\varepsilon > 0$ with

$$\left(\frac{2\varepsilon}{b_1}\right)^{\frac{1}{\rho_1}} \leq c_1, \text{ hence, } 0 < \varepsilon \leq \frac{b_1}{2}{c_1}^{\rho_2},$$

we obtain the inclusion

$$\left\{f \leq f^* + \left(\frac{2\varepsilon}{b_1}\right)^{\frac{1}{\rho_1}}\right\} \supset K(x^*, 2\varepsilon) \supset \overline{K(x^*, \varepsilon)}.$$

This yields

$$\left\{y \in D : \mid y - x^* \mid > \varepsilon\right\} \supset \left\{f > f^* + \left(\frac{2\varepsilon}{b_1}\right)^{\frac{1}{\rho_1}}\right\}.$$

Using part (b), for $0 < \varepsilon \leq \min\left\{f(x_0) - f^*, \frac{b_1}{2}{c_1}^{\rho_2}\right\}$ we obtain

$$P\left(\mid X_n - x^* \mid > \varepsilon\right) \geq P\left(f(X_n) - f^* > \left(\frac{2\varepsilon}{b_1}\right)^{\frac{1}{\rho_1}}\right)$$
$$\geq \left(1 - bmb_2{}^d\left(\frac{2\varepsilon}{b_1}\right)^{\frac{d\rho_2}{\rho_1}}\right)^n .$$

(d) Let now $N \in \mathbb{N}$ be arbitrary and $\alpha > \frac{2}{d\rho_2}$.
Let $K \in \mathbb{N}$ be chosen such that $bmb_2{}^d(Nn^{-\alpha})^{d\rho_2} \leq 1$ and $Nn^{-\alpha} \leq f(x_0) - f^*$, if $n \geq K$.
Then part (b) and the Bernoulli inequality yield

$$P\left(f(X_n) - f^* \leq Nn^{-\alpha}\right) \leq 1 - \left(1 - bmb_2{}^d(Nn^{-\alpha})^{d\rho_2}\right)^n$$
$$\leq bmb_2{}^d(Nn^{-\alpha})^{d\rho_2} n = bmb_2{}^d N^{d\rho_2} n^{1-\alpha d\rho_2} .$$

Because of $\alpha > \frac{2}{d\rho_2}$ it follows that $1 - \alpha d\rho_2 < -1$ and therefore

$$\sum_{n=1}^{\infty} P\left(n^\alpha\left(f(X_n) - f^*\right) \leq N\right) \leq K + bmb_2{}^d N^{d\rho_2} \sum_{n=K}^{\infty} n^{1-\alpha d\rho_2} < \infty.$$

From Borel–Cantelli's lemma it follows therefore that

$$P\left(n^\alpha\left(f(X_n) - f^*\right) \leq N \text{ for infinite many } n \in \mathbb{N}\right) = 0 ,$$
$$P\left(n^\alpha\left(f(X_n) - f^*\right) > N \text{ for finally all } n \in \mathbb{N}\right) = 1, \text{ respectively.}$$

Because this is true for all $n \in \mathbb{N}$, the assertion (i) follows.
Similarly, Part (c) is used to show assertion (ii).

(e) Let again $x_0 \in D_{f_0} \backslash D^*$.
Define $F_n(\varepsilon) := P\left(f(X_n) - f^* \leq \varepsilon\right)$.
From part (b) for all $\varepsilon \in (0, f(x_0) - f^*]$ we get the inequality

$$F_n(\varepsilon) \leq 1 - \left(1 - bmb_2{}^d \varepsilon^{d\rho_2}\right)^n .$$

The continuity of F_n from the right guarantees this inequality also for $\varepsilon = 0$.
Let

$$G_n(\varepsilon) := \begin{cases} 0 & \text{if } \varepsilon < 0 \\ 1 - \left(1 - \left(\frac{\varepsilon}{f(x_0) - f^*}\right)^{d\rho_2}\right)^n & \text{if } 0 \leq \varepsilon \leq f(x_0) - f^* \\ 1 & \text{if } \varepsilon > f(x_0) - f^* . \end{cases}$$

Because of the assumption $b\lambda^d(K(x^*, b_2c^{\rho_2})) < 1$ (see part (a)), we get

$$f(x_0) - f^* \leq f_0 - f^* = c < \left(bmb_2{}^d\right)^{-\frac{1}{d\rho_2}},$$

hence,

$$\left(\frac{1}{f(x_0) - f^*}\right)^{d\rho_2} \geq bmb_2{}^d$$

and therefore $F_n(\varepsilon) \leq G_n(\varepsilon)$ for all $\varepsilon \in \mathbb{R}$.

Therefore,

$$\begin{aligned} n^{\frac{1}{d\rho_2}} E\left(f(X_n) - f^*\right) &= n^{\frac{1}{d\rho_2}} \int_0^\infty (1 - F_n(x))dx \\ &\geq n^{\frac{1}{d\rho_2}} \int_0^{f(x_0)-f^*} \left(1 - \left(\frac{x}{f(x_0) - f^*}\right)^{d\rho_2}\right)^n dx \\ &= n^{\frac{1}{d\rho_2}} \left(f(x_0) - f^*\right) \int_0^1 (1 - u^{d\rho_2})^n du. \end{aligned}$$

Now, Lemma B.16 yields assertion (iii).

(f) Select an arbitrary $x_0 \in D \backslash D^*$.

Let

$$T := \begin{cases} \inf\{n \in \mathbb{N}_0 : X_n \in (D_{f_0} \backslash D^*)\} & \text{if}\{n \in \mathbb{N}_0 : X_n \in (D_{f_0} \backslash D^*)\} \neq \emptyset \\ \infty & \text{else} \end{cases}.$$

Because of the monotonicity of the sequence $E(f(X_n) - f^*)$ and due to $E(f(X_n) - f^*) \geq 0$, there exists $b \geq 0$ with

$$\lim_{n\to\infty} E(f(X_n) - f^*) = b.$$

If $b > 0$, the last assertion of Theorem 14.2 follows because of $n^{\frac{1}{d\rho_2}} \to \infty$.

If $b = 0$, the theorem of monotone convergence yields

$$E\left(\lim_{n\to\infty} (f(X_n) - f^*)\right) = 0.$$

Because of $\lim_{n\to\infty} (f(X_n) - f^*) \geq 0 \quad P-$a.s., we therefore have

$$\lim_{n\to\infty} (f(X_n) - f^*) = 0 \qquad P-\text{a.s.}.$$

From this and on account of

$$P(X_n \in D^*) = P(f(X_n) - f^* = 0) = F_n(0) = 0 \text{ (cf. e))}$$

it follows that

$$P(T < \infty) = 1.$$

Thus, we have

$$\begin{aligned} E\left(f(X_n) - f^*\right) &\geq E\left(f(X_{n+T}) - f^*\right) \\ &= \int E\left(f(X_{n+T}) - f^* \mid X_T = x\right) P_{X_T}(dx). \end{aligned}$$

Since the stationary case is present, (cf. [2]), we have

$$E\big(f(X_{n+T}) - f^* \mid X_T = x\big) = E\big(f(X_n) - f^* \mid X_0 = x\big) \qquad P_{X_T} - \text{a.s..}$$

For $x_0 = x$, according to part (e) we get

$$\begin{aligned} n^{\frac{1}{d\rho_2}} E\left(f(X_n) - f^* \mid X_0 = x\right) &= n^{\frac{1}{d\rho_2}} E\left(f(X_n) - f^*\right) \\ &\geq \left[n^{\frac{1}{d\rho_2}} \int_0^1 \left(1 - u^{d\rho_2}\right)^n du\right] (f(x) - f^*). \end{aligned}$$

Because of $P\left(X_T \in (D_{f_0} \setminus D^*)\right) = 1$, this implies that

$$\begin{aligned} & n^{\frac{1}{d\rho_2}} E\left(f(X_n) - f^*\right) \\ & \geq n^{\frac{1}{d\rho_2}} \int_{D_{f_0} \setminus D^*} E\left(f(X_{n+T}) - f^* \mid X_T = x\right) P_{X_T}(dx) \\ & = n^{\frac{1}{d\rho_2}} \int_{D_{f_0} \setminus D^*} E\left(f(X_n) - f^* \mid X_0 = x\right) P_{X_T}(dx) \\ & \geq \left[n^{\frac{1}{d\rho_2}} \int_0^1 \left(1 - u^{d\rho_2}\right)^n du\right] \int_{D_{f_0} \setminus D^*} (f(x) - f^*) P_{X_T}(dx). \end{aligned}$$

Now, it is $f(x) - f^* > 0$ on $D_{f_0} \setminus D^*$ and therefore

$$\int_{D_{f_0} \setminus D^*} (f(x) - f^*) P_{X_T}(dx) > 0.$$

With $h(x) := \Gamma\left(1 + \frac{1}{d\rho_2}\right) \int_{D_{f_0} \setminus D^*} (f(x) - f^*) P_{X_T}(dx)$, the last assertion Theorem 14.2 follows from Lemma B.16.

□

Remark 14.5 Suppose that the mutation sequence $(p_n : n \in \mathbb{N})$ is absolutely continuous with the sequence of densities $(g_n : n \in \mathbb{N})$. Furthermore, assume

$$\begin{cases} \text{There exist constants } b_0 \geq 0 \text{ and } R \in \mathbb{R}_+ \cup \{\infty\} \text{with} \\ g_n(x_0, \ldots, x_{n-1}, y_n) \leq b_0 \text{ for all } n \in \mathbb{N}, x_0, \ldots, x_{n-1} \in D \setminus D^* \\ \qquad \text{and } y_n \in D \cap K(x_{n-1}, R). \end{cases} \tag{14.8}$$

Then condition (14.7) is satisfied.

Condition (14.8) is satisfied in particular in the stationary case, D is bounded, and g is continuous on $\bar{D} \times \bar{D}$

This can be shown analogously to the proof of Remark 14.4.

Remark 14.6 Let the RSM $(X_n : n \in \mathbb{N}_0)$ be stationary with the mutation transition probability p, which is assumed to be absolutely continuous with the continuous and positive Lebesgue density g.

Furthermore, let the conditions of Remark 14.3 be fulfilled.

Then all assumptions of Theorem 14.1 and Theorem 14.2 hold (with $\rho_1 = \rho_2 = \frac{1}{2}$). This follows from the Remarks 14.3 and 14.5.

In particular, for each starting point $x_0 \in D \setminus D^*$ there exist constants $c_1 > 0$ and $c_2 < \infty$ with

$$\begin{aligned} c_1 &\leq \liminf_{n \to \infty} \left(n^{\frac{2}{d}} E(f(X_n) - f^*) \right) \\ &\leq \limsup_{n \to \infty} \left(n^{\frac{2}{d}} E(f(X_n) - f^*) \right) \leq c_2. \end{aligned}$$

The first inequality means that rate of convergence of the sequence $E(f(X_n) - f^*)$ is at most of order $O\left(n^{-\frac{2}{d}}\right)$, while the last inequality means that the rate of convergence is at least of this order.

Hence, the rate of convergence is exactly of order $O\left(n^{-\frac{2}{d}}\right)$.

Example 14.1 Let $f : D \to \mathbb{R}$, $D = \mathbb{R}$, be continuously differentiable and convex.

Furthermore, let D^* be a singleton, hence, $D^* = \{x^*\}$.

Let denote $h : \mathbb{R} \to \mathbb{R}$ a positive, continuous, and bounded probability density. Let

$$g_n(x_0, \ldots, x_{n-1}, y_n) = g(x_{n-1}, y_n) = \frac{1}{\sigma(x_{n-1})} h\left(\frac{y_n - x_{n-1}}{\sigma(x_{n-1})}\right)$$

(cf. Example 12.1).

(Note: If the random variable X has the density $f_X(x)$, then the random variable $Y = a + bX (b > 0)$ has the density $f_Y(x) = \frac{f_X(\frac{x-a}{b})}{b}$).

Let us consider the following situation:

(i) *uncontrolled* case: $\sigma(x) \equiv \sigma > 0$.
(ii) *controlled* case: $\sigma(x) =\mid \nabla f(x) \mid=\mid f'(x) \mid$.

In the uncontrolled case, obviously condition (14.8) is satisfied because of $g(x, y) = \frac{1}{\sigma} \sup\{h(z) : z \in D\}$.

Since D^* is bounded and f is convex, any D_a $(a > f^*)$ is bounded (cf. Lemma B.3), and Remark 14.4 implies that condition (14.1) is satisfied too.

If in the controlled case $x_0 = x^* + 2$ is chosen for the starting point and $B = [x^* + 1, x^* + 2]$, then

$$p(x, B) = \int_B g(x, y)\lambda^1(dy) = \frac{1}{\sigma(x)} \int_{x^*+1}^{x^*+2} h\left(\frac{y-x}{\sigma(x)}\right) dy$$

$$= \int_{\frac{(x^*+1-x)}{\sigma(x)}}^{\frac{(x^*+2-x)}{\sigma(x)}} h(u)du \le \int_{\frac{(x^*+1-x)}{\sigma(x)}}^{\infty} h(u)du \underset{x \to x^*}{\longrightarrow} 0$$

$$\text{since } \frac{1}{\sigma(x)} \to \infty \text{, if } x \to x^*.$$

If it is taken into consideration that because of the convexity of f the set B is included $D_{f(x_0)}$, then this indicates that in the controlled case condition (14.1) is violated. The same is also true in case of condition(14.7). Indeed, if we select

$$B_x := \left[x^* - \sqrt{\sigma(x)}, x^* + \sqrt{\sigma(x)}\right],$$

then on one hand we have

$$\lambda^1(B_x) = 2\sqrt{\sigma(x)} \to 0 \text{ for } x^* \to x.$$

But on the other hand it is

$$p(x, B_x) = \frac{1}{\sigma(x)} \int_{x^*-\sqrt{\sigma(x)}}^{x^*+\sqrt{\sigma(x)}} h\left(\frac{y-x}{(x)}\right) dy$$

$$= \int_{\frac{(x^*-x-\sqrt{\sigma(x)})}{\sigma(x)}}^{\frac{(x^*-x+\sqrt{\sigma(x)})}{\sigma(x)}} h(u)du \underset{x \to x^*}{\longrightarrow} \int_{-\infty}^{+\infty} h(u)du = 1.$$

Therefore, condition (14.7) cannot be satisfied.

With the methods of Chap. 14 it is therefore not possible to make a statement about the rate of convergence in the controlled case. This gap will be closed in the next chapter.

Additionally, it should be noted that in the controlled case it is possible to prove the validity of condition (12.1) with Remark 12.3. Thus, from Remark 12.2 and Theorem 12.1 at least the convergence of the controlled random search procedure can be obtained.

References

1. Gänssler, P., Stute, W.: Wahrscheinlichkeitstheorie. Springer, Berlin (1977)
2. Neveu, J.: Mathematische Grundlagen der Wahrscheinlichkeitstheorie. Oldenburg Verlag, Munich (1969)

Chapter 15
Random Search Methods with a Linear Rate of Convergence

15.1 Methods with a Rate of Convergence that Is at Least Linear

We return now once again to the example of Chap. 14. In this example we have

$$D = \mathbb{R} \quad , \quad f(x) = | x |^s \qquad (s > 0 \text{ arbitrary}), \text{ and} \quad p(x, \cdot)$$

denotes the uniform distribution on the interval $[x - r(x), x + r(x)]$.

We know that in the uncontrolled case

$$r(x) = r > 0 \quad \text{for any} \quad x \in \mathbb{R}$$

the rate of convergence of the sequence $E(f(X_n) - f^*)$ is of the order $O(n^{-s})$.

Of course, we may choose a different function $r(x)$ in order to improve the type of the convergence rate. However, what is an appropriate choice of $r(x)$? To find an answer to this question, we remember that for the process history $x_0, x_1, \ldots, x_n$ the next mutation point y_{n+1} is the realization of the uniform distribution on the interval $[x_n - r(x_n), x_n + r(x_n)]$. If $r(x)$ is chosen as a constant function, then the probability of a search success (i.e., $f(y_{n+1}) < f(x_n)$) becomes very small if x_n is close to x^*. It even becomes all the smaller, the closer x_n approaches x^*. Obviously, this has a negative influence on the rate of convergence. Hence, to remedy this, we have to adjust the length of the mutation interval $[x_n - r(x_n), x_n + r(x_n)]$ to the distance between x_n and x^*. In the case of an optimal point x^*, this is not a problem. But how to obtain information about $| x - x^* |$, if x^* is not known? In general this is very difficult. However, if some information about the form of f are known, things may be different. For example, for quadratic functions f we have the equation

K. Marti, *Optimization Under Stochastic Uncertainty*, International Series in Operations Research & Management Science 296,
https://doi.org/10.1007/978-3-030-55662-4_15

$$x - x^* = H_f{}^{-1}(x)\nabla f(x) \qquad (H_f = \text{Hessian})$$

If f is convex and twice continuously differentiable with a regular Hessian at x^*, then for each $R > 0$ at least constants $c_1, c_2 > 0$ can be found such that on $K(x^*, R)\backslash\{x^*\}$ the estimate

$$c_1 \leq \frac{\mid \nabla f(x) \mid}{\mid x - x^* \mid} \leq c_2$$

holds (cf. Lemma B.13).

Now we will return again to our example and adapt the length of the interval $[x - r(x), x + r(x)]$ to the distance between x and x^* by setting

$$r(x) = \mid x - x^* \mid = \mid x \mid .$$

Furthermore, with this control of the interval length we examine whether the convergence rate $O(n^{-s})$ of the uncontrolled case is exceeded under the above control law.

According to Lemma 9.1, for $x > 0$ we have

$$\begin{aligned}
&\int (f(y) - f^*)q(x, dy) = f(x) - f^* - \int (f(x) - f(y))q(x, dy) \\
&= f(x) - f^* - \int_{\{f<f(x)\}} (f(x) - f(y))p(x, dy) \\
&= \mid x \mid^s - \int_{-|x|}^{|x|} \left(\mid x \mid^s - \mid y \mid^s\right) p(x, dy) \\
&= \mid x \mid^s - \int_{-|x|}^{|x|} \left(\mid x \mid^s - \mid y \mid^s\right) \frac{1}{2 \mid x \mid} 1_{[x-|x|, x+|x|]}(y)dy \\
&= \mid x \mid^s - \frac{1}{2 \mid x \mid} \int_0^{|x|} \left(\mid x \mid^s - \mid y \mid^s\right) dy \\
&= \mid x \mid^s - \frac{1}{2 \mid x \mid} \left(\mid x \mid^{s+1} - \frac{1}{s+1} \mid x \mid^{s+1}\right) = \frac{1}{2}\frac{s+2}{s+1} \mid x \mid^s \\
&= \frac{1}{2}\frac{s+2}{s+1}(f(x) - f^*)
\end{aligned}$$

It is easy to see that this formula is also valid for $x < 0$. Then, we get

$$\begin{aligned}
&E(f(X_n) - f^*) \\
&= \int \ldots \int (f(x_n) - f^*)q(x_{n-1}, dx_n) \ldots q(x_0, dx_1)
\end{aligned}$$

$$= \frac{1}{2}\frac{s+2}{s+1}\int \dots \int (f(x_{n-1}) - f^*)q(x_{n-2}, dx_{n-1}) \dots q(x_0, dx_1)$$

$$= \dots = \left(\frac{1}{2}\frac{s+2}{s+1}\right)^n (f(x_0) - f^*)$$

Because of $s > 0$, it follows that $\frac{1}{2}\frac{s+2}{s+1} < 1$.
With $\beta := \frac{1}{2}\frac{s+2}{s+1}$ we then obtain

$$E(f(X_n) - f^*) = O(\beta^n).$$

Comparing this result with the uncontrolled case, a considerable improvement of the rate of convergence is found. The reason for this is the fact that the step length parameter $r(x)$ has been adapted to the distance between x and x^*. As already mentioned, such an adjustment is possible only if specific demands are made on the shape of f and D.

Hence, in the following we consider the following assumptions:

$$\begin{cases} D^* \text{ is assumed to be bounded and } D^* \cap \dot{D} \neq \emptyset\text{. Furthermore, let there exist an} \\ \alpha > f^* \text{ with a convex } D_\alpha\text{, such that } f \text{ is continuous and convex on } D_\alpha. \end{cases} \tag{15.1}$$

$$\begin{cases} \text{There exist an } x^* \in D^* \cap \dot{D} \text{ as well as } \tilde{a} > f^* \text{ and } \tilde{b} > 0 \text{ with} \\ \inf\{|\, y - x^* \,| : y \in D, f(y) = f(x)\} \geq \\ \tilde{b} \cdot \sup\{|\, y - x^* \,| : y \in D, f(y) = f(x)\} \text{ for all } x \in D_{\tilde{a}} \backslash D^* \end{cases} \tag{15.2}$$

Remark 15.1 Condition (15.2) ensures, roughly spoken, that the level sets $\{x \in D : f(x) = c\}$ do not get too small for values of c close to f^*.

This will be illustrated by means of two simple examples.

If $D = \mathbb{R}^d$ and $f(x, y) := ax^2 + by^2$ (a, b fixed with $0 < a < b$), then the level sets $\{x \in D : f(x) = c\}$ $(c > 0)$ are ellipses with the half axis $\sqrt{\frac{c}{a}}$ and $\sqrt{\frac{c}{b}}$. Therefore,

$$\inf\{|\, z - x^* \,| : f(z) = c\} = \sqrt{\frac{c}{b}} = \sqrt{\frac{a}{b}}\sqrt{\frac{c}{a}}$$

$$= \sqrt{\frac{a}{b}} \sup\{|\, z - x^* \,| : f(z) = c\},$$

i.e. condition (15.2) holds with $\tilde{b} := \sqrt{\frac{a}{b}}$.

Conversely, for the function $f(x, y) := x^2+y^4$ condition (15.2) violated. Indeed, for $0 < c \leq 1$ we have

$$\frac{\inf\{| z - x^* |: f(z) = c\}}{\sup\{| z - x^* |: f(z) = c\}} = \frac{c^{\frac{1}{2}}}{c^{\frac{1}{4}}} = c^{\frac{1}{4}} \underset{c \to 0}{\longrightarrow} 0.$$

Remark 15.2 Let $D^* = \{x^*\} \subset \dot{D}$, and let f be twice continuously differentiable in a neighborhood of x^* with a positive definite Hessian at x^*. Furthermore, assume that f is continuous on D.

If D is compact or D and f are convex, then the conditions (12.1) and (12.2) are satisfied.

Indeed, by Lemma B.13 we know that all assumptions of Lemma B.12 are satisfied. According to Remark B1, the assumptions (15.1) and (15.2) are then satisfied.

Definition 15.1

(a) A random search procedure $(X_n : n \in \mathbb{N}_0)$ starting in $x_0 \in D$ is called *convergent in the mean, linearly convergent P-a.s.*, resp., with respect to the starting point $x_0 \in D$ if a $q > 1$ exists such that

$$q^n \left(E(f(X_n) - f^*)\right) \underset{n\to\infty}{\longrightarrow} 0 \quad ,$$

$$q^n \left(f(X_n) - f^*\right) \underset{n\to\infty}{\longrightarrow} 0 \qquad P\text{-a.s. , respectively.}$$

(b) A random search procedure $(X_n : n \in \mathbb{N}_0)$ is called *convergent in the mean, globally linearly convergent* P – a.s., provided that for each starting point $x_0 \in D$ the procedure $(X_n : n \in \mathbb{N}_0)$ is convergent in the mean, linearly convergent P-a.s., respectively.

For the proof of a linear rate of convergence the following two lemmas are of crucial importance.

Instead of $\{x \in D : f(x) < c\}$ we will use again the shorter notation $\{f < c\}$.

Lemma 15.1 *Let denote* $(X_n : n \in \mathbb{N}_0)$ *the random search procedure starting in* $x_0 \in D\backslash D^*$ *and having the mutation sequence* $(p_n : n \in \mathbb{N})$.

Furthermore, let there exist a $c > 0$ *with*

$$\int_{\{f<f(x_{n-1})\}} (f(x_{n-1}) - f(y_n))p_n(x_0, x_1, \ldots, x_{n-1}, dy_n)$$
$$\geq c(f(x_{n-1}) - f^*)$$

for any $n \in \mathbb{N}, x_1, \ldots, x_{n-1} \in D_{f(x_0)}\backslash D^*$.

Then,

$$E(f(X_n) - f^*) \leq (1 - c)^n(f(x_0) - f^*)$$

for any $n \in \mathbb{N}$.

Hence, the procedure $(X_n : n \in \mathbb{N}_0)$ *is in the mean linearly convergent for the starting point* x_0.

Proof Because of

$$\begin{aligned}
&\int_{\mathbb{R}^d} (f(x_n) - f^*) q_n(x_0, x_1, \ldots, x_{n-1}, dx_n) \\
&\quad = f(x_{n-1}) - f^* - \int_{\mathbb{R}^d} (f(x_{n-1}) - f(x_n)) q_n(x_0, x_1, \ldots, x_{n-1}, dx_n) \\
&\quad = f(x_{n-1}) - f^* \int_{\{f < f(x_{n-1})\}} (f(x_{n-1}) - f(y_n)) p_n(x_0, x_1, \ldots, x_{n-1}, dy_n) \\
&\quad \le (1-c)(f(x_{n-1}) - f^*),
\end{aligned}$$

from Lemma 15.1 it follows that

$$\begin{aligned}
&E(f(X_n) - f^*) \\
&\quad = \int \ldots \int (f(x_n) - f^*) q_n(x_0, x_1, \ldots, x_{n-1}, dx_n) \ldots q_1(x_0, dx_1) \\
&\quad \le (1-c)^n (f(x_0) - f^*).
\end{aligned}$$

□

Lemma 15.2 *Let denote* $(X_n : n \in \mathbb{N}_0)$ *the random search procedure starting in* $x_0 \in D \backslash D^*$ *and having the mutation sequence* $(p_n : n \in \mathbb{N})$. *Moreover, suppose that there exist* $c_1 \in [0, 1)$, $c_2 > 0$ *such that*

$$p_n\left(x_0, x_1, \ldots, x_{n-1}, \left\{y \in D : f(y) - f^* \le c_1(f(x_{n-1}) - f^*)\right\}\right) \ge c_2$$
$$\text{for all } n \in \mathbb{N}, x_1, \ldots, x_{n-1} \in D_{f(x_0)} \backslash D^*.$$

Then the procedure $(X_n : n \in \mathbb{N}_0)$ *is linearly convergent for the starting point* $x_0 \quad P-a.s.$.

Proof Let $x_1, \ldots, x_{n-1} \in D_{f(x_0)}$ be arbitrary points.

If $x_{n-1} \in D^*$, then because of $f(X_n) \le f(X_{n-1}) \quad P$-a.s., we have

$$P\left(f(X_n) = f^* \mid X_{n-1} = x_{n-1}, \ldots, X_1 = x_1, X_0 = x_0\right) = 1$$

and therefore

$$\begin{aligned}
P(f(X_n) - f^* \le c_1(f(X_{n-1}) - f^*) \mid X_{n-1} = x_{n-1}, \ldots, X_1 = x_1, X_0 = x_0) \\
= 1 \ge c_2.
\end{aligned}$$

If x_{n-1} does not lie in D^*, then, because of $c_1 < 1$, x_{n-1} is also not included in the set $\{y \in D : f(y) - f^* \leq c_1(f(x_{n-1}) - f^*)\}$.

Thus, (cf. (9.5)), we have

$$q_n\left(x_0, x_1, \ldots, x_{n-1}, \left\{y \in D : f(y) - f^* \leq c_1(f(x_{n-1}) - f^*)\right\}\right)$$
$$= p_n\left(x_0, x_1, \ldots, x_{n-1}, \left\{y \in D : f(y) - f^* \leq c_1(f(x_{n-1}) - f^*)\right\}\right).$$

This yields

$$P(f(X_n) - f^* \leq c_1(f(X_{n-1}) - f^*) \mid X_{n-1} = x_{n-1}, \ldots, X_1 = x_1, X_0 = x_0)$$
$$= P\Big(X_n \in \left\{y \in D : f(y) - f^* \leq c_1(f(x_{n-1}) - f^*)\right\} \mid$$
$$X_{n-1} = x_{n-1}, \ldots\ldots, X_0 = x_0\Big)$$
$$= q_n\left(x_0, x_1, \ldots, x_{n-1}, \left\{y \in D : f(y) - f^*\right.\right.$$
$$\left.\left.\leq c_1(f(x_{n-1}) - f^*)\right\}\right) \geq c_2.$$

Consequently, for arbitrary points $x_1, \ldots, x_{n-1} \in D_{f(x_0)}$ the following inequality

$$P\Big(f(X_n) - f^* \leq c_1(f(X_{n-1}) - f^*) \mid X_{n-1} = x_{n-1}, \ldots$$
$$\ldots, X_1 = x_1, X_0 = x_0\Big) \geq c_2$$

holds. Because of the definition of the factorized conditional expectation (cf. [1, p.191]) we have

$$P\left(f(X_n) - f^* \leq c_1(f(X_{n-1}) - f^*) \mid X_{n-1}, \ldots, X_1, X_0\right)(\omega)$$
$$= P\Big(f(X_n) - f^* \leq c_1(f(X_{n-1}) - f^*) \mid X_{n-1} = X_{n-1}(\omega), \ldots$$
$$X_{n-1}(\omega), \ldots, X_0 = X_0(\omega)\Big).$$

Hence, the above inequality is equivalent to

$$P\left(f(X_n) - f^* \leq c_1(f(X_{n-1}) - f^*) \mid X_{n-1}, \ldots, X_1, X_0\right) \geq c_2$$
$$P - \text{a.s. on} \bigcap_{i=1}^{n-1} \left\{X_i \in D_{f(x_0)}\right\}.$$

However, we have (cf. Lemma 9.1)

$$P\left(\bigcap_{i=1}^{n-1}\{X_i \in D_{f(x_0)}\}\right) = 1.$$

Hence,

$$P\left(f(X_n) - f^* \le c_1(f(X_{n-1}) - f^*) \mid \mathcal{A}_{n-1}\right) \ge c_2 \qquad P - \text{a.s.},$$

where $\mathcal{A}_k := \sigma(X_0, X_1, \ldots, X_k)$ denotes the σ-algebra generated by $X_0, X_1, \ldots, X_k$.

Define $Z_n := 1_{\{f(X_n)-f^* \le c_1(f(X_{n-1})-f^*)\}}$.

Since the σ-algebra generated by $Z_1, \ldots, Z_n$ is included in $\mathcal{A}_n$, iteration of the conditional expectations (cf. [1], p. 171) yields

$$\begin{aligned} &E\left(Z_{n+1} \mid Z_n, \ldots, Z_1\right) = E\left(E\left(Z_{n+1} \mid \mathcal{A}_n\right) \mid Z_n, \ldots, Z_1\right) \\ &\quad = E\left(P\left(f(X_n) - f^* \le c_1(f(X_{n-1}) - f^*) \mid \mathcal{A}_n\right) \mid Z_n, \ldots, Z_1\right) \ge c_2 \\ &\quad P - \text{a.s.}. \end{aligned}$$

Lemma C.1 yields then

$$\liminf_{n\to\infty} \frac{1}{n}\sum_{i=1}^{n} Z_i \ge c_2 \qquad P - \text{a.s.}.$$

Thus, for all $\varepsilon > 0$ we get

$$P\left(\frac{1}{n}\sum_{i=1}^{n} 1_{\{f(X_i)-f^* \le c_1(f(X_{i-1})-f^*)\}} \ge c_2 - \varepsilon \quad \text{finally for all } n \in \mathbb{N}\right) = 1$$

and therefore

$$P\left(\sum_{i=1}^{n} 1_{\{f(X_i)-f^* \le c_1(f(X_{i-1})-f^*)\}} \ge n(c_2 - \varepsilon) \quad \text{finally for all } n \in \mathbb{N}\right) = 1.$$

Because of $f(X_i) - f^* \leq f(X_{i-1}) - f^* \quad P-$a.s. for all $i \in \mathbb{N}$, for arbitrary $k \in \mathbb{N}$ we get

$$\left\{\sum_{i=1}^{n} 1_{\{f(X_i)-f^* \leq c_1(f(X_{i-1})-f^*)\}} \geq k\right\} \subset$$
$$\left\{f(X_n) - f^* \leq {c_1}^k (f(x_0) - f^*)\right\} \qquad P-\text{a.s..}$$

For any $\varepsilon > 0$ this yields

$$P\Big(f(X_n) - f^* \leq {c_1}^{n(c_2-\varepsilon)}(f(x_0) - f^*)$$
$$\text{finally for all } n \in \mathbb{N}\Big) = 1.$$

This yields the assertion. □

The following theorem contains sufficient conditions for the linear convergence of a random search procedure $(X_n : n \in \mathbb{N}_0)$ fulfilling the integration property (10.4).

Since in particular random direction methods and procedures with an absolutely continuous mutation sequence have the property (10.4), (cf. Lemma 10.1), it is possible to prove—under suitable conditions—also for these methods a linear rate of convergence. This will be done in Theorems 15.2 and 15.3.

Theorem 15.1 *Let $(X_n : n \in \mathbb{N}_0)$ be a random search procedure with the mutation sequence $(p_n : n \in \mathbb{N})$.*

Suppose that there exist transition probabilities ρ_n from $\left(\mathbb{R}^d\right)^n$ to $S := \left\{x \in \mathbb{R}^d :\mid x \mid= 1\right\}$ and transition probabilities μ_n from $\left(\mathbb{R}^d\right)^n \times S$ to $\mathbb{R}_+$ fulfilling the integration property (10.4).

Furthermore, let the conditions (15.1) and (15.2) be satisfied.

Then:

(i) Suppose that for an $a > f^*$ the following conditions hold:

$$\begin{cases} \text{There exists } \varepsilon > 0 \text{ with} \\ \int_{\left[0, \frac{1}{2}|x_{n-1}-x^*|\right]} r\mu_n(x_0, x_1, \ldots, x_{n-1}, s, dr) \geq \varepsilon_0 \mid x_{n-1} - x^* \mid \\ \text{for all } n \in \mathbb{N}, x_0, \ldots, x_{n-1} \in D_a \backslash D^*, s \in S \end{cases} \tag{15.3}$$

and

$$\begin{cases} \textit{There exists } \tau > 0 \textit{ with} \\ \rho_n(x_0, x_1, \ldots, x_{n-1}, A) \geq \tau\gamma(A) \\ \textit{for all } n \in \mathbb{N}, x_0, \ldots, x_{n-1} \in D_a \backslash D^*, A \subset S \textit{ measurable} . \end{cases} \tag{15.4}$$

Then there is $a_0 > f^*$ such that the random search procedure $(X_n : n \in \mathbb{N}_0)$ is linearly convergent in the mean for any starting point $x_0 \in D_{a_0}$.

(ii) If for an $a > f^*$ condition (12.4) as well as the following condition

$$\begin{cases} \textit{There exist } \varepsilon_1 \in \left(0, \frac{1}{2}\right) \textit{ and } \sigma > 0 \textit{ such that} \\ \mu_n\left(x_0, x_1, \ldots, x_{n-1}, s, \left[\varepsilon_1 \mid x_{n-1} - x^* \mid, \frac{1}{2} \mid x_{n-1} - x^* \mid\right]\right) \geq \sigma \\ \textit{for any } n \in \mathbb{N}, x_0, \ldots, x_{n-1} \in D_a \backslash D^*, s \in S \end{cases} \tag{15.5}$$

is satisfied, then there is $a_1 > f^*$ such that the procedure $(X_n : n \in \mathbb{N}_0)$ is linearly convergent for all starting points $x_0 \in D_{a_1}$ P-a.s.

(iii) Suppose that D is convex and f is convex and continuous. If for each $a > f^*$ the conditions (12.4) and (15.4), the conditions (12.4) and (15.5), are satisfied, then the procedure $(X_n : n \in \mathbb{N}_0)$ is globally linearly convergent in the mean, P-a.s., respectively.

Proof

(1) Suppose that the conditions (15.3) and (15.4) are satisfied for an $a > f^*$.

If in Lemma B.11 the set D is replaced by D_α (α according to assumption (15.1)), then, together with assumption (15.2), it follows that there is a $\phi \in \left(0, \frac{\pi}{2}\right]$ with

$$\begin{aligned} &x + C\left(\phi, x^* - x, \frac{1}{2} \mid x^* - x \mid\right) \\ &\subset \left\{y \in D : f(x) - f(y) \geq \frac{\mid y - x \mid}{3 \mid x^* - x \mid}(f(x) - f^*)\right\} \\ &\subset \{f \leq f(x)\} \text{ for all } x \in D_\alpha \backslash D^*. \end{aligned}$$

For the definition of $C\left(\phi, x^* - x, \frac{1}{2} \mid x^* - x \mid\right)$, see Definition B.1.

Let $a_0 := \min\{a, \alpha\}$, and let denote x_0 an arbitrary starting point from D_{a_0}.

Without loss of generality we may assume that $f(x_0) > f^*$, because for $x_0 \in D^*$ the assertion is trivial.

Select any $n \in \mathbb{N}$ and arbitrary $x_1, \ldots, x_{n-1} \in D_{f(x_0)} \backslash D^*$.

Because of

$$D_{f(x_0)} \backslash D^* \subset D_{a_0} \backslash D^* \subset D_\alpha \backslash D^*$$

we have

$$\int_{x_{n-1}+C\left(\phi,x^*-x_{n-1},\frac{1}{2}|x_{n-1}-x^*|\right)}(f(x_{n-1})-f(y_n)p_n(x_0,x_1,\dots,x_{n-1},dy_n)$$

$$\geq \frac{f(x_{n-1})-f^*}{3\mid x_{n-1}-x^*\mid}\int_{\mathbb{R}^d}\mid y_n-x_{n-1}\mid 1_{C\left(\phi,x^*-x_{n-1},\frac{1}{2}|x_{n-1}-x^*|\right)}\cdot$$

$$\cdot\,(y_n-x_{n-1})p_n(x_0,x_1,\dots,x_{n-1},dy_n)$$

$$=\frac{f(x_{n-1})-f^*}{3\mid x_{n-1}-x^*\mid}\int_S\int_{\mathbb{R}_+}\mid rs\mid 1_{C\left(\phi,x^*-x_{n-1},\frac{1}{2}|x_{n-1}-x^*|\right)}(rs)\cdot$$

$$\cdot\,\mu_n(x_0,x_1,\dots,x_{n-1},s,dr)\rho_n(x_0,x_1,\dots,x_{n-1},ds)$$

$$=\frac{f(x_{n-1})-f^*}{3\mid x_{n-1}-x^*\mid}\int_S\int_{\mathbb{R}_+} r1_{C(\phi,x^*-x_{n-1})}(s)1_{\left[0,\frac{1}{2}|x_{n-1}-x^*|\right]}(r)\cdot$$

$$\cdot\,\mu_n(x_0,x_1,\dots,x_{n-1},s,dr)\rho_n(x_0,x_1,\dots,x_{n-1},ds)$$

$$=\frac{f(x_{n-1})-f^*}{3\mid x_{n-1}-x^*\mid}\int_{S\cap C(\phi,x^*-x_{n-1})}\int_{\left[0,\frac{1}{2}|x_{n-1}-x^*|\right]} r\cdot$$

$$\cdot\,\mu_n(x_0,x_1,\dots,x_{n-1},s,dr)\rho_n(x_0,x_1,\dots,x_{n-1},ds)$$

$$\geq\frac{\varepsilon_0}{3}(f(x_{n-1})-f^*)\rho_n\left(x_0,x_1,\dots,x_{n-1},S\cap C\left(\phi,x^*-x_{n-1}\right)\right)$$

(due to (15.3))

$$\geq\frac{\varepsilon_0}{3}(f(x_{n-1})-f^*)\tau\gamma\left(S\cap C\left(\phi,x^*-x_{n-1}\right)\right).$$

(because of (15.4))

We know that $\gamma\,(S\cap C\,(\phi,x^*-x_{n-1}))$ is positive and does not depend on x_{n-1} (cf. Lemma A.3). Considering the inclusion $\{f<f(x_{n-1})\}\subset x_{n-1}+C\,(\phi,x^*-x_{n-1})$, Lemma 15.1 yields the statement of part (i).

(2) Suppose that for a certain $a>f^*$ conditions (15.4) and (15.5) are fulfilled. Let α be selected according to assumption (15.1), and let $a_1:=\min\{a,\alpha\}$. Select $x_0\in D_{a_1}\backslash D^*$ and $x_1,\dots,x_{n-1}\in D_{f(x_0)}\backslash D^*$ arbitrarily. Applying Lemma B.11 to D_α, we obtain

$$x_{n-1} + C\left(\phi, x^* - x_{n-1}, \varepsilon_1 \mid x^* - x_{n-1} \mid, \frac{1}{2} \mid x^* - x_{n-1} \mid\right)$$

(cf. Definition B.1))

$$\subset \left\{y \in D : f(x_{n-1}) - f(y) \geq (f(x_{n-1}) - f^*)\frac{\varepsilon_1}{3}\right\}$$

$$\subset \left\{y \in D : f(y) - f^* \leq \left(1 - \frac{\varepsilon_1}{3}\right)(f(x_{n-1}) - f^*)\right\}.$$

Consequently, we have

$$p_n\left(x_0, x_1, \ldots, x_{n-1}, \left\{y \in D : f(y) - f^* \leq \left(1 - \frac{\varepsilon_1}{3}\right)(f(x_{n-1}) - f^*)\right\}\right)$$

$$\geq p_n\Bigg(x_0, x_1, \ldots, x_{n-1}, x_{n-1}+$$

$$+ C\left(\phi, x^* - x_{n-1}, \varepsilon_1 \mid x^* - x_{n-1} \mid, \frac{1}{2} \mid x^* - x_{n-1} \mid\right)\Bigg)$$

$$= \int_{S \cap C(\phi, x^* - x_{n-1})} \int_{\left[\varepsilon_1 |x^* - x_{n-1}|, \frac{1}{2}|x^* - x_{n-1}|\right]} \mu_n(x_0, x_1, \ldots, x_{n-1}, s, dr)\cdot$$

$$\cdot\, \rho_n(x_0, x_1, \ldots, x_{n-1}, ds)$$

$$\geq \sigma \rho_n\left(x_0, x_1, \ldots, x_{n-1}, S \cap C\left(\phi, x^* - x_{n-1}\right)\right)$$

$$\geq \sigma \tau \gamma\left(S \cap C\left(\phi, x^* - x_{n-1}\right)\right).$$

Lemma 15.2 yields now the assertion in part (ii).

(3) Assume that D is convex and f is convex and continuous.

Let $x_0 \in D \backslash D^*$ be selected arbitrarily.

According to the assumptions, for $a := f(x_0)$ (15.3) and (15.4) are satisfied. Due to Lemma B.11 there exists a $\phi > 0$ with

$$x + C\left(\phi, x^* - x, \frac{1}{2} \mid x^* - x \mid\right)$$

$$\subset \left\{y \in D : f(x) - f(y) \geq \frac{\mid x - y \mid}{3 \mid x - x^* \mid}(f(x) - f^*)\right\}$$

$$\subset \{f \leq f(x)\} \qquad \text{for all } x \in D_a \backslash D^*.$$

The assertions analogous to (i), (ii), resp., follow now exactly in the same way as in part (1), part (2), respectively.

□

Theorem 15.2 *Suppose that the assumptions (15.1) and (15.2) are satisfied.*

Let $(X_n : n \in \mathbb{N}_0)$ be a random direction method with the step length distribution λ_n. Let h denote a bounded probability density on $\mathbb{R}_+$ and let α_n be a positive, measurable function with

$$\lambda_n(x_0, x_1, \ldots, x_{n-1}, B) = (\alpha_n(x_0, \ldots, x_{n-1}))^{-1} \int_B h\left(\frac{r}{\alpha_n(x_0, \ldots, x_{n-1})}\right) dr$$

$$\text{for any } x_0, \ldots, x_{n-1} \in D\backslash D^*, \ B \subset \mathbb{R}_+ \text{ measurable}$$

(cf. Example 14.1).

Furthermore, let h be positive on the interval $(0, c)$ $\quad (c > 0)$.
Then:

(i) If for an $a > f^*$ the following condition

$$\begin{cases} \text{There exist } \alpha_0, \alpha_1 > 0 \text{ with} \\ \alpha_0 \leq \frac{\alpha_n(x_0,\ldots,x_{n-1})}{|x_{n-1}-x^*|} \leq \alpha_1 \\ \text{for any } n \in \mathbb{N}, x_0, \ldots, x_{n-1} \in D_a\backslash D^*. \end{cases} \tag{15.6}$$

holds, then there is an $a_0 > f^*$ such that $(X_n : n \in \mathbb{N}_0)$ converges linearly for any starting point $x_0 \in D_{a_0}$ in the mean and P-a.s.

(ii) Provided that D is convex, the function f is continuous and convex, and for each $a > f^*$ Condition (15.6) holds, then $(X_n : n \in \mathbb{N}_0)$ converges globally linear in the mean and P-a.s.

Proof According to Lemma 10.1 the integration property (10.4) is satisfied with

$$\mu_n(x_0, x_1, \ldots, x_{n-1}, s, \cdot) = \lambda_n(x_0, x_1, \ldots, x_{n-1}, \cdot)$$

and

$$\rho_n(x_0, x_1, \ldots, x_{n-1}, \cdot) = \gamma.$$

Obviously, for each $a > f^*$ condition (15.4) holds. Hence, it remains to show that—under the assumptions of Theorem 15.2—assumption (15.6) yields conditions (15.3) and (15.5).

Let (15.6) be satisfied for $a > f^*$.

Moreover, let $x_0, \ldots, x_{n-1} \in D_a\backslash D^*$ and $s \in S$ be arbitrary points.

Then:

$$\int_{\left[0, \frac{1}{2}|x_{n-1}-x^*|\right]} r\mu_n(x_0, x_1, \ldots, x_{n-1}, s, dr)$$

$$= \int_{\left[0, \frac{1}{2}|x_{n-1}-x^*|\right]} r\lambda_n(x_0, x_1, \ldots, x_{n-1}, dr)$$

$$= \int_0^{\frac{1}{2}|x_{n-1}-x^*|} \frac{r}{\alpha_n(x_0, x_1, \ldots, x_{n-1})} h\left(\frac{r}{\alpha_n(x_0, x_1, \ldots, x_{n-1})}\right) dr$$

$$= \alpha_n(x_0, x_1, \ldots, x_{n-1}) \cdot \int_0^{\frac{1}{2}|x_{n-1}-x^*|\alpha_n{}^{-1}} rh(r)dr$$

$$\geq | x_{n-1} - x^* | \left(\alpha_0 \cdot \int_0^{\frac{1}{2}\alpha_1{}^{-1}} rh(r)dr\right).$$

Since h is positive on the interval $(0, c)$, also the constant in the round brackets is positive.

This yields (15.3).

If $\varepsilon \in \left(0, \frac{1}{2}\right)$ is chosen so small that $\frac{\varepsilon_1}{\alpha_0} < c$ and $\frac{\varepsilon_1}{\alpha_0} < \frac{1}{2}\alpha_1{}^{-1}$, then under the above assumptions the constant

$$\sigma := \int_{\varepsilon_1\alpha_0{}^{-1}}^{\frac{1}{2}\alpha_1{}^{-1}} h(r)dr$$

is positive.

A simple calculation shows that

$$\mu_n\left(x_0, x_1, \ldots, x_{n-1}, s, \left[\varepsilon_1 \mid x_{n-1} - x^* \mid, \frac{1}{2} \mid x_{n-1} - x^* \mid\right]\right) \geq \sigma$$

Thus, also (15.5) holds.

Now, Theorem 15.1 yields the assertion. □

Remark 15.3 Suppose that $D^* = \{x^*\} \subset \mathring{D}$, D are compact, and the function f is continuous and twice continuous differentiable on a neighborhood of x^* with a positive definite Hessian at x^*.

Assume that

(1) $\alpha_n(x_0, x_1, \ldots, x_{n-1}) =| A(x_{n-1})\nabla f(x_{n-1}) |$
with a continuous matrix function $A : \mathbb{R}^d \to \mathbb{R}^{d\times d}$, where A is invertible at x^*.

(2) $\alpha_n(x_0, x_1, \ldots, x_{n-1}) =| \nabla f(x_{n-1}) | \eta(x_{n-1})$
with a continuous function $\eta : \mathbb{R}^d \to (0, \infty)$.

Then condition (15.6) is satisfied for an $a > f^*$.

In addition, according to Remark 15.1, also the assumptions (15.1) and (15.2) hold.

Proof In the first case the regularity of $A(x^*)$ and the continuity of A yield that $A(x)$ is invertible for each x contained in a certain neighborhood of $K(x^*, r)$. Hence, the matrices $(A(x))^T A(x)$ are positive definite (and of course symmetric) for any $x \in K(x^*, r)$.

Define now

$$\lambda_1 := \max \left\{ \lambda(x) : \lambda(x) \text{ eigenvalue of } (A(x))^T A(x), x \in K(x^*, r) \right\}$$

and

$$\lambda_0 := \min \left\{ \lambda(x) : \lambda(x) \text{ eigenvalue of } (A(x))^T A(x), x \in K(x^*, r) \right\}.$$

For $x \in K(x^*, r)$ we have then the following estimate (cf. Lemma B.7):

$$| \nabla f(x) | \sqrt{\lambda_0} \leq | A(x) \nabla f(x) | \leq | \nabla f(x) | \sqrt{\lambda_1}.$$

In the first case we get the assertion due to $\lambda_0 > 0$ and $\lambda_1 < \infty$, since according to Lemma B.13, for an $a > f^*$ the set D_a is included in $K(x^*, r)$.

In the second case it follows from Lemma B.13 und Lemma B.10 that there exist an $a > f^*$ and positive constants b_1, b_2 with

$$b_1 \leq \frac{| \nabla f(x) |}{| x - x^* |} \leq b_2 \qquad \text{for any } x \in D_a \backslash D^*.$$

Furthermore, due to the continuity of η and the compactness of D_a, constants $c_1, c_2 > 0$ exist with

$$c_1 \leq \eta(x) \leq c_2 \qquad \text{for all } x \in D_a,$$

cf. Lemma B.4. Therefore,

$$b_1 c_1 \leq \frac{| \nabla f(x) | \eta(x)}{| x - x^* |} \leq b_2 c_2 \qquad \text{for any } x \in D_a \backslash D^*$$

which yields the assertion in the second case. □

Remark 15.4 Dropping in Remark 15.2 the requirement for compactness of D and demands instead that $D := \mathbb{R}^d$ and f is a convex, twice continuously differentiable function, then according to Remark 15.1 assumptions (15.1) and (15.2) are satisfied again.

Let α_n be defined as in (1) and (2), respectively, of Remark 15.2, where in the first case $A(x)$ is assumed to be invertible for any $x \in \mathbb{R}^d$.

Then condition (15.6) holds for all $a > f^*$.

Taking into account that all sets D_a $(a \geq f^*)$ are compact (cf. Lemma B.4), the proof follows similarly to that of Remark 15.2.

Example 15.1 Let $(X_n : n \in \mathbb{N}_0)$ be a stationary random search procedure having the step length distribution λ.

Let $\lambda(x, \cdot)$ denote the uniform distribution on the interval $(0, \alpha(x))$.

Hence, $\lambda(x, \cdot)$ has the density

$$g(x, r) = \frac{1}{\alpha(x)} 1_{(0,\alpha(x))}(r) = \frac{1}{\alpha(x)} 1_{(0,1)} \left(\frac{r}{\alpha(x)} \right)$$

$$= \frac{1}{\alpha(x)} h \left(\frac{r}{\alpha(x)} \right) \quad \text{with } h(r) = 1_{(0,1)}(r).$$

Suppose that $D = \mathbb{R}^d$, $D^* = \{x^*\}$, and let the function f be convex and twice continuously differentiable with a positive definite Hessian at x^*. From Remark 15.3 and Theorem 15.2 it follows then that

$$\alpha(x) = \mid \nabla f(x) \mid$$

yields a procedure $(X_n : n \in \mathbb{N}_0)$ being globally linear convergent in the mean and P-a.s.

Analogously to Theorem 15.2, for methods with an absolutely continuous mutation sequence we have the following result, where its proof is again primarily based on Theorem 15.1.

Theorem 15.3 *Assume that the assumptions (15.1) and (15.2) are satisfied.*

Let $(X_n : n \in \mathbb{N}_0)$ denote a random search procedure with an absolutely continuous mutation sequence $(p_n : n \in \mathbb{N})$ and associated sequence of densities $(g_n : n \in \mathbb{N})$.

Let h_1, h_2 be measurable, nonnegative functions on $\mathbb{R}_+$.

Suppose that h_1 is positive on the interval $(0, c)$ $(c > 0)$.

Furthermore, assume that $\int_{\mathbb{R}_+} r^{d-1} h_2(r) \lambda^1(dr) < \infty$.

Then:

(i) Provided that for an $a > f^*$ the following condition

$$\begin{cases} \textit{There are sequences of measurable, positive functions} \\ (\alpha_n : n \in \mathbb{N}) \textit{ und } (\tau_n : n \in \mathbb{N}) \textit{ as well as positive constants} \\ \alpha_0, \alpha_1 \textit{ such that} \\ \alpha_0 \le \frac{\alpha_n(x_0,\ldots,x_{n-1})}{|x_{n-1}-x^*|} \le \alpha_1 \qquad \textit{and} \\ (\alpha_n(x_0, \ldots, x_{n-1}))^{-d} h_1 \left(\frac{|z|}{\alpha_n(x_0,\ldots,x_{n-1})} \right) \\ \quad \le g_n(x_0, \ldots, x_{n-1}, x_{n-1} + z) \\ \quad \le (\tau_n(x_0, \ldots, x_{n-1}))^{-d} h_2 \left(\frac{|z|}{\tau_n(x_0,\ldots,x_{n-1})} \right) \\ \textit{for any } n \in \mathbb{N}, x_0, \ldots, x_{n-1} \in D_a \backslash D^*, z \in \mathbb{R}^d \end{cases} \tag{15.7}$$

is satisfied, then an $a_0 > f^*$ *exists such that* $(X_n : n \in \mathbb{N}_0)$ *converges linearly in the mean and* $P-a.s.$ *for any starting point* $x_0 \in D_{a_0}$.

(ii) If D *is convex and* f *is continuous and convex, and if furthermore condition (15.7) holds for any* $a > f^*$*, then* $(X_n : n \in \mathbb{N}_0)$ *is globally linearly convergent in the mean and* P*-a.s.*

Proof According to Lemma 10.1, for μ_n and ρ_n defined according to (10.3), the integration property (10.4) is satisfied.

Theorem 15.1 yields then the assertion, provided that the conditions (15.3), (15.4) and (15.5) are satisfied for an $a > f^*$.

In the following it will be shown that this is true.

According to (15.3) we have

$$\rho_n(x_0, \ldots, x_{n-1}, A) = \frac{2\pi^{\frac{d}{2}}}{\Gamma(\frac{d}{2})} \int_A \int_{\mathbb{R}_+} r^{d-1} g_n(x_0, \ldots, x_{n-1}, x_{n-1} + rs) \cdot \\ \cdot \lambda^1(dr)\gamma(ds).$$

From (15.7) we get then

$$\rho_n(x_0, \ldots, x_{n-1}, A) \geq \frac{2\pi^{\frac{d}{2}}}{\Gamma(\frac{d}{2})} \int_A \int_{\mathbb{R}_+} \left[\frac{r}{\alpha_n(x_0, \ldots, x_{n-1})}\right]^d \frac{1}{r} \cdot \\ \cdot h_1\left(\frac{r}{\alpha_n(x_0, \ldots, x_{n-1})}\right) dr\gamma(ds) \\ = \left(\frac{2\pi^{\frac{d}{2}}}{\Gamma(\frac{d}{2})} \int_{\mathbb{R}_+} u^{d-1} h_1(u) du\right) \gamma(A).$$

According to the assumption, the constant in the brackets is positive. This implies condition (15.4).

The term occurring in (10.3) at μ_n, hence,

$$\int_{\mathbb{R}_+} r^{d-1} g_n(x_0, \ldots, x_{n-1}, x_{n-1} + rs)\lambda^1(dr)$$

can be estimated from below—as just shown—by $\int_{\mathbb{R}_+} u^{d-1} h_1(u) du > 0$.

Moreover, this term can be estimated from above by $I := \int_{\mathbb{R}_+} u^{d-1} h_2(u) du$. Since by assumption $I < \infty$, we get

$$\int_{\left[0, \frac{1}{2}|x_{n-1} - x^*|\right]} r\mu_n(x_0, x_1, \ldots, x_{n-1}, s, dr)$$

$$\geq \frac{1}{I} \int_{\left[0, \frac{1}{2}|x_{n-1}-x^*|\right]} r \cdot r^{d-1} g_n(x_0, \ldots, x_{n-1}, x_{n-1} + rs) \lambda^1(dr)$$

$$\geq \frac{1}{I} \int_{\left[0, \frac{1}{2}|x_{n-1}-x^*|\right]} \left(\frac{r}{\alpha_n(x_0, \ldots, x_{n-1})} \right)^d h_1 \left(\frac{r}{\alpha_n(x_0, \ldots, x_{n-1})} \right) dr$$

$$= \frac{\alpha_n(x_0, \ldots, x_{n-1})}{I} \int_0^{\frac{|x_{n-1}-x^*|}{2\alpha_n(x_0, \ldots, x_{n-1})}} u^d h_1(u) du$$

$$\geq | x_{n-1} - x^* | \left(\frac{\alpha_0}{I} \int_0^{\frac{1}{2}\alpha_1^{-1}} u^d h_1(u) du \right).$$

This implies (15.3).

Condition (15.5) can be shown in the same way (cf. also the proof of Theorem 15.2). □

Example 15.2

(1) Let $r : \mathbb{R}^d \to \mathbb{R}_+$ be measurable.

Moreover, let $p_n(x_0, \ldots, x_{n-1}, \cdot)$ denote the uniform distribution on the d-dimensional rectangle

$$R_{x_{n-1}} := x_{n-1} + (-r(x_{n-1}), r(x_{n-1}))^d .$$

Here,

$$g_n(x_0, \ldots, x_{n-1}, y_n) = (2r(x_{n-1}))^{-d} 1_{R_{x_{n-1}}}(y_n),$$

hence,

$$g_n(x_0, \ldots, x_{n-1}, x_{n-1} + z) = (2r(x_{n-1}))^{-d} 1_{(-r(x_{n-1}), r(x_{n-1}))^d}(z)$$

$$= (2r(x_{n-1}))^{-d} 1_{(-1,1)^d} \left(\frac{z}{r(x_{n-1})} \right).$$

Because of $K(0, 1) \subset (-1, 1)^d \subset K(0, \sqrt{d})$, we have

$$(2r(x_{n-1}))^{-d} 1_{K(0,1)} \left(\frac{z}{r(x_{n-1})} \right) \leq g_n(x_0, \ldots, x_{n-1}, x_{n-1} + z)$$

$$\leq (2r(x_{n-1}))^{-d} 1_{K(0,\sqrt{d})} \left(\frac{z}{r(x_{n-1})} \right)$$

and therefore

$$(2r(x_{n-1}))^{-d}\, 1_{(0,1)}\left(\frac{|\, z\,|}{r(x_{n-1})}\right) \le g_n(x_0, \ldots, x_{n-1}, x_{n-1} + z)$$
$$\le (2r(x_{n-1}))^{-d}\, 1_{(0,\sqrt{d})}\left(\frac{|\, z\,|}{r(x_{n-1})}\right).$$

Choosing

$$r(x) :=|\, \nabla f(x)\,|,$$

then with

$$\alpha_n(x_0, \ldots, x_{n-1}) := r(x_{n-1}) \quad h_1 := 1_{(0,1)}$$
$$\text{and} \quad h_2 := 1_{(0,\sqrt{d})}$$

and under suitable assumptions on D and f (cf. Remark 15.2, Remark 15.3), resp., condition (15.7) is satisfied for an (for all, resp.) $a > f^*$.

(2) Let $p_n(x_0, \ldots, x_{n-1}, \cdot)$ denote a d-dimensional normal distribution with mean x_{n-1} and regular covariance matrix $C = C(x_{n-1})$.

Then,

$$g_n(x_0, \ldots, x_{n-1}, x_{n-1} + z) = (2\pi)^{-\frac{d}{2}}|\, C(x_{n-1})\,|^{-\frac{1}{2}} \exp\,(-\frac{1}{2} z^T C(x_{n-1})^{-1} z).$$

With $\lambda_1 := \max\{\lambda : \lambda \text{ eigenvalue of } C\}$ and $\lambda_0 := \min\{\lambda : \lambda \text{ eigenvalue of } C\}$ we get

$$\lambda_1^{-\frac{d}{2}} \le |\, C\,|^{-\frac{1}{2}} \le \lambda_0^{-\frac{d}{2}}$$

and

$$\left(|\, z\,|\, \lambda_1^{-\frac{1}{2}}\right)^2 \le z^T C^{-1} z \le \left(|\, z\,|\, \lambda_0^{-\frac{1}{2}}\right)^2.$$

This yields the estimate

$$(2\pi)^{-\frac{d}{2}} \left(\frac{\lambda_0}{\lambda_1}\right)^{\frac{d}{2}} \lambda_0^{-\frac{d}{2}} \exp\left(-\frac{1}{2}\left(|\, z\,|\, \lambda_0^{-\frac{1}{2}}\right)^2\right)$$
$$\le g_n(x_0, \ldots, x_{n-1}, x_{n-1} + z)$$
$$\le (2\pi)^{-\frac{d}{2}} \left(\frac{\lambda_1}{\lambda_0}\right)^{\frac{d}{2}} \lambda_1^{-\frac{d}{2}} \exp\left(-\frac{1}{2}\left(|\, z\,|\, \lambda_1^{-\frac{1}{2}}\right)^2\right).$$

Supposing that for a given $a > f^*$ the following condition

$$\begin{cases} \text{There are } \alpha_0, \alpha_1 > 0 & \text{with} \\ \alpha_0 \le \frac{\sqrt{\lambda}}{|x-x^*|} \le \alpha_1 & \text{for all } x \in D_a \backslash D^* \text{ and all} \\ & \text{eigenvalues } \lambda \text{ of } C(x) \end{cases} \tag{15.8}$$

holds, then for this given $a > f^*$ condition (15.7) holds (with $\alpha_n(x_0, \ldots, x_{n-1}) = \sqrt{\lambda_0(x_{n-1})}$, $\tau_n(x_0, \ldots, x_{n-1}) = \sqrt{\lambda_1(x_{n-1})}$, $h_1(r) := (2\pi)^{-\frac{d}{2}} \left(\alpha_0 \alpha_1{}^{-1}\right)^d \exp\left(-\frac{r^2}{2}\right)$ and $h_2(r) := (2\pi)^{-\frac{d}{2}} \left(\alpha_1 \alpha_0{}^{-1}\right)^d \exp\left(-\frac{r^2}{2}\right)$).

If $D = \mathbb{R}^d$, $D^* = \{x^*\}$ and if f is a twice continuously differentiable function with a Hessian being positive definite at all points, then (15.1) and (15.2) are satisfied.

This follows from Remark 15.1 taking into account that the positive definiteness of the Hessian implies the convexity of f (cf. Lemma B.4).

Choosing

$$C(x) = |\nabla f(x)|^2 I \qquad (I = \text{d-dimensional unit matrix}),$$

then the Remark 15.3 implies the validity of (15.8).

Choosing (cf. [3])

$$C(x) = \left(\nabla f(x)^T H^{-1}(x) \nabla f(x)\right) H^{-1}(x)$$
$$(H = \text{Hessian of } f),$$

then the supposition (15.8) is satisfied as well. This follows from Lemma B.14.

Hence, according to Theorem 15.3, in both cases we know that the corresponding procedure converges globally linear in the mean and P-a.s.

Conversely, choosing

$$C(x) \equiv C \qquad (C \text{ regular}),$$

according to Remark 14.4 and Theorem 14.1, for all starting points $x_0 \in \mathbb{R}^d$ we have the considerably smaller convergence rate

$$O\left(n^{-\frac{2}{d}}\right).$$

The above two examples have two important properties in common. Firstly, the stationary case applies. Secondly, in both examples the mutation distribution $p(x, \cdot)$ has the same structure. For if we choose in the first, second example, resp.

$$A_x := r(x)I$$

and for μ the uniform distribution on $(-1, 1)^d$,

$$A_x := (C(x))^{\frac{1}{2}} \qquad \left(\text{i.e. } A_x {A_x}^T = C(x)\right)$$

and for μ the d-dimensional standard normal distribution,

then in both cases we have

$$p(x, \cdot) = \tau_x(\mu) \qquad \text{with} \qquad \tau_x(z) := x + A_x z.$$

Let us consider now this situation in more detail.

According to the transformation rule for densities, $p(x, \cdot)$ has the Lebesgue density

$$g(x, y) = \frac{1}{\mid \det A_x \mid} h\left({A_x}^{-1}(y - x)\right)$$

provided that μ has the density h and A_x is regular.

Furthermore, if functions $h_1, h_2 : \mathbb{R}_+ \to \mathbb{R}_+$ exist with

$$h_1(\mid z \mid) \le h(z) \le h_2(\mid z \mid) \qquad \text{for all } z \neq 0$$

(this holds in both examples), then—analogously to the second example—we get the following chain of inequalities:

$$\begin{aligned}\left(\frac{\lambda_0}{\lambda_1}\right)^{\frac{d}{2}} \lambda_0^{-\frac{d}{2}} h_1\left(\mid z \mid \lambda_0^{-\frac{1}{2}}\right) &\le g(x, x + z) \\ &\le \left(\frac{\lambda_1}{\lambda_0}\right)^{\frac{d}{2}} \lambda_1^{-\frac{d}{2}} h_2\left(\mid z \mid \lambda_1^{-\frac{1}{2}}\right),\end{aligned}$$

where $\lambda_0 = \lambda_0(x)$ ($\lambda_1 = \lambda_1(x)$) denotes the smallest (the largest, resp.) eigenvalue of $A_x {A_x}^T$.

Note that under suitable additional conditions by means of Theorem 15.3 it is possible to show—just as in the second example—a linear rate of convergence.

However, the next theorem shows that in case

$$p(x, \cdot) = \tau_x(\mu)$$

we don't need very restrictive requirements concerning the shape of h.

Theorem 15.4 *Let the assumptions (15.1) and (15.2) be satisfied.*

Let μ be a probability measure on $(\mathbb{R}^d, \mathcal{B}^d)$ having the Lebesgue density h.

Moreover, suppose that there are positive constants σ_1, σ_2 such that $h(x) \geq \sigma_1$ for any $x \in K(0, \sigma_2)$.

Let denote $(A_x : x \in D)$ a family of regular $d \times d$-matrices, and assume that the mapping $x \to A_x$ is continuous.

Consider a stationary random search procedure $(X_n : n \in \mathbb{N}_0)$ having the mutation transition probability p given by

$$p(x, \cdot) = \tau_x(\mu) \qquad \text{, where } \tau_x(y) := x + A_x y.$$

Then,

(i) Assuming that for an $a > f^*$ the following condition

$$\begin{cases} \text{There are } \alpha_0, \alpha_1 > 0 & \text{with} \\ \alpha_0 \leq \frac{\sqrt{\lambda}}{|x-x^*|} \leq \alpha_1 & \text{for any } x \in D_a \backslash D^* \text{ and all} \\ & \text{eigenvalues } \lambda \text{ of } {A_x}^T A_x \end{cases} \tag{15.9}$$

is satisfied, then there is an $a_0 > f^*$ such that for any starting point $x_0 \in D_{a_0}$ the algorithm $(X_n : n \in \mathbb{N}_0)$ converges linearly in the mean and P-a.s.

(ii) Let D be convex and f be continuous and convex. Furthermore, if for any $a > f^*$ the condition (15.9) is satisfied, then $(X_n : n \in \mathbb{N}_0)$ is globally linearly convergent in the mean and P-a.s.

Proof

(1) Let (15.9) be satisfied for an $a > f^*$.

If in Lemma B.9 the set D is replaced by D_α (α selected according to assumption (15.1)), then together with assumption (15.2) we have the existence of a $\phi \in \left(0, \frac{\pi}{2}\right]$ with

$$\begin{aligned} & x + C\left(\phi, x^* - x, \frac{1}{2} \mid x^* - x \mid\right) \\ & \subset \left\{y \in D : f(x) - f(y) \geq \frac{\mid y - x \mid}{3 \mid x^* - x \mid}\left(f(x) - f^*\right)\right\} \\ & \subset \{f \leq f(x)\} \qquad \text{for any } x \in D_a \backslash D^*. \end{aligned}$$

Let $a_0 := \min\{a, \alpha\}$,
and let denote x_0 an arbitrary starting point from D_{a_0}.

Without loss of generality we may assume that $f(x_0) > f^*$.

Furthermore, let $x \in D_{f(x_0)} \backslash D^*$ be an arbitrary point.

Because of

$$D_{f(x_0)} \backslash D^* \subset D_{a_0} \backslash D^* \subset D_\alpha \backslash D^*$$

we have

$$\int_{\{f \le f(x)\}} (f(x) - f(y))\, p(x, dy)$$
$$\ge \frac{f(x) - f^*}{3 \mid x - x^* \mid} \int_{x + C\left(\phi, x^* - x, \frac{1}{2}|x - x^*|\right)} \mid x - y \mid \tau_x(\mu)(dy)$$
$$= \frac{f(x) - f^*}{3 \mid x - x^* \mid} \int_{\tau_x{}^{-1}\left(x + C\left(\phi, x^* - x, \frac{1}{2}|x^* - x|\right)\right)} \mid A_x y \mid \mu(dy)$$
$$= \frac{f(x) - f^*}{3 \mid x - x^* \mid} \int_{A_x{}^{-1}\left(C\left(\phi, x^* - x, \frac{1}{2}|x^* - x|\right)\right)} \mid A_x y \mid \mu(dy).$$

According to Lemma B.9, there is $\varphi > 0$ with

$$A_x{}^{-1}\left(C\left(\phi, x^* - x, \frac{1}{2} \mid x^* - x \mid\right)\right)$$
$$\supset C\left(\varphi, A_x{}^{-1}(x^* - x), \frac{1}{2} \mid x^* - x \mid \lambda_1{}^{-\frac{1}{2}}\right)$$

where $\lambda_1 := \max\left\{\lambda : \lambda \text{ eigenvalue of } A_x{}^T A_x\right\}$.

Note that $D_{f(x_0)}$ is compact—due to Lemma B.4.

Because of

$$\frac{1}{2} \mid x - x^* \mid \lambda_1{}^{-\frac{1}{2}} \ge \frac{1}{2}\alpha_1{}^{-1}$$

with $z := A_x{}^{-1}(x^* - x)$ it follows that

$$\int_{\{f \le f(x)\}} (f(x) - f(y))\, p(x, dy)$$
$$\ge \frac{f(x) - f^*}{3 \mid x - x^* \mid} \int_{C\left(\varphi, z, \frac{1}{2}\alpha_1{}^{-1}\right)} \mid A_x y \mid h(y)\lambda^d(dy)$$
$$\ge \frac{f(x) - f^*}{3 \mid x - x^* \mid}\lambda_0{}^{\frac{1}{2}} \int_{C\left(\varphi, z, \frac{1}{2}\alpha_1{}^{-1}\right)} \mid y \mid h(y)\lambda^d(dy),$$

where $\lambda_0 := \min\left\{\lambda : \lambda \text{ eigenvalue of } A_x{}^T A_x\right\}$.

From (15.9) we get

$$\frac{f(x) - f^*}{3 \mid x - x^* \mid}\lambda_0{}^{\frac{1}{2}} \ge \frac{1}{3}\alpha_0\left(f(x) - f^*\right).$$

Let $\sigma_3 := \frac{1}{2} \cdot \min\{\frac{1}{2}{\alpha_1}^{-1}, \sigma_2\}$.

Because of $C\left(\varphi, z, \frac{1}{2}{\alpha_1}^{-1}\right) \supset C\left(\varphi, z, \sigma_3, \frac{1}{2}{\alpha_1}^{-1}\right)$ and due to $h(y) \geq \sigma_1$ for any $y \in C\left(\varphi, z, \sigma_3, \frac{1}{2}{\alpha_1}^{-1}\right)$, we find

$$\int_{C\left(\varphi, z, \frac{1}{2}{\alpha_1}^{-1}\right)} \mid y \mid h(y)\lambda^d(dy) \geq \sigma_3\sigma_1\lambda^d\left(C\left(\varphi, z, \sigma_3, \frac{1}{2}{\alpha_1}^{-1}\right)\right).$$

From the rotation invariance of the Lebesgue measure (cf. also Appendix A in the appendix) it follows that

$$\lambda^d\left(C\left(\varphi, z, \sigma_3, \frac{1}{2}{\alpha_1}^{-1}\right)\right)$$

does not depend on z.

Therefore, a $c > 0$ exists such that for any $x \in D_{f(x_0)} \backslash D^*$ we have

$$\int_{\{f \leq f(x)\}} (f(x) - f(y))\, p(x, dy) \geq c\left(f(x) - f^*\right).$$

Hence, Lemma 15.1 yields then the linear convergence in the mean.

(2) Select again an arbitrary point $x \in D_{f(x_0)} \backslash D^*$.

Furthermore, consider an arbitrary $\varepsilon \in \left(0, \frac{1}{2}\right)$.

According to Lemma B.9, there exists a $\varphi > 0$ with

$$\begin{aligned} &{A_x}^{-1}\left(C\left(\phi, x^* - x, \varepsilon \mid x^* - x \mid, \frac{1}{2} \mid x^* - x \mid\right)\right) \\ &\supset C\left(\varphi, {A_x}^{-1}(x^* - x), \varepsilon \mid x^* - x \mid {\lambda_0}^{-\frac{1}{2}}, \frac{1}{2} \mid x^* - x \mid {\lambda_1}^{-\frac{1}{2}}\right) \end{aligned}$$

(λ_0, λ_1 as in (1))

$$\supset C\left(\varphi, {A_x}^{-1}(x^* - x), \varepsilon{\alpha_0}^{-1}, \frac{1}{2}{\alpha_1}^{-1}\right)$$

(according to (15.9)).

If now $\varepsilon \in \left(0, \frac{1}{2}\right)$ is chosen as small that $\frac{\varepsilon}{\alpha_0} < \frac{1}{2}{\alpha_1}^{-1}$, then it follows that

$$\begin{aligned} &p\left(x, \left\{y \in D : f(y) - f^* \leq \left(1 - \frac{\varepsilon}{3}\right)(f(x) - f^*)\right\}\right) \\ &\geq p\left(x, \left\{y \in D : f(x) - f(y) \geq \frac{\varepsilon}{3}(f(x) - f^*)\right\}\right) \end{aligned}$$

$$
\begin{aligned}
&\geq p\left(x, x + C\left(\phi, x^* - x, \varepsilon \mid x^* - x \mid, \frac{1}{2} \mid x^* - x \mid\right)\right) \\
&= \mu\left(A_x{}^{-1}\left(C\left(\phi, x^* - x, \varepsilon \mid x^* - x \mid, \frac{1}{2} \mid x^* - x \mid\right)\right)\right) \\
&\geq \mu\left(C\left(\varphi, A_x{}^{-1}(x^* - x), \varepsilon \alpha_0{}^{-1}, \frac{1}{2}\alpha_1{}^{-1}\right)\right).
\end{aligned}
$$

The second inequality follows from the inclusion

$$
\begin{aligned}
&x + C\left(\phi, x^* - x, \frac{1}{2} \mid x^* - x \mid\right) \\
&\quad \subset \left\{y \in D : f(x) - f(y) \geq \frac{\mid y - x \mid}{3 \mid x^* - x \mid}\left(f(x) - f^*\right)\right\}
\end{aligned}
$$

and the definition of $C\left(\phi, x^* - x, \varepsilon \mid x^* - x \mid, \frac{1}{2} \mid x^* - x \mid\right)$ (cf. Definition B.1).

Analogously to part (1), the linear convergence P-a.s. follows now from Lemma 15.2.

(3) Suppose that D is convex and f is a convex and continuous function.
Consider an arbitrary point $x_0 \in D \backslash D^*$.
According to the assumption, condition (15.9) is satisfied for $a = f(x_0)$.
According to Lemma B.11, there is again a $\phi > 0$ with

$$
\begin{aligned}
&x + C\left(\phi, x^* - x, \frac{1}{2} \mid x^* - x \mid\right) \\
&\quad \subset \left\{y \in D : f(x) - f(y) \geq \frac{\mid y - x \mid}{3 \mid x^* - x \mid}\left(f(x) - f^*\right)\right\} \\
&\quad \subset \{f \leq f(x)\} \qquad \text{for any } x \in D_a \backslash D^*.
\end{aligned}
$$

Now the statements analogously to (i) follow exactly as in part (1) and part (2). □

15.2 Methods with a Rate of Convergence that Is at Most Linear

Under suitable assumptions, the preceding theorems provide statements about a linear rate of convergence, i.e. it was shown the existence of a constant $q > 1$ such that

$$q^n \left(E\left(f(X_n) - f^*\right)\right) \to 0 \qquad \text{and}$$
$$q^n \left(f(X_n) - f^*\right) \to 0 \qquad P\text{-a.s.}$$

At least theoretically, a procedure with a linear rate of convergence could converge still faster, e.g. superlinear.

Hence, we are interested therefore to make also estimates of the convergence rate in the other direction.

It will turn out (cf. Theorems 15.5, 15.6, 15.7, and 15.8) that under assumptions implying a linear rate of convergence (cf. Theorems 15.2 and 15.3), stronger results are not possible. Indeed, we will show the existence of an $r > 1$ with

$$r^n \left(E\left(f(X_n) - f^*\right)\right) \to \infty \qquad \text{and}$$
$$r^n \left(f(X_n) - f^*\right) \to \infty \qquad P\text{-a.s.}$$

Thus, the rate of convergence cannot be superlinear: In the other case, i.e. if, e.g., the sequence $E\left(f(X_n) - f^*\right)$ would have a superlinear rate of convergence, hence,

$$\lim_{n\to\infty} \frac{E\left(f(X_{n+1}) - f^*\right)}{E\left(f(X_n) - f^*\right)} = 0,$$

then—as it is easy to see—we would have

$$r^n \left(E\left(f(X_n) - f^*\right)\right) \to 0 \quad \text{for any } r > 1$$

The same holds true of course also for the sequence $f(X_n) - f^*$.

The rates of convergence of the sequences $E\left(f(X_n) - f^*\right)$ and $f(X_n) - f^*$ are not necessarily always identical. As the next example shows, they may differ considerably. It is therefore not sufficient to limit the examination of the rate of convergence to one of the two sequences.

Example 15.3 Let $D = \mathbb{R}$ and consider the function

$$f(x) = \begin{cases} 0 & \text{if} \quad x \in [-1, 1] \\ |\, x \,| - 1 & \text{else.} \end{cases}$$

Let $(X_n : n \in \mathbb{N}_0)$ denote a stationary random search procedure having a mutation distribution $p(x, \cdot)$ being normal with mean x and fixed variance $\sigma^2 > 0$.

Because of $D^* = [-1, 1]$, it follows in particular that $\lambda^1(D^*) > 0$.

Therefore assumption (14.4) is satisfied.

From the Remarks 14.4, 14.2, resp., we obtain the validity of the assumptions (14.1) and (14.2).

Theorem 14.1, part (vii) firstly shows now that the sequence $E\left(f(X_n) - f^*\right)$ is globally linear convergent, and, secondly, that the sequence $f(X_n) - f^*$ is P-a.s. convergent of any order, i.e.

$$a_n\left(f(X_n) - f^*\right) \to 0 \qquad P\text{-a.s.}$$

for each sequence $(a_n : n \in \mathbb{N})$ in $\mathbb{R}$.

Theorem 15.5 *Suppose that $D^* \neq \emptyset$, and D and f are convex.*
Let $(X_n : n \in \mathbb{N}_0)$ be a random direction procedure. Then

$$p_n\left(x_0, \ldots, x_{n-1}, \{f < f(x_{n-1})\}\right) \leq \frac{1}{2}$$

for any $n \in \mathbb{N}$, $x_0, \ldots, x_{n-1} \in D \backslash D^*$

and

$$E\left(f(X_n) - f^*\right) \geq 2^{-n}\left(f(x_0) - f^*\right)$$

for any $n \in \mathbb{N}$ and all starting points $x_0 \in D$.

Proof For $z \in \mathbb{R}^d \backslash \{0\}$ define $H_z := \left\{y \in \mathbb{R}^d : y^T z \leq 0\right\}$.
For arbitrary $n \in \mathbb{N}$, $x_0, \ldots, x_{n-1} \in D$ and $z \in \mathbb{R}^d \backslash \{0\}$ we have then:

$$\begin{aligned}
&p_n\left(x_0, \ldots, x_{n-1}, x_{n-1} + H_z\right) \\
&= \int_{\{(s,r) \in S \times \mathbb{R}_+ : sr \in H_z\}} (\gamma \otimes \lambda_n(x_0, \ldots, x_{n-1}, \cdot))\, d(s, r) \\
&= \int_{\{s \in S : s^T z \leq 0\}} \int_{\mathbb{R}_+} \gamma(ds) \lambda_n(x_0, \ldots, x_{n-1}, dr) \\
&= \gamma\left(\left\{s \in S : s^T z \leq 0\right\}\right) = \frac{1}{2}.
\end{aligned}$$

Since f is convex, also the level set $\{f \leq f(x)\}$ is convex.

Furthermore, according to Lemma B.6 we have $\{f = f(x)\} \subset rd\{f \leq f(x)\}$ for $x \in D \backslash D^*$.

Hence, the set $\{f \leq f(x)\}$ has a supporting hyperplane at x.

Consequently, there is a $z \neq 0$ with (cf. Lemma B.5)

$$\{f \leq f(x)\} \subset x + H_z.$$

This yields $p_n\left(x_0, \ldots, x_{n-1}, \{f < f(x_{n-1})\}\right) \leq \frac{1}{2}$ and therefore

$$\int_{\{f < f(x_{n-1})\}} \left(f(x_{n-1}) - f(y_n)\right) p_n(x_0, \ldots, x_{n-1}, dy_n) \leq \frac{1}{2}\left(f(x_{n-1}) - f^*\right).$$

Exchanging in the proof of 15.1 the signs of inequality, the assertion follows. □

Theorem 15.6 *Suppose again that $D^* \neq \emptyset$, and assume that D and f are convex.*

Let μ denote a probability measure on $\left(\mathbb{R}^d, \mathcal{B}^d\right)$ having the der Lebesgue density h.

Moreover, we assume that there exist positive constants σ_1, σ_2 with $h(x) \geq \sigma_1$ for any $x \in K(0, \sigma_2)$.

Let $(A_x : x \in D)$ be a family of regular $d \times d$-matrices.

Consider the stationary random search procedure $(X_n : n \in \mathbb{N}_0)$ with the mutation transition probability p given by $p(x, \cdot) = \tau_x(\mu)$, where $\tau_x(y) := x + A_x y$.

Then there is a $c \in (0, 1)$ such that

$$E\left(f(X_n) - f^*\right) \geq c^n \left(f(x_0) - f^*\right)$$

for all starting points $x_0 \in D$ and any $n \in \mathbb{N}$.

Proof Let $x \in D, z \in \mathbb{R}^d \setminus \{0\}$ be arbitrary. Then:

$$\begin{aligned}
{\tau_x}^{-1}(x + H_z) &= \left\{y \in \mathbb{R}^d : A_x y \in H_z\right\} \\
&= \left\{y \in \mathbb{R}^d : (A_x y)^T z \leq 0\right\} = \left\{y \in \mathbb{R}^d : y^T ({A_x}^T z) \leq 0\right\} = H_{{A_x}^T z}.
\end{aligned}$$

From the convexity of f it follows again for any $x \in D$ the existence of a $z \neq 0$ with $\{f < f(x)\} \subset x + H_z$.

This yields

$$\begin{aligned}
p\left(x, \{f < f(x)\}\right) &\leq p\left(x, x + H_z\right) = \left(\tau_x(\mu)\right)(x + H_z) \\
&= \mu\left({\tau_x}^{-1}(x + H_z)\right) = \mu\left(H_{{A_x}^T z}\right).
\end{aligned}$$

From the absolute continuity of μ we get

$$\mu\left(\left\{y \in \mathbb{R}^d : y^T z = 0\right\}\right) = 0$$

and therefore

$$\mu\left(H_z\right) + \mu\left(H_{-z}\right) = 1 \qquad \text{for any } z \neq 0.$$

Consequently, for $z \neq 0$ we find

$$\begin{aligned}\mu\left(H_z\right) &= 1 - \mu\left(H_{-z}\right) \leq 1 - \mu\left(H_{-z} \cap K(0, \sigma_2)\right) \\ &= 1 - \int_{H_{-z} \cap K(0,\sigma_2)} h(x)\lambda^d(dx) \leq 1 - \sigma_1 \lambda^d\left(H_{-z} \cap K(0, \sigma_2)\right) \\ &= 1 - \frac{\sigma_1}{2}\lambda^d\left(K(0, \sigma_2)\right).\end{aligned}$$

Thus,

$$p\left(x, \{f < f(x)\}\right) \leq 1 - \frac{\sigma_1}{2}\lambda^d\left(K(0, \sigma_2)\right) < 1.$$

The rest of the proof follows now as in the proof of Theorem 15.5. □

Example 15.4 Let $p(x, \cdot)$ denote a d-dimensional normal distribution with mean x and a regular covariance matrix $C(x)$.

Then, a matrix A_x exists with $A_x A_x{}^T = C(x)$.

Choosing τ_x as in Theorem 15.5, and μ is the standard normal distribution, then

$$p(x, \cdot) = \tau_x(\mu).$$

Because of $\mu\left(H_z\right) = \frac{1}{2}$ it follows that (cf. the proof of Theorem 15.6)

$$p(x, x + H_z) = \frac{1}{2} \qquad \text{for any } x \in D \text{ and } z \neq 0.$$

For a convex f we obtain now

$$E\left(f(X_n) - f^*\right) \geq 2^{-n}\left(f(x_0) - f^*\right) \qquad \text{for any } n \in \mathbb{N}.$$

In the rest of this section we show that under the assumptions of Theorems 15.2 and 15.3 the rate of convergence is at most linear. This relates to the P-a.s. of the sequence $f(X_n) - f^*$. A crucial point in the proof is to find an estimate from above of the distribution function F_n of the variable $f(X_n) - f^*$. This is the purpose of the next lemma.

Lemma 15.3 *Let $(X_n : n \in \mathbb{N}_0)$ be a random search procedure starting at $x_0 \in D \backslash D^*$ and having the mutation sequence $(p_n : n \in \mathbb{N})$. Moreover, assume that there are constants $a > 0$ and $c \in (0, 1]$ such that*

$$p_n\left(x_0, \ldots, x_{n-1}, \left\{y \in D : f(y) - f^* \leq \varepsilon\left(f(x_{n-1}) - f^*\right)\right\}\right) \leq \left(\frac{\varepsilon}{c}\right)^a$$

for any $\varepsilon \in (0, c]$, $n \in \mathbb{N}$ and $x_1, \ldots, x_{n-1} \in D_{f(x_0)} \backslash D^*$.

For $n \in \mathbb{N}$ define $Z_n := \left(\frac{f(X_n) - f^*}{c^n (f(x_0) - f^*)}\right)^a$.

Then for any $n \in \mathbb{N}$ and $x \in (0, 1]$ we have

$$P(Z_n \le x) \le x \sum_{i=0}^{n-1} \frac{1}{i!} \left(\ln \frac{1}{x} \right)^i .$$

Proof

(1) Next we show that for $x \in (0, c(f(x_0) - f^*)]$ and $n \in \mathbb{N}$ we have

$$P\left(f(X_n) - f^* \le x \mid f(X_{n-1}) - f^* = y\right)$$
$$\le \left(\frac{x}{cy}\right)^a 1_{(\frac{x}{c}, f(x_0)-f^*]}(y) + 1_{[0,\frac{x}{c}]}(y)$$
$$P_{f(X_{n-1})-f^*} - a.s.$$

Obviously, this holds for $y = 0$.

Hence, to prove the assertion, the following inequality must be shown for arbitrary measurable $B \subset (0, f(x_0) - f^*]$. Note that $P_{f(X_{n-1})-f^*}$ $([0, f(x_0) - f^*]) = 1$:

$$P\left(\{f(X_n) - f^* \le x\} \cap \{f(X_{n-1}) - f^* \in B\}\right)$$
$$\le \int_B \left(\left(\frac{x}{cy}\right)^a 1_{(\frac{x}{c}, f(x_0)-f^*]}(y) + 1_{[0,\frac{x}{c}]}(y) \right) P_{f(X_{n-1})-f^*}(dy).$$

Thus, let $B \subset (0, f(x_0) - f^*]$ and take $x \in (0, c(f(x_0) - f^*)]$.
Then:

$$P\left(\{f(X_n) - f^* \le x\} \cap \{f(X_{n-1}) - f^* \in B\}\right)$$
$$= \int_{\{f(X_{n-1})-f^* \in B\}} P\left(f(X_n) - f^* \le x \mid X_{n-1}, \ldots, X_0\right) dP$$
$$= \int_{\{f(X_{n-1})-f^* \in B\}} q_n\left(X_0, \ldots, X_{n-1}, \left\{ y \in D : f(y) - f^* \le \right.\right.$$
$$\left.\left. \le \frac{x}{f(X_{n-1}) - f^*} \left(f(X_{n-1}) - f^*\right) \right\}\right) dP$$
$$= \int_{\{f(X_{n-1})-f^* \in B\} \cap \left\{\frac{x}{f(X_{n-1})-f^*} \in (0,c)\right\}} p_n(\ldots) dP +$$
$$+ \int_{\{f(X_{n-1})-f^* \in B\} \cap \left\{\frac{x}{f(X_{n-1})-f^*} \ge c\right\}} q_n(\ldots) dP$$
$$\le \int_{\{f(X_{n-1})-f^* \in B\} \cap \{f(X_{n-1})-f^* \in (\frac{x}{c}, f(x_0)-f^*]\}} \left(\frac{x}{c\left(f(X_{n-1}) - f^*\right)} \right)^a .$$

$$\cdot\, dP + \int_{\{f(X_{n-1})-f^* \in B\}} 1_{[0,\frac{x}{c}]}\left(f(X_{n-1}) - f^*\right) dP$$

$$= \int_{\{f(X_{n-1})-f^* \in B\}} \left(\left(\frac{x}{c\,(f(X_{n-1}) - f^*)} \right)^a 1_{(\frac{x}{c}, f(x_0)-f^*]} \cdot \right.$$

$$\left. \cdot \left(f(X_{n-1}) - f^*\right) + 1_{[0,\frac{x}{c}]}\left(f(X_{n-1}) - f^*\right) \right) dP$$

$$= \int_B \left(\left(\frac{x}{cy} \right)^a 1_{(\frac{x}{c}, f(x_0)-f^*]}(y) + 1_{[0,\frac{x}{c}]}(y) \right) P_{f(X_{n-1})-f^*}(dy).$$

(2) For $n \in \mathbb{N}$ and $x \geq 0$ define

$$F_n(x) := P\left(f(X_n) - f^* \leq x\right) \qquad \text{and}$$

$$G_n(x) := \left[\left(\frac{x}{c^n\,(f(x_0) - f^*)} \right)^a \sum_{i=0}^{n-1} \frac{1}{i!} \left(\ln \left(\frac{c^n\,(f(x_0) - f^*)}{x} \right)^a \right)^i \right] \cdot$$

$$\cdot\, 1_{(0, c^n(f(x_0)-f^*)]}(x) + 1_{(c^n(f(x_0)-f^*),\infty)}(x).$$

Furthermore, for $x \geq 0$ and a fixed $n \in \mathbb{N}$ let

$$G(x) := \left(x \sum_{i=0}^{n-1} \frac{1}{i!} (-\ln x)^i \right) 1_{(0,1]}(x) + 1_{(1,\infty)}(x).$$

For $x \in (0, 1)$ we have then

$$G'(x) = \sum_{i=0}^{n-1} \frac{1}{i!} (-\ln x)^i - \sum_{i=1}^{n-1} \frac{1}{(i-1)!} (-\ln x)^{i-1}$$

$$= \frac{(-\ln x)^{n-1}}{(n-1)!} > 0,$$

which means that G is a distribution function.

Hence, because of $G_n(x) = G\left(\left(\frac{x}{c^n(f(x_0)-f^*)}\right)^a\right)$, also G_n is a distribution function having the Lebesgue density

$$g_n(x) := \left[\frac{a x^{a-1}}{(n-1)!\,(c^n\,(f(x_0) - f^*))^a} \left[\ln \left[\frac{c^n\,(f(x_0) - f^*)}{x} \right]^a \right]^{n-1} \right] \cdot$$

$$\cdot\, 1_{[0, c^n(f(x_0)-f^*)]}(x).$$

According to the definition of Z_n, the assertion follows by showing for any $x \in (0, c^n\,(f(x_0) - f^*)]$ that the inequality

$$P\left(f(X_n) - f^* \leq x\right) \leq G_n(x)$$

holds.

This will be done by means of complete induction.

(3) Let $n = 1$.

Because of $P(f(X_0) - f^* = f(x_0) - f^*) = 1$, for arbitrary $x \in (0, c(f(x_0) - f^*))$ we have

$$\begin{aligned} P\left(f(X_1) - f^* \leq x\right) &= P\left(f(X_1) - f^* \leq x \mid f(X_0) - f^* = f(x_0) - f^*\right) \\ &\leq \left(\frac{x}{c(f(x_0) - f^*)}\right)^a 1_{(\frac{x}{c}, f(x_0)-f^*]}\left(f(x_0) - f^*\right) + \\ &\quad + 1_{[0,\frac{x}{c}]}\left(f(x_0) - f^*\right) = \left(\frac{x}{c(f(x_0) - f^*)}\right)^a. \end{aligned}$$

Due to $c(f(x_0) - f^*) \leq f(x_0) - f^*$ and
$P(f(X_1) - f^* \leq f(x_0) - f^*) = 1$,
obviously, the above inequality holds also for $x = c(f(x_0) - f^*)$.

Consider the step $n \to n + 1$:
Select an arbitrary $x \in \left[0, c^{n+1}(f(x_0) - f^*)\right]$.
With $b := f(x_0) - f^*$ and because of $F_n(b) = 1$ we get

$$\begin{aligned} &P\left(f(X_{n+1}) - f^* \leq x\right) \\ &= \int_{[0,b]} P\left(f(X_{n+1}) - f^* \leq x \mid f(X_n) - f^* = y\right) F_n(dy) \\ &\leq \int_{[0,b]} \left(\left(\frac{x}{cy}\right)^a 1_{(\frac{x}{c},b]}(y) + 1_{[0,\frac{x}{c}]}(y)\right) F_n(dy) \\ &= F_n\left(\frac{x}{c}\right) + \left(\frac{x}{c}\right)^a \int_{(\frac{x}{c},b]} y^{-a} F_n(dy) \\ &= F_n\left(\frac{x}{c}\right) + \left(\frac{x}{c}\right)^a \left(y^{-a} F_n(y) \mid_{\frac{x}{c}}^{b} + a \cdot \int_{(\frac{x}{c},b]} y^{-a-1} F_n(y) dy\right) \\ &\leq F_n\left(\frac{x}{c}\right) + \left(\frac{x}{c}\right)^a \cdot \\ &\quad \cdot \left(b^{-a} F_n(b) - \left(\frac{x}{c}\right)^{-a} F_n\left(\frac{x}{c}\right) + a \cdot \int_{(\frac{x}{c},b]} y^{-a-1} G_n(y) dy\right), \end{aligned}$$

where the last inequality follows from induction assumption, and the last equation follows from the rule of integration by parts.

Note that the use of this rule is not obvious, because here we do not have a Riemann integral, but a Lebesgue integral.

However, replacing in the corresponding integral above the term y^{-a} by

$$\left(\frac{x}{c}\right)^{-a} - a \cdot \int_{(\frac{x}{c},y]} t^{-a-1} dt,$$

then we obtain a double integral. Applying Fubini's theorem, i.e. by interchanging the integrals, the rule of integration by parts follows in the present case.

Applying again the rule of the integration by parts to the integral $\int_{(\frac{x}{c},b]} y^{-a-1} G_n(y)dy$, we find

$$\begin{aligned}
&P\left(f(X_{n+1}) - f^* \le x\right)\\
&\le \left(\frac{x}{bc}\right)^a + \left(\frac{x}{c}\right)^a \left(-y^{-a} G_n(y) \mid_{\frac{x}{c}}^{b} + \int_{(\frac{x}{c},b]} y^{-a} g_n(y)dy\right)\\
&= \left(\frac{x}{bc}\right)^a + \left(\frac{x}{c}\right)^a \left(-b^{-a} + \left(\frac{x}{c}\right)^{-a} G_n\left(\frac{x}{c}\right) + \int_{\frac{x}{c}}^{b} y^{-a} g_n(y)dy\right)\\
&= G_n\left(\frac{x}{c}\right) + \left(\frac{x}{c}\right)^a \int_{\frac{x}{c}}^{c^n b} y^{-a} g_n(y)dy\\
&= G_n\left(\frac{x}{c}\right) + \frac{a}{(n-1)!(c^n b)^a} \left(\frac{x}{c}\right)^a \int_{\frac{x}{c}}^{c^n b} y^{a-1} y^{-a}.\\
&\quad \cdot \left(\ln(c^n b)^a - a \ln y\right)^{n-1} dy\\
&= G_n\left(\frac{x}{c}\right) + \frac{a x^a}{(n-1)!(c^{n+1} b)^a} \left(\frac{-1}{an} \left(\ln(c^n b)^a - a \ln y\right)^n\right)\Big|_{\frac{x}{c}}^{c^n b}\\
&= G_n\left(\frac{x}{c}\right) + \left(\frac{x}{c^{n+1} b}\right)^a \frac{1}{n!} \left(\ln(c^n b)^a - a \ln \frac{x}{c}\right)^n\\
&= G_n\left(\frac{x}{c}\right) + \left(\frac{x}{c^{n+1} b}\right)^a \frac{1}{n!} \left[\ln\left(\frac{c^{n+1} b}{x}\right)^a\right]^n = G_{n+1}(x),
\end{aligned}$$

since $\frac{x}{c} \in [0, c^n b]$.

□

Lemma 15.4 *Let the assumptions of Lemma 15.3 be satisfied. Then there is a* $q = q(x_0) > 1$ *with*

$$q^n \left(f(X_n) - f^*\right) \to \infty \qquad P\text{-a.s.}$$

Proof

(1) Defining $Z_n := \left(\frac{f(X_n)-f^*}{c^n(f(x_0)-f^*)}\right)^a$,
for $x \in [0, 1]$ we get

$$P\left(Z_n \le x\right) = P\left(f(X_n) - f^* \le x^{\frac{1}{a}} c^n \left(f(x_0) - f^*\right)\right)$$
$$\le x \sum_{i=0}^{n-1} \frac{1}{i!} \left(\ln \frac{1}{x}\right)^i .$$

To prove the assertion, it is sufficient to show that

$$q^n Z_n \to \infty \qquad P\text{-a.s.}$$

holds for a $q > 1$. Indeed, in this case we get

$$\left(\frac{q^{\frac{1}{a}}}{c}\right)^n \left(f(X_n) - f^*\right) \to \infty \qquad P\text{-a.s.}$$

(2) A sufficient, but not necessary condition for
$q^n Z_n \to \infty \quad P-$ a.s. reads:

$$\sum_{n=0}^{\infty} P\left(Z_n \le \frac{K}{q^n}\right) < \infty \qquad \text{for any } K \in \mathbb{N}.$$

Namely, in this case, the Lemma of Borel–Cantelli yields

$$P\left(q^n Z_n \le K \text{ for } \infty\text{-many } n \in \mathbb{N}\right) = 0 \text{ , hence}$$
$$P\left(q^n Z_n > K \text{ for finally all } n \in \mathbb{N}\right) = 1 \text{ for any } K \in \mathbb{N}$$

and therefore

$$P\left(\liminf_{n\to\infty} q^n Z_n \ge K\right) = 1 \text{ for any } K \in \mathbb{N}.$$

(3) Let $q > 1$ be chosen such that the inequality

$$4e \ln q < q - 1$$

is satisfied (e.g. $q \ge 41, 404$).
Let $K \in \mathbb{N}$ be arbitrary.
Because of $q^n \to \infty$, there is an $N \in \mathbb{N}$ with $Kq^{-n} \le 1$ for any $n \ge N$.

Then:

$$
\begin{aligned}
\sum_{n=0}^{\infty} P\left(Z_n \leq K q^{-n}\right) &\leq N + \sum_{n=N}^{\infty} P\left(Z_n \leq K q^{-n}\right) \\
&\leq N + K \cdot \sum_{n=N}^{\infty} q^{-n} \sum_{i=0}^{n-1} \frac{1}{i!} \left(\ln q^n - \ln K\right)^i \\
&\leq N + K \cdot \sum_{n=1}^{\infty} q^{-n} \sum_{i=0}^{n-1} \frac{1}{i!} \left(2 \ln q^n\right)^i \\
&= N + K \cdot \sum_{n=1}^{\infty} q^{-n} \sum_{i=0}^{n-1} \frac{(2n)^i}{i!} (\ln q)^i \\
&= N + K \cdot \sum_{i=0}^{\infty} \sum_{n=i+1}^{\infty} \frac{(\ln q^2)^i}{i!} n^i q^{-n} \\
&= N + K \cdot \sum_{i=0}^{\infty} \frac{(\ln q^2)^i}{i!} \sum_{n=0}^{\infty} (n+1+i)^i q^{-(n+i+1)} \\
&= N + K \cdot \sum_{i=0}^{\infty} \frac{(\ln q^2)^i}{i!} q^{-i-1} \sum_{n=0}^{\infty} (n+1+i)^i q^{-n}.
\end{aligned}
$$

According to Remark B.5 we have $(n+1+i)^i \leq 2^i (n+1)^i + (2i)^i$. Furthermore, due to Remark B.4 it is

$$
\sum_{n=0}^{\infty} (n+1) \ldots (n+i) q^{-n} = \frac{i!}{\left(1 - \frac{1}{q}\right)^{i+1}}.
$$

This yields the estimate

$$
\begin{aligned}
\sum_{n=0}^{\infty} (n+1+i)^i q^{-n} &\leq (2i)^i \left(1 - \frac{1}{q}\right)^{-1} + 2^i \sum_{n=0}^{\infty} (n+1)^i q^{-n} \\
&\leq (2i)^i \left(1 - \frac{1}{q}\right)^{-1} + 2^i \sum_{n=0}^{\infty} (n+1) \ldots (n+i) q^{-n} \\
&= (2i)^i \left(1 - \frac{1}{q}\right)^{-1} + 2^i i! \left(1 - \frac{1}{q}\right)^{-i-1} \\
&\leq 2^i e^i i! \left(1 - \frac{1}{q}\right)^{-1} + 2^i i! \left(1 - \frac{1}{q}\right)^{-i-1} \qquad \text{according to Remark B.6.}
\end{aligned}
$$

From this it follows that

$$\frac{(\ln q^2)^i}{i!} q^{-i-1} \sum_{n=0}^{\infty} (n+1+i)^i q^{-n}$$

$$\leq \left(\frac{4e\ln q}{q}\right)^i \frac{1}{q-1} + \frac{1}{q-1}\left(\frac{4\ln q}{q-1}\right)^i \leq \left(\frac{4e\ln q}{q-1}\right)^i \frac{2}{q-1},$$

hence,

$$\sum_{n=0}^{\infty} P\left(Z_n \leq Kq^{-n}\right) \leq N + \frac{2K}{q-1} \sum_{i=0}^{\infty} \left(\frac{4e\ln q}{q-1}\right)^i < \infty.$$

This implies the assertion.

□

Remark 15.5 Suppose that the assumptions of Lemma 15.3 are satisfied.

According to the proof of Lemma 15.4, for $q \geq 42$ it follows then that

$$\left(\frac{q^{\frac{1}{a}}}{c}\right)^n \left(f(X_n) - f^*\right) \to \infty \qquad P\text{-a.s.}$$

The stochastic convergence towards ∞ is obtained for much smaller values of q. Here, it follows already for $q > e$ that

$$\left(\frac{q^{\frac{1}{a}}}{c}\right)^n \left(f(X_n) - f^*\right) \to \infty \qquad P - \text{stoch.}$$

Proof Consider a sequence $(Y_n : n \in \mathbb{N})$ of random variables having values in $[0, 1]$ such that

$$P\left(Y_n \leq x\right) = x \sum_{i=0}^{n-1} \frac{1}{i!} \left(\ln \frac{1}{x}\right)^i$$

for $x \in [0, 1]$. Let then $Q_n := \ln \frac{1}{Y_n}$ for $n \in \mathbb{N}$.

A simple calculation shows that Q_n Γ-distributed with the parameters 1 and n. The characteristic function φ_n of Q_n is then given by

$$\varphi_n(t) = (1 - it)^{-n},$$

hence, the for characteristic function ϕ_n of $\frac{1}{n} Q_n$ we get

$$\phi_n(t) = \varphi_n\left(\frac{t}{n}\right) = \left(1 - \frac{it}{n}\right)^{-n} \to e^{it}.$$

However, e^{it} is the characteristic function of the one-point probability distribution at the point 1.

Because of the convergence of the characteristic function we obtain the convergence of $\frac{1}{n}Q_n$ towards constant 1 in distribution and therefore also P-stochastic.

Consequently,

$$-n\left(\frac{1}{n}Q_n - s\right) = -Q_n + ns \overset{P-st.}{\longrightarrow} \infty \quad \text{if } s > 1$$

and therefore

$$\left(e^s\right)^n Y_n = \left(e^s\right)^n \exp\left(-Q_n\right) \overset{P-st.}{\longrightarrow} \infty \quad \text{for } s > 1,$$

hence

$$r^n Y_n \overset{P-stoch.}{\longrightarrow} \infty \quad \text{if } r > e.$$

Defining Z_n as in Lemma 15.3, and because of

$$P\left(Z_n \leq x\right) \leq P\left(Y_n \leq x\right),$$

for any $K > 0$ we get the inequality

$$P\left(Z_n > K\right) \geq P\left(Y_n > K\right).$$

Now, if $r^n Y_n$ converges stochastically towards infinity, i.e. if

$$P\left(r^n Y_n > K\right) \to 1 \quad \text{for any } K > 0,$$

then this also applies to the sequence $r^n Z_n$.

Thus, the assertion is shown. □

Lemma 15.5 *Let $(X_n : n \in \mathbb{N}_0)$ be a random search procedure with the mutation sequence*
$(p_n : n \in \mathbb{N})$.

Suppose that there are transition probabilities ρ_n from $(\mathbb{R}^d)^n$ to S and transition probabilities μ_n from $(\mathbb{R}^d)^n \times S$ to $\mathbb{R}_+$ fulfilling the integration property (10.4). Moreover, let the following assumptions be satisfied:

(1) D is compact or D and f are convex
(2) $D^* = \{x^*\}$, where $x^* \in \overset{\circ}{D}$

(3) f is twice continuously differentiable in a neighborhood of x^* with a Hessian matrix that is positively definite at x^*.
Furthermore, let f be continuous on D.
(4) For any $x_0 \in D\backslash D^*$ there are constants $a > 0$ and $c \in (0, 1]$ with $\mu_n\left(x_0, \ldots, x_{n-1}, s, \left[(1-\varepsilon) \mid x_{n-1} - x^* \mid, (1+\varepsilon) \mid x_{n-1} - x^* \mid\right]\right) \leq \left(\frac{\varepsilon}{c}\right)^a$ for all $n \in \mathbb{N}$, $x_1, \ldots, x_{n-1} \in D_{f(x_0)}\backslash D^*$, $\varepsilon \in (0, c]$, $s \in S$.

Then, for any starting point $x_0 \in D\backslash D^*$, a constant $q > 1$ exists with

$$q^n\left(f(X_n) - f^*\right) \to \infty \qquad P\text{-a.s.}$$

Proof According to Lemma B.13, the assumptions (1), (2), and (3) imply all assumptions of Lemma B.12. Hence, due to Corollary B.3, for any $x_0 \in D\backslash D^*$ there exist constants $d_1 \in (0, 1]$ and $d_2 > 0$ with

$$\{y \in D : f(y) - f^* \leq \varepsilon\left(f(x) - f^*\right)\} \subset K\left(x^*, d_2\sqrt{\varepsilon} \mid x - x^* \mid\right)$$
$$\text{for any } x \in D_{f(x_0)} \text{ and any } \varepsilon \in (0, d_1].$$

Without loss of generality we may assume that $c \leq d_1$, and let $x_0 \in D\backslash D^*$ be arbitrary.

For any $n \in \mathbb{N}$, $x_1, \ldots, x_{n-1} \in D_{f(x_0)}\backslash D^*$ and $\varepsilon \in (0, c]$ we have

$$\begin{aligned}
&p_n\left(x_0, \ldots, x_{n-1}, \left\{y \in D : f(y) - f^* \leq \varepsilon\left(f(x_{n-1}) - f^*\right)\right\}\right)\\
&\leq p_n\left(x_0, \ldots, x_{n-1}, K\left(x^*, d_2\sqrt{\varepsilon} \mid x_{n-1} - x^* \mid\right)\right)\\
&= \int_{\mathbb{R}^d} 1_{K(x^*-x_{n-1}, d_2\sqrt{\varepsilon}|x_{n-1}-x^*|)}\left(y_n - x_{n-1}\right) p_n(x_0, \ldots, x_{n-1}, dy_n)\\
&= \int_S \int_{\mathbb{R}_+} 1_{K(x^*-x_{n-1}, d_2\sqrt{\varepsilon}|x_{n-1}-x^*|)}(rs)\mu_n(x_0, \ldots, x_{n-1}, s, dr)\cdot\\
&\quad \cdot \rho_n(x_0, \ldots, x_{n-1}, ds)\\
&= \int_S \int_{\mathbb{R}_+} 1_{K(x^*, d_2\sqrt{\varepsilon}|x_{n-1}-x^*|)}(x_{n-1} + rs)\mu_n(x_0, \ldots, x_{n-1}, s, dr)\cdot\\
&\quad \cdot \rho_n(x_0, \ldots, x_{n-1}, ds).
\end{aligned}$$

If $|r- \mid x_{n-1} - x^* \mid| \geq \rho$, then

$$\mid x_{n-1} + rs - x^* \mid \geq \left|\mid rs \mid - \mid x_{n-1} - x^* \mid\right| = \left|r - \mid x_{n-1} - x^* \mid\right| \geq \rho.$$

Thus, $\mid x_{n-1} + rs - x^* \mid < \rho$ implies the inequality

$$\left|r - \mid x_{n-1} - x^* \mid\right| < \rho.$$

This yields

$$1_{K(x^*, d_2\sqrt{\varepsilon}|x_{n-1}-x^*|)}(x_{n-1}+rs) \le 1_{((1-d_2\sqrt{\varepsilon})|x_{n-1}-x^*|,(1+d_2\sqrt{\varepsilon})|x_{n-1}-x^*|)}(r)$$

and therefore

$$\begin{aligned} &p_n\left(x_0, \ldots, x_{n-1}, \left\{y \in D : f(y) - f^* \le \varepsilon\left(f(x_{n-1}) - f^*\right)\right\}\right) \\ &\le \int_S \mu_n\Bigg(x_0, \ldots, x_{n-1}, s, \Big(\left(1 - d_2\sqrt{\varepsilon}\right) \mid x_{n-1} - x^* \mid, \\ &\left(1 + d_2\sqrt{\varepsilon}\right) \mid x_{n-1} - x^* \mid \Big)\Bigg) \cdot \\ &\cdot \rho_n(x_0, \ldots, x_{n-1}, ds). \end{aligned}$$

Let $\tau := \min\left\{c, c^2 {d_2}^{-2}\right\}$.
Thus, if $\varepsilon \in (0, \tau]$, then $d_2\sqrt{\varepsilon} \le c$, and according to Assumption (4) we have

$$\begin{aligned} &p_n\left(x_0, \ldots, x_{n-1}, \left\{y \in D : f(y) - f^* \le \varepsilon\left(f(x_{n-1}) - f^*\right)\right\}\right) \\ &\le \left(\frac{d_2\sqrt{\varepsilon}}{c}\right)^a = \left(\varepsilon c^{-2} {d_2}^2\right)^{\frac{a}{2}} \le \left(\frac{\varepsilon}{\tau}\right)^{\frac{a}{2}} \\ &\text{for all } \varepsilon \in (0, \tau]. \end{aligned}$$

The assertion follows now from Lemma 15.4. □

Theorem 15.7 *Let $(X_n : n \in \mathbb{N}_0)$ denote a random direction procedure with the step length distribution λ_n.*

Let h be a bounded probability density on $\mathbb{R}_+$, and let α_n denote positive, measurable functions with

$$\lambda_n\left(x_0, \ldots, x_{n-1}, B\right) = \left(\alpha_n(x_0, \ldots, x_{n-1})\right)^{-1} \int_B h\left(\frac{r}{\alpha_n(x_0, \ldots, x_{n-1})}\right) dr$$

$$\text{for any } n \in \mathbb{N}, x_0, \ldots, x_{n-1} \in D\backslash D^*, B \subset \mathbb{R}_+ \text{ measurable}.$$

Moreover, assume that the assumptions (1), (2), and (3) of Lemma 15.5 are satisfied.

Provided that for all $x_0 \in D\backslash D^*$ there exists a constant $K < \infty$ such that

$$\frac{\mid x_{n-1} - x^* \mid}{\alpha_n(x_0, \ldots, x_{n-1})} \le K \qquad \text{for all } n \in \mathbb{N}, x_1, \ldots, x_{n-1} \in D_{f(x_0)}\backslash D^*,$$

then for each starting point $x_0 \in D\backslash D^*$ there is a constant $q > 1$ with

$$q^n\left(f(X_n) - f^*\right) \to \infty \qquad P\text{-a.s.}$$

Proof Since $(X_n : n \in \mathbb{N}_0)$ is a random direction procedure, according to Lemma 10.1 we have

$$\mu_n(x_0, \ldots, x_{n-1}, s, \cdot) = \lambda_n(x_0, \ldots, x_{n-1}, \cdot).$$

For arbitrary $n \in \mathbb{N}$, $x_1, \ldots, x_{n-1} \in D_{f(x_0)} \backslash D^*$ and $\varepsilon \in [0, 1]$ this yields

$$\begin{aligned}
&\mu_n\left(x_0, \ldots, x_{n-1}, s, \left((1-\varepsilon) \mid x_{n-1} - x^* \mid, (1+\varepsilon) \mid x_{n-1} - x^* \mid\right)\right) \\
&= (\alpha_n(x_0, \ldots, x_{n-1}))^{-1} \int_{(1-\varepsilon)|x_{n-1}-x^*|}^{(1+\varepsilon)|x_{n-1}-x^*|} h\left(\frac{r}{\alpha_n(x_0, \ldots, x_{n-1})}\right) dr \\
&= \int_{(1-\varepsilon)|x_{n-1}-x^*|(\alpha_n(x_0,\ldots,x_{n-1}))^{-1}}^{(1+\varepsilon)|x_{n-1}-x^*|(\alpha_n(x_0,\ldots,x_{n-1}))^{-1}} h(u)du \\
&\leq \text{const} \cdot 2\varepsilon \frac{\mid x_{n-1} - x^* \mid}{\alpha_n(x_0, \ldots, x_{n-1})} \leq \text{const} \cdot \varepsilon.
\end{aligned}$$

The assertion follows now directly from Lemma 15.5. □

Theorem 15.8 *Let $(X_n : n \in \mathbb{N}_0)$ be a random search procedure with an absolutely continuous mutation sequence $(p_n : n \in \mathbb{N})$ and associated density sequence $(g_n : n \in \mathbb{N})$.*

Let h_1, h_2 denote measurable, nonnegative functions on $\mathbb{R}_+$.

Assume that h_1 is positive on the interval $(0, c)$ $(c > 0)$.

Furthermore, suppose that h_2 is bounded.

Let $(\alpha_n : n \in \mathbb{N})$ and $(\tau_n : n \in \mathbb{N})$ denote sequences of measurable, positive functions with

$$\begin{aligned}
&(\alpha_n(x_0, \ldots, x_{n-1}))^{-d} h_1\left(\frac{\mid z \mid}{\alpha_n(x_0, \ldots, x_{n-1})}\right) \leq g_n(x_0, \ldots, x_{n-1}, x_{n-1} + z) \\
&\leq (\tau_n(x_0, \ldots, x_{n-1}))^{-d} h_2\left(\frac{\mid z \mid}{\tau_n(x_0, \ldots, x_{n-1})}\right)
\end{aligned}$$

for any $n \in \mathbb{N}$, $x_0, \ldots, x_{n-1} \in D \backslash D^*$, $z \in \mathbb{R}^d$.

Furthermore, let there exist for all $x_0 \in D \backslash D^*$ a positive constant $K < \infty$ with

$$\frac{\mid x_{n-1} - x^* \mid}{\tau_n(x_0, \ldots, x_{n-1})} \leq K \qquad \text{for any } n \in \mathbb{N}, x_1, \ldots, x_{n-1} \in D_{f(x_0)} \backslash D^*.$$

If f and D satisfy assumptions (1), (2), and (3) of Lemma 15.5, then there is for each starting point $x_0 \in D \backslash D^*$ a $q > 1$ such that

$$q^n\left(f(X_n) - f^*\right) \to \infty \qquad P\text{-a.s.}$$

Proof Let $x_0 \in D \backslash D^*$ be an arbitrary point,
and consider $n \in \mathbb{N}$, $x_1, \ldots, x_{n-1} \in D_{f(x_0)} \backslash D^*$, $\varepsilon \in [0, 1]$ and an arbitrary $s \in S$.

According to (10.3) and with $I := \int_{\mathbb{R}_+} r^{d-1} g_n(x_0, \ldots, x_{n-1}, x_{n-1} + rs)dr$ we have

$$\mu_n \left(x_0, \ldots, x_{n-1}, s, \left((1-\varepsilon) \mid x_{n-1} - x^* \mid, (1+\varepsilon) \mid x_{n-1} - x^* \mid\right)\right)$$

$$= \frac{1}{I} \int_{(1-\varepsilon)|x_{n-1}-x^*|}^{(1+\varepsilon)|x_{n-1}-x^*|} r^{d-1} g_n(x_0, \ldots, x_{n-1}, x_{n-1} + rs)dr$$

$$\leq \frac{1}{I} \frac{1}{\tau_n(x_0, \ldots, x_{n-1})} \int_{(1-\varepsilon)|x_{n-1}-x^*|}^{(1+\varepsilon)|x_{n-1}-x^*|} \left(\frac{r}{\tau_n(x_0, \ldots, x_{n-1})}\right)^{d-1} \cdot$$

$$\cdot h_2 \left(\frac{r}{\tau_n(x_0, \ldots, x_{n-1})}\right) dr$$

$$= \frac{1}{I} \int_{(1-\varepsilon)|x_{n-1}-x^*|(\tau_n(x_0,\ldots,x_{n-1}))^{-1}}^{(1+\varepsilon)|x_{n-1}-x^*|(\tau_n(x_0,\ldots,x_{n-1}))^{-1}} u^{d-1} h_2(u)du$$

$$\leq \frac{1}{I} 2\varepsilon \mid x_{n-1} - x^* \mid (\tau_n(x_0, \ldots, x_{n-1}))^{-1} \cdot$$

$$\cdot \sup \left\{u^{d-1} h_2(u) : 0 \leq u \leq \frac{2 \mid x_{n-1} - x^* \mid}{\tau_n(x_0, \ldots, x_{n-1})}\right\}$$

$$\leq \frac{1}{I} 2\varepsilon K (2K)^{d-1} \tilde{c},$$

where $\tilde{c} := \sup \{h_2(u) : u \in \mathbb{R}_+\}$.

Since, according to the assumption, I can be estimated from below by

$$\int_{\mathbb{R}_+} u^{d-1} h_1(u)du > 0,$$

we have $\frac{1}{I} < \infty$.

The assertion follows now from Lemma 15.5. □

15.3 Linear Convergence for Positive Probability of Success

The considerations at the beginning of this section have shown that a decreasing probability of success, in particular near x^*, has negative consequences for the rate of convergence. The next example shows that the same holds true if the probability of success becomes too large.

Example 15.5 Let $D = \mathbb{R}^2$ and $f(x) = \mid x \mid^2$.

Consider a stationary random direction method $(X_n : n \in \mathbb{N}_0)$ having the step length distribution $\lambda(x, \cdot) = \varepsilon_\rho$ (with arbitrary $\rho > 0$).

The probability of success $p(x, \{f < f(x)\})$ is then given by

$$\begin{aligned}
p(x, \{f < f(x)\}) &= p(x, \{y :\mid y \mid<\mid x \mid\}) \\
&= (\gamma \otimes \lambda(x, \cdot))(\{(s, r) \in S \times \mathbb{R}_+ :\mid x + rs \mid<\mid x \mid\}) \\
&= \gamma(\{s \in S :\mid x + s\rho \mid<\mid x \mid\}) = \frac{1}{\pi} \arccos \frac{\rho}{2 \mid x \mid} \\
&\quad \text{for } 0 \leq \rho \leq 2 \mid x \mid,
\end{aligned}$$

where the last equation results from a simple geometric consideration.

For any $x \neq 0$, the probability $p(x, \{f < f(x)\})$ converges therefore monotonously increasing towards $\frac{1}{2}$, provided that ρ convergences monotonously decreasing towards zero. Thus, the probability of success becomes the greater, the smaller ρ is chosen. However, on one hand, if ρ is chosen very small, then this has of course a negative effect on the rate of convergence, since together with ρ, the step length will become extremely small. If, on the other hand, ρ is chosen very large, then also the step lengths will be very large, while the probability of success will be very small.

This example shows that the step lengths, the parameters of the step length distribution, resp., are closely linked to the probability of success and vice versa. Hence, one could try to control the step length parameters by means of of the probability of success. Attempts into this direction can be found in [2, 4–6]. However, with exception of the last paper, no practicable algorithms can be found there, but only rules of the thumb, such as Rechenberg's $\frac{1}{5}$-*success rule*, cf. [4]. The goal of the algorithm given in [2] is that, in course of the search, the probability of success is near to a given fixed value $p > 0$.

However, there is given no answer to the question whether this goal is actually achieved. Also the question, whether the algorithm under consideration yields a convergent method, is not answered. From the mathematical viewpoint, the paper of Tarasenko is more useful. There, the current step length is controlled taking into account the successes and failures in the preceding steps. A detailed discussion of this topic can be found then in Chap. 12.

Considering now a random search procedure with the property that the probability of success lies always within given, fixed limits, we ask now for the corresponding rate of convergence in this case. According to Chap. 13 we already know that the method converges under certain conditions. In the following, the question concerning the convergence rate can be answered for random direction procedures only.

Theorem 15.9 *Let $(X_n : n \in \mathbb{N}_0)$ denote a random direction procedure having the step length distribution λ_n.*

Suppose that $D = \mathbb{R}^d$ and $f : D \to \mathbb{R}$ are convex.

Let $D^* = \{x^*\}$, and assume that f is twice continuously differentiable with a Hessian matrix which is positively definite at x^*.

Let h be a bounded probability density on $\mathbb{R}_+$, and let α_n denote a positive, measurable function with

$$\lambda_n(x_0, \ldots, x_{n-1}, B) = \frac{1}{\alpha_n(x_0, \ldots, x_{n-1})} \int_B h\left(\frac{r}{\alpha_n(x_0, \ldots, x_{n-1})}\right) dr$$

$$\text{for any } x_0, \ldots, x_{n-1} \in D\backslash D^*, B \subset \mathbb{R}_+ \text{ measurable}.$$

Furthermore, assume that h is positive on an interval $(0, c)$ $(c > 0)$.

Then:

If for all $a > f^*$ there are constants $c_1, c_2 \in \left(0, \frac{1}{2}\right)$ with

$$p_n\left(x_0, \ldots, x_{n-1}, \{f < f(x_{n-1})\}\right) \in [c_1, c_2]$$

$$\text{for any } n \in \mathbb{N}, x_0, \ldots, x_{n-1} \in D_a\backslash D^*,$$

then $(X_n : n \in \mathbb{N}_0)$ is globally linear convergent in the mean and P-a.s.

Proof

(1) From Remark 15.1 and Lemma B.2 it follows that the assumptions (15.1) and (12.2) are satisfied. The assertion follows then from Theorem 15.2, part (ii), provided that condition (15.6) holds for all $a > f^*$. In the following we show that this holds true.

For this purpose let $a > f^*$ be arbitrary.

(2) For $x \in \mathbb{R}^d$ and $r > 0$ define

$$A_{x,r} := \{s \in S : f(x + rs) - f(x) < 0\}.$$

From Corollary B.1 we know that for any $x, y \in \mathbb{R}^d$ we have

$$(\nabla f(x))^T (x - y) \geq f(x) - f(y).$$

Hence, if $f(x + rs) - f(x) < 0$, then $(\nabla f(x))^T (-rs) > 0$ and therefore $(\nabla f(x))^T s < 0$.

This yields

$$A_{x,r} \subset \left\{s \in S : (\nabla f(x))^T s < 0\right\}.$$

From the compactness of the set D_a (cf. Lemma B.2 and Lemma B.4), and because f is twice continuously differentiable, we obtain the Lipschitz continuity of ∇f on the set D_a.

Hence, there exists a constant $L > 0$ with

$$| \nabla f(x) - \nabla f(y) | \leq L \mid x - y \mid \qquad \text{for any } x, y \in D_a.$$

Moreover, the mean value theorem of calculus yields for all $x, y \in D_a$, with a certain $\tau \in [0, 1]$, the equation

$$\begin{aligned} f(x) - f(y) &= (\nabla f\,(y + \tau(x - y)))^T\,(x - y) \\ &= (\nabla f(y))^T\,(x - y) + (\nabla f\,(y + \tau(x - y)) - \nabla f(y))^T\,(x - y). \end{aligned}$$

Since D_a is convex, we have $y + \tau(x - y) \in D_a$.
Thus, from the inequality of Schwarz we obtain

$$\begin{aligned} f(x) - f(y) &\leq (\nabla f(y))^T\,(x - y) + \mid \nabla f\,(y + \tau(x - y)) - \nabla f(y) \mid\mid x - y \mid \\ &\leq (\nabla f(y))^T\,(x - y) + L \mid x - y \mid^2. \end{aligned}$$

Consequently, for all $x \in D_a \backslash D^*$, $r > 0$, and $s \in A_{x,r}$ we have

$$f(x + rs) - f(x) \leq (\nabla f(x))^T\,rs + Lr^2$$

and therefore

$$\begin{aligned} &\left\{ s \in S : \left(\frac{f(x)}{\mid \nabla f(x) \mid} \right)^T s < -\frac{Lr}{\mid \nabla f(x) \mid} \right\} \\ &\quad = \left\{ s \in S : (\nabla f(x))^T\,(rs) + Lr^2 < 0 \right\} \subset A_{x,r}. \end{aligned}$$

Because of $f(x) - f^* \leq (\nabla f(x))^T\,(x - x^*) \leq \mid \nabla f(x) \mid\mid x - x^* \mid$ (cf. Corollary B.1), according to Corollary B.4 with an $m = m(a) > 0$ we have

$$\mid \nabla f(x) \mid \geq \frac{f(x) - f^*}{\mid x - x^* \mid} \geq m \frac{\mid x - x^* \mid^2}{\mid x - x^* \mid} = m \mid x - x^* \mid$$

and therefore

$$A_{x,r} \supset \left\{ s \in S : \left(\frac{f(x)}{\mid \nabla f(x) \mid} \right)^T s < -\frac{Lr}{m \mid x - x^* \mid} \right\}.$$

Again from Corollary B.4, with a certain $M = M(a) < \infty$, for arbitrary $x, y \in D_a$ such that $f(y) < f(x)$ we have the inequality

$$\mid y - x^* \mid^2 \leq \frac{f(y) - f^*}{m} < \frac{f(x) - f^*}{m} \leq \frac{M}{m} \mid x - x^* \mid^2$$

and therefore

$$| y - x | \leq | y - x^* | + | x - x^* | < \left(1 + \sqrt{\frac{M}{m}}\right) | x - x^* | .$$

Thus, for all $x \in D_a$ we get

$$\{f < f(x)\} \subset K\left(x, \left(1 + \sqrt{\frac{M}{m}}\right) | x - x^* |\right).$$

(3) For any $x_0, \ldots, x_{n-1} \in D_a \backslash D^*$ it follows now that

$$\begin{aligned}
& p_n\left(x_0, \ldots, x_{n-1}, \{f < f(x_{n-1})\}\right) \\
& \leq p_n\left(x_0, \ldots, x_{n-1}, K\left(x_{n-1}, \left(1 + \sqrt{\frac{M}{m}}\right) | x_{n-1} - x^* |\right)\right) \\
& = \lambda_n\left(x_0, \ldots, x_{n-1}, \left[0, \left(1 + \sqrt{\frac{M}{m}}\right) | x_{n-1} - x^* |\right]\right) \\
& = \frac{1}{\alpha_n\left(x_0, \ldots, x_{n-1}\right)} \int_0^{\left(1+\sqrt{\frac{M}{m}}\right)|x_{n-1}-x^*|} h\left(\frac{r}{\alpha_n\left(x_0, \ldots, x_{n-1}\right)}\right) dr \\
& = \frac{| x_{n-1} - x^* |}{\alpha_n\left(x_0, \ldots, x_{n-1}\right)} \int_0^{1+\sqrt{\frac{M}{m}}} h\left(\frac{| x_{n-1} - x^* |}{\alpha_n\left(x_0, \ldots, x_{n-1}\right)} u\right) du \\
& \leq \frac{| x_{n-1} - x^* |}{\alpha_n\left(x_0, \ldots, x_{n-1}\right)} K\left(1 + \sqrt{\frac{M}{m}}\right),
\end{aligned}$$

where $K > 0$ is chosen such that $h \leq K$, which is possible according to the assumption.

This yields

$$\frac{\alpha_n\left(x_0, \ldots, x_{n-1}\right)}{| x_{n-1} - x^* |} \leq \frac{K\left(1 + \sqrt{\frac{M}{m}}\right)}{p_n\left(x_0, \ldots, x_{n-1}, \{f < f(x_{n-1})\}\right)}.$$

If $p_n\left(x_0, \ldots, x_{n-1}, \{f < f(x_{n-1})\}\right) \geq c_1$, it follows therefore that

$$\frac{\alpha_n\left(x_0, \ldots, x_{n-1}\right)}{| x_{n-1} - x^* |} \leq \frac{K\left(1 + \sqrt{\frac{M}{m}}\right)}{c_1} =: d_2.$$

(4) Let μ denote the probability measure on $(\mathbb{R}_+, \mathcal{B}^l\mid_{\mathbb{R}_+})$ having the density h. Then for arbitrary $x_0, \ldots, x_{n-1} \in D_a \backslash D^*$ we have again

$$
\begin{aligned}
p_n & (x_0, \ldots, x_{n-1}, \{f < f(x_{n-1})\}) \\
&= \int_{\{(s,r)\in S\times\mathbb{R}_+ : f(x_{n-1}+rs)<f(x_{n-1})\}} (\gamma \otimes \lambda_n (x_0, \ldots, x_{n-1}, \cdot)) (d(s, r)) \\
&= \int_{\mathbb{R}_+} \gamma \left(A_{x_{n-1},r}\right) \lambda_n (x_0, \ldots, x_{n-1}, dr) \\
&\geq \frac{1}{\alpha_n (x_0, \ldots, x_{n-1})} \in_{\mathbb{R}_+} \gamma \cdot \\
&\quad \cdot \left(\left\{s \in S : \left(\frac{\nabla f(x_{n-1})}{\mid \nabla f(x_{n-1}) \mid}\right)^T s < -\frac{Lr}{m \mid x_{n-1} - x^* \mid}\right\}\right) \cdot \\
&\quad \cdot h\left(\frac{r}{\alpha_n (x_0, \ldots, x_{n-1})}\right) dr \\
&= \int_{\mathbb{R}_+} \gamma \left(\left\{s \in S : \left(\frac{\nabla f(x_{n-1})}{\mid \nabla f(x_{n-1}) \mid}\right)^T s < -\frac{L}{m} u \frac{\alpha_n (x_0, \ldots, x_{n-1})}{\mid x_{n-1} - x^* \mid}\right\}\right) \cdot \\
&\quad \cdot \mu(du) \\
&= \int_{\mathbb{R}_+} \gamma \left(\left\{s \in S : z^T s < -\frac{L}{m} u \frac{\alpha_n (x_0, \ldots, x_{n-1})}{\mid x_{n-1} - x^* \mid}\right\}\right) \mu(du)
\end{aligned}
$$

for any $z \in S$ (cf. Lemma A.3).

From the theorem on majorized convergence we get

$$
\begin{aligned}
\lim_{\tau\to 0} & \int_{\mathbb{R}_+} \gamma \left(\left\{s \in S : z^T s < -\frac{L}{m} u\tau\right\}\right) \mu(du) \\
&= \int_{\mathbb{R}_+} \gamma \left(\left\{s \in S : z^T s < 0\right\}\right) \mu(du) = \frac{1}{2}.
\end{aligned}
$$

Thus, for any $\varepsilon > 0$ there exists $\sigma > 0$ such that

$$
p_n (x_0, \ldots, x_{n-1}, \{f < f(x_{n-1})\}) \geq \frac{1}{2} - \varepsilon,
$$

provided that

$$
\frac{\alpha_n (x_0, \ldots, x_{n-1})}{\mid x_{n-1} - x^* \mid} \leq \sigma.
$$

According to the assumption, to $a > f^*$ there exists $\varepsilon_0 > 0$ with

$$p_n\left(x_0, \ldots, x_{n-1}, \{f < f(x_{n-1})\}\right) < \frac{1}{2} - \varepsilon_0$$

for any $x_0, \ldots, x_{n-1} \in D_a \backslash D^*$.

Hence, there is a $\sigma > 0$ with

$$\frac{\alpha_n\left(x_0, \ldots, x_{n-1}\right)}{\mid x_{n-1} - x^* \mid} \geq \sigma \quad \text{for any } x_0, \ldots, x_{n-1} \in D_a \backslash D^*.$$

Thus, Condition (15.6) holds for all $a > f^*$.

As already mentioned, the assertion follows now from Theorem 15.2.

□

References

1. Gänssler, P., Stute, W.: Wahrscheinlichkeitstheorie. Springer, Berlin (1977)
2. Lawson, G., Lavi, A.: Monte Carlo programming: random searching in the solution of parameter optimization problems. In: Conference on Systems Science and Cybernetics. Pittsburgh (1969)
3. Marti, K.: Optimierungsverfahren. vorlesung an der hochschule der bundeswehr münchen (1983)
4. Richter, H.: Wahrscheinlichkeitstheorie. Springer, Berlin (1966)
5. Schumer M.A.; Steiglitz, K.: Adaptive step size random search. IEEE Trans. Autom. control **AC** **(13)**, 270–276 (1968)
6. Tarasenko, G.: Über die konvergenzgeschwindigkeit der adaptiven zufälligen suche. Russisch), Problemy slucajnogo poiska (1980)

Chapter 16
Success/Failure-Driven Random Direction Procedures

In this section we consider a random direction procedure, cf. G.S. Tarasenko [2], based on a very simple step length control. At any iteration step n the step length 1_n is chosen deterministically according to the following algorithm:

$$\ell_1 = \ell > 0$$

$$\ell_{n+1} = \begin{cases} \gamma_1 \ell_n & \text{in case of success at the } n\text{-th step} \\ \gamma_2 \ell_n & \text{in case of failure in the } n\text{-th step,} \end{cases}$$

$$\text{where} \quad \gamma_1 > \ell, \gamma_2 \in (0, \ell)$$

are fixed parameters.

Hence, after successes the step lengths will be increased, and after failures they will be reduced. If the function f is convex, which is generally assumed in this section, this type of step size control appears to be quite plausible. Indeed, under not too restrictive assumptions, in Chap. 16 we derive now global linear (even P-a.s.) for a rather broad class of random direction methods.

Definition 16.1 (Model) Suppose that $D = \mathbb{R}^d$ and $f : D \to \mathbb{R}$ is convex. Furthermore, let $D^* = \{x^*\}$. Choose coefficients $\gamma_1 > \ell, \gamma_2 \in (0, \ell)$ and $0 < \ell_0 \le L_0 < \infty$, and let λ_0 denote a probability measure on $[\ell_0, L_0]$ (initial distribution).

Let the sequence of mappings $\ell_n : D^n \to \mathbb{R}_+$ be defined as follows:

$$\ell_1(x_0) := 1 \text{ for all } x_0 \in D$$

$$\ell_{n+1}(x_0, \ldots, x_n) := \begin{cases} \gamma_1 \ell_n(x_0, \ldots, x_{n-1}) & \text{if } f(x_n) < f(x_{n-1}) \\ \gamma_2 \ell_n(x_0, \ldots, x_{n-1}) & \text{if } f(x_n) \ge f(x_{n-1}). \end{cases}$$

K. Marti, *Optimization Under Stochastic Uncertainty*, International Series in Operations Research & Management Science 296,
https://doi.org/10.1007/978-3-030-55662-4_16

For $1 \in \mathbb{R}_+$ and $x \in D$, let $T_1(x) := 1 \cdot x$.

Furthermore, for $n \in \mathbb{N}$ and $x_0, \ldots, x_{n-1} \in D$, define

$$\lambda_n(x_0, \ldots, x_{n-1}, \cdot) := T_{1_n}(x_0, \ldots, x_{n-1})^{(\lambda_0)}$$

(in particular we have $\lambda_1(x_0, \cdot) = \lambda_0$).

From the recursive definition of ℓ_n it follows that

$$\ell_{n+k}(x_0, \ldots, x_{n-1}, x_n, \ldots, x_{n+k-1}) = \gamma_1^{e_{nk}} \cdot \gamma_2^{k-e_{nk}} \cdot \ell_n(x_0, \ldots, x_{n-1}),$$

where the quantity

$$e_{nk} := \sum_{i=n}^{n+k-1} \ell_{\{(x,y):f(x)>f(y)\}}(x_{i-1}, x_i)$$

represents the number of "successes" in the time interval $[n, n+k-1]$. Let denote $(X_n : n \in \mathbb{N}_0)$ the random direction procedure with the just defined step length distribution λ_n.

Therefore,

$$p_n(x_0, \ldots, x_{n-1}, B) = \Big(\gamma \otimes \lambda_n(x_0, \ldots, x_{n-1}, \cdot)\Big) \times \Big(\Big\{(x, r) \in S \times \mathbb{R}_+ : x_{n-1} + rs \in B\Big\}\Big),$$

where γ is the uniform distribution on the d-dimensional unit sphere S.

Let $\mathcal{A}_n := \sigma(X_0, \ldots, X_n)$ denote the σ-algebra generated by $X_0, \ldots, X_n$. Furthermore, let $\gamma_{x,r}$ be the uniform distribution on the sphere with center x and radius r (cf. Appendix A). For a stopping time with respect to $(\mathcal{A}_n : n \in \mathbb{N}_0)$, let

$$\mathcal{A}_T := \Big\{A \in \mathcal{A} : A \cap \{T = n\} \in \mathcal{A}_n \text{ for all } n \in \mathbb{N}_0\Big\}.$$

Sets of the form $\Big\{x \in D : f(x) \leq c\}$ are denoted again shortly by $\{f \leq c\}$.

The following theorem represents the basic results of the present Chap. 16. Its proof is very complicated and requires several preparations. For simplification, we start with an outline of the proof illustrating the most important ideas of the proof. This outline can be found after Theorem 16.3.

Theorem 16.1 *Let the following conditions be satisfied:*

$$\left.\begin{array}{l} \textit{for all } x_0 \neq x^* \textit{ there exists } b = b(x_0) > 0 \textit{ with} \\ \inf\Big\{|y - x^*| : f(y) = \alpha\Big\} \leq b \sup\Big\{|y - x^*| : f(y) = \alpha\Big\} \\ \qquad \textit{for all } \alpha \in \Big(f^*, f(x_0)\Big] \end{array}\right\} \qquad (16.1)$$

$$\left.\begin{array}{l} \textit{for all } x_0 \neq x^* \textit{ there exists } p > 0 \textit{ with} \\ \quad (i)\ \gamma_1^p \gamma_2^{1-p} > \ell \textit{ and} \\ \quad (ii) \textit{ for all } \delta > 0 \textit{ there is } \varepsilon > 0 \textit{ such that the inequality} \\ \qquad \gamma_{x,r}\left\{f < f(x)\right\} \geq p - \delta \textit{ is satisfied for all} \\ \qquad r > 0 \textit{ and } x \in D_{f(x_0)} > D^* \textit{ with } \dfrac{r}{|x - x^*|} < \varepsilon. \end{array}\right\} \tag{16.2}$$

Then, $(X_n : n \in \mathbb{N}_0)$ is P-a.s. globally linear convergent.

Remark 16.1

(I) The assumption (16.1) is already satisfied, provided that an $a_1 > f^*$ and a $b_1 > 0$ exist such that

$$\inf\left\{|y - x^*| : f(y) = \alpha\right\} \geq b_1 \sup\left\{|y - x^*| : f(y) = \alpha\right\} \text{ for all } \alpha \in (f^*, a_1].$$

(II) If assumption (16.1) is satisfied, then even

$$\inf\left\{|y - x^*| : f(y) = \alpha\right\} \leq b \sup\left\{|y - x^*| : f(y) \leq \alpha\right\} \text{ for all } \alpha \in \left(f^*, f(x_0)\right].$$

Proof

(I) Let $x_0 \neq x^*$ be arbitrary.
If $f(x_0) \leq a_1$, then the assertion is clear. Hence, let $f(x_0) > a_1$. From the boundedness of D^*, with Lemmas B.2 and B.4 we obtain the boundedness of the set $D_{f(x_0)}$.
Thus, $R := \sup\left\{|y - x^*| : f(y) \leq f(x_0)\right\} < \infty$.
Now, for $\alpha \in \left[a_1, f(x_0)\right]$ we have

$$\frac{\inf\left\{|y - x^*| : f(y) = \alpha\right\}}{\sup\left\{|y - x^*| : f(y) = \alpha\right\}} \leq \frac{\inf\left\{|y - x^*| : f(y) \leq a_1\right\}}{R} =: \tilde{b} > 0.$$

For $\alpha \in (f^*, a_1]$, according to the assumption it is

$$\inf\left\{|y - x^*| : f(y) = \alpha\right\} \geq b_1 \sup\left\{|y - x^*| : f(y) = \alpha\right\}.$$

The assertion follows now with $b := \min\{b_1, b\}$.

(II) Select $\alpha \in \left(f^*, f(x_0)\right]$ and $y \in D\backslash D^*$ with $f(y) \leq \alpha$ arbitrarily. For $t \geq 0$, define $g(t) := f\left(x^* + t(y - x^*)\right)$.

From the boundedness of D_α and the continuity of g, we obtain the existence of a point $t_0 > 0$ with $g(t_0) = \alpha$.

If $t_0 < 1$, then $x^* + t_0(y - x^*)$ lies between x^* and y. Lemma B.1 yields then

$$\frac{f(y) - f^*}{|y - x^*|} \leq \frac{f\Big(x^* + t_0(y - x^*)\Big) - f^*}{t_0|y - x^*|},$$

hence, $f(y) \geq f^* + \dfrac{1}{t_0}(\alpha - f^*) > \alpha$ in contradiction to $f(y) \leq \alpha$. Thus, $t_0 \geq 1$ and therefore

$$\begin{aligned}|y - x^*| \leq t_0|y - x^*| &= \Big|x^* + t_0(y - x^*) - x^*\Big| \\ &\leq \sup\Big\{|z - x^*| : f(z) = \alpha\Big\}.\end{aligned}$$

Since y was arbitrary, it follows that

$$\sup\Big\{|y - x^*| : f(y) \leq \alpha\Big\} \leq \sup\Big\{|z - x^*| : f(z) = \alpha\Big\}.$$

This inequality and assumption (16.1) yield the assertion.

□

Remark 16.2 Suppose that condition (16.1) is fulfilled. Then, for all $x_0 \neq x^*$ there exists $q = q(x_0) > 0$ with

$$\frac{f(y) - f^*}{f(x) - f^*} \leq q\frac{|y - x^*|}{|x - x^*|} \quad \text{for all } x \in D_{f(x_0)}\backslash D^* \text{ and } y \in D_{f(x)}.$$

Proof Let $x_0 \neq x^*$ be arbitrary.

Choose $b = b(x_0)$ according to assumption (16.1). Select arbitrary vectors $x \in D_{f(x_0)}\backslash D^*$ and $y \in D_{f(x)}$.

Moreover, let $t \in [0, 1]$ be chosen such that $f\Big(x^* + t(x - x^*)\Big) = f(y)$. This is possible because of $f(x^*) = f^* \leq f(y)$ and $f(x) \geq f(y)$. Define $z := x^* + t(x - x^*)$. Because of $f(z) = f(y)$ we have

$$\begin{aligned}|z - x^*| &\leq \sup\Big\{|z - x^*| : f(z) = f(y)\Big\} \leq \\ &= \frac{1}{b}\inf\Big\{|z - x^*| : f(z) = f(y)\Big\} \leq \frac{|y - x^*|}{b}.\end{aligned}$$

Now it follows from Lemma B.1 that

$$b\frac{f(y)-f^*}{|y-x^*|} \le \frac{f(z)-f^*}{|z-x^*|} \le \frac{f(x)-f^*}{|x-x^*|}$$

and therefore

$$\frac{f(y)-f^*}{f(x)-f^*} \le \frac{1}{b}\frac{|y-x^*|}{|x-x^*|}.$$

The assertion follows then with $q = 1/b$. □

Theorem 16.2 *Let $\gamma_1\gamma_2 > 1$, and assume that ∇f is Lipschitz continuous on compact sets, i.e., for all compact set $M \subset D$ there exists $L > 0$ with*

$$\left|\nabla f(x) - \nabla f(y)\right| \le L|x-y| \text{ for all } x, y \in M.$$

Furthermore:

$$\begin{cases} \text{for all } R > 0 \text{ there exist } m_R > 0,\, M_R > 0 \text{ and } s_R \in (1,2] \\ \text{with } m_R|x-x^*|^{s_R} \le f(x) - f^* \le M_R|x-x^*|^{s_R} \text{ for all } x \in K(x^*, R). \end{cases} \tag{16.3}$$

Then, $(X_n : n \in \mathbb{N}_0)$ is P-a.s. globally linear convergent.

Remark 16.3 Note that assumption (16.1) holds (with $s_R = 2$), provided that f is twice continuously differentiable in neighborhood of x^* with a positive definite Hessian at x^*. This follows from Corollary B.3.

Remark 16.4 If f is twice continuously differentiable, then ∇f is Lipschitz continuous on compact sets. This follows from the mean value theorem of differential calculus and from Schwarz's inequality.

Remark 16.5 Assumption (16.3) implies assumption (16.1).

Proof Let (16.3) be satisfied and take an arbitrary $x_0 \ne x^*$. If $R > 0$ is chosen such that $D_{f(x_0)} \subset K(x^*, R)$, then (16.1) implies for all $\alpha \in \left(f^*, f(x_0)\right]$ the inclusions

$$\{f = \alpha\} \subset D\backslash K\left(x^*, \left(\frac{\alpha - f^*}{M_R}\right)^{1/s_R}\right) \text{ and}$$

$$\{f = \alpha\} \subset K\left(x^*, \left(\frac{\alpha - f^*}{m_R}\right)^{1/s_R}\right).$$

Consequently,

$$\inf\Big\{|y-x^*| : f(y)=\alpha\Big\} \geq \left(\frac{\alpha-f^*}{M_R}\right)^{1/s_R} =$$

$$\left(\frac{m_R}{M_R}\right)^{1/s_R}\left(\frac{\alpha-f^*}{m_R}\right)^{1/s_R} \leq \left(\frac{m_R}{M_R}\right)^{1/s_R} \sup\Big\{|y-x^*| : f(y)=\alpha\Big\}.$$

Hence, (16.1) follows with $b = \left(\frac{m_R}{M_R}\right)^{1/s_R}$. □

Proof of Theorem 16.2 In order to apply Theorem 16.1, according to Remark 16.5 we have only to show that assumption (16.3) holds. For this purpose, consider an arbitrary $x_0 \neq x^*$. For $x \in D_{f(x_0)} \backslash D^*$ and $r > 0$, define $A_{x,r} := \Big\{s \in S : f(x+rs) < f(x)\Big\}$. Then, $\gamma_{x,r}\Big\{f < f(x)\Big\} = \gamma(A_{x,r})$.

From the compactness of $D_{f(x_0)}$ (cf. Lemma B.2 and Lemma B.4) we get the existence of a constant $L > 0$ with

$$\Big|\nabla f(x) - \nabla f(y)\Big| \leq L|x-y| \quad \text{for all } x, y \in D_{f(x_0)}.$$

In the same way as in the proof of Theorem 15.9, the following inclusion can be shown:

$$\left\{s \in S : \left(\frac{\nabla f(x)}{\Big|\nabla f(x)\Big|}\right)^{T_s} < -\frac{Lr}{\Big|\nabla f(x)\Big|}\right\} \subset A_{x,r}.$$

Because of

$$f(x) - f^* \leq \Big(\nabla f(x)\Big)^T (x-x^*) \leq \Big|\nabla f(x)\Big||x-x^*| \text{ (cf. Corollary B.1)},$$

according to assumption (16.3) we have the inequality

$$\Big|\nabla f(x)\Big| \geq \frac{f(x)-f^*}{|x-x^*|} \geq m\frac{|x-x^*|^s}{|x-x^*|} = m|x-x^*|^{s-1}$$

with a suitable $m > 0$ and a certain $s \in (1, 2]$. Hence,

$$A_{x,r} \supset \left\{s \in S : \left(\frac{\nabla f(x)}{\Big|\nabla f(x)\Big|}\right)^{T_s} < -\frac{Lr}{m \cdot |x-x^*|}|x-x^*|^{2-s}\right\}$$

$$\supset \left\{ s \in S : \left(\frac{\nabla f(x)}{\left|\nabla f(x)\right|} \right)^{T_s} < -\frac{LR^{2-s}}{m} \frac{r}{|x - x^*|} \right\},$$

where $R > 0$ is chosen such that $D_{f(x_0)} \subset \overline{K(x^*, R)}$. Thus, if $\frac{r}{|x - x^*|} < \varepsilon$, then

$$\gamma_{x,r}\Big\{f < f(x)\Big\} \geq \gamma \left\{ s \in S : \left(\frac{\nabla f(x)}{\left|\nabla f(x)\right|} \right)^{T_s} < -\frac{LR^{2-s}}{m} \varepsilon \right\}$$

$$= \gamma \left\{ s \in S : Z^{T_s} < -\frac{LR^{2-s}}{m} \varepsilon \right\}$$

for all $z \in S$ (cf. Lemma A.3).

The equation $\gamma\{s \in S : z^{T_s} < 0\} = \frac{1}{2}$ yields therefore (16.2) with $p = 1/2$. Global linear convergence is now provided by Theorem 16.1. □

Remark 16.6 Theorem 16.2 yields for all starting points $x_0 \in D$ the existence of a constant $\beta > 1$ with

$$\beta^n (f(X_n) - f^*) \to 0 \quad P - \text{a.s.}$$

Moreover, for each starting point $x_0 \in D$ there is also a constant $\alpha > 1$ with

$$\alpha^n |X_n - x^*| \to 0 \quad P - \text{a.s.}$$

Proof For each starting point $x_0 \in D$ there is a constant $R > 0$ with $D_{f(x_0)} \subset K(x^*, R)$. Choosing $\alpha = \beta^{1/s_R}$ then (16.1) yields

$$\alpha_n |X_n - x^*| \leq \left(\frac{1}{m_R} \beta^n \Big(f(X_n) - f^* \Big) \right)^{1/s_R} \quad P\text{-a.s.}$$

Of course, $\beta > 1$ implies $\alpha > 1$. □

Theorem 16.3 *Suppose that the function f is twice continuously differentiable with a Hessian which is positive definite at x^*.*

If $\gamma_1 \gamma_2 > \ell$, then for each starting point $x_0 \in D$ there is a $\beta > 1$ such that

$$\beta^n \Big(f(X_n) - f^* \Big) \to 0 \ P\text{-a.s.}$$

and

$$\beta^n |X_n - x^*| \to 0 \ P\text{-}a.s.$$

The proof of this assertion follows directly from Theorem 16.2 taking into account the remarks 16.3, 16.4, and 16.6.

Outline of the proof of Theorem 16.1

Let us consider the special case $\ell_0 = L_0$, i.e. $\lambda_0 = \varepsilon_{\ell_0}$. The step length distribution λ_n is then given by

$$\lambda_n(x_0, \ldots, x_{n-1},) = \varepsilon_{\ell_n}(x_0, \ldots, x_{n-1}).$$

Once the selection points $x_0, \ldots, x_{n-1}$ are generated, then, according to the construction of a random direction procedure, in the present case the mutation points y_n are obtained as realizations of the uniform distribution on the sphere with center x_{n-1} and radius $r = \ell_n(x_0, \ldots, x_{n-1})$. Therefore

$$x_n = x_{n-1} \text{ or } |x_{n-1} - x_n| = \ell_n(x_0, \ldots, x_{n.1}).$$

Of crucial importance to the proof of Theorem 16.1 are then the quotients

$$Q_n = \frac{\ell_{n+1}(X_0, \ldots, X_n)}{|X_n - x^*|}$$

and the angles

$$Z_n = \varphi(X_{n+1} - X_n, x^* - X_n) \qquad \text{cf. Definition B.1.}$$

According to Lemma B.9 we know that for all starting points $x_0 \in D \backslash D^*$ there exists an angle $\psi > 0$ such that for all $a \in \left(0, \frac{1}{2}\right)$ we have

$$\left\{Q_n \in \left[a, \frac{1}{2}\right]\right\} \cap \left\{Z_n \in [0, \Psi]\right\} =$$

$$\left\{a|X_n - x^*| \le |X_{n+1} - X_n| \le \frac{1}{2}|X_n - x^*|\right\} \cap \left\{Z_n \in [0, \psi]\right\} =$$

$$\left\{(X_{n+1} - X_n) \in C\left(\psi, x^* - X_n, a|X_n - x^*|\frac{1}{2}|X_n - x^*|\right)\right\} \subset$$

$$\left\{f(X_n) - f(X_{n+1}) \ge \frac{|X_{n+1} - X_n|}{3|x^* - X_n|}\left(f(X_n) - f^*\right)\right\} \cap \left\{\frac{|X_{n+1} - X_n|}{|x^* - X_n|} \ge a\right\} \subset$$

(according to Lemma B.11, part (2)

$$\left\{f(X_n) - f(X_{n+1}) \geq \frac{a}{3}\Big(f(X_n) - f^*\Big)\right\} =$$
$$\left\{f(X_{n+1}) - f^* \leq \Big(1 - \frac{a}{3}\Big)\Big(f(X_n) - f^*\Big)\right\}.$$

To obtain, firstly, the convergence of the random search procedure $(X_n : n \in \mathbb{N}_0)$, it is sufficient to show that for infinite many $n \in \mathbb{N}$ the event

$$\left\{Q_n \in \left[a, \frac{1}{2}\right]\right\} \cap \left\{Z_n \in [0, \psi]\right\}$$

occurs. Then it is also true that for infinite many $n \in \mathbb{N}$ the event

$$\left\{f(X_{n+1}) - f^* \leq \Big(1 - \frac{a}{3}\Big)\Big(f(X_n) - f^*\Big)\right\}$$

occurs. Since for the remaining iteration stages n the inequality

$$F(X_{n+1}) - f^* \leq f(X_n) - f^*$$

holds, we get then the convergence of $f(X_n)$ towards f^*. For the event

$$\left\{Q_n \in \left[a, \frac{1}{2}\right]\right\} \cap \left\{Z_n \in [0, \psi]\right\}$$

to occur for infinite many n, it is sufficient that $\left\{Q_n \in \left[a, \frac{1}{2}\right]\right\}$ infinite often occurs. Indeed, there exists a $p > 0$ with

$$p\Big(\Big|Z_n \in [0, \psi]\Big|\mathcal{A}_n\Big) \geq p \ P\text{-a.s. on } \left\{Q_n \in \left[a, \frac{1}{2}\right]\right\}$$

(cf. part 2 of the proof of Lemma 16.10). For the meaning of the above inequality, cf. Definition C.1, Lemma C.5, respectively.

From this the desired result can be derived. Hence, it is sufficient to show that for an $a \in \left(0, \frac{1}{2}\right)$ the event

$$\left\{Q_n \in \left[a, \frac{1}{2}\right]\right\}$$

will occur for infinite many $n \in \mathbb{N}$.

However, the proof of this assertion is rather difficult, since the stochastic behavior of the sequence $(Q_n : n \in \mathbb{N}_0)$ cannot be described easily. In this context the following properties are of crucial importance:

(i)

$$P\left(Q_{n+1} = \gamma_2 Q_n | \mathcal{A}_n\right) \geq \frac{1}{2}.$$

This follows from the fact that in a random direction procedure the probability of failure is at least $\frac{1}{2}$ (cf. Theorem 15.5 and Lemma 16.1; note that f is convex) and a failure in the $n+1$-ten step implies the event $\{Q_{n+1} = \gamma_2 Q_n\}$.

(ii) If c is chosen sufficiently large, a failure will occur in the $n+1$-th step (i.e. $X_{n+1} = X_n$ and therefore $Q_{n+1} = \gamma_2 Q_n$), provided that $Q_n > c$.

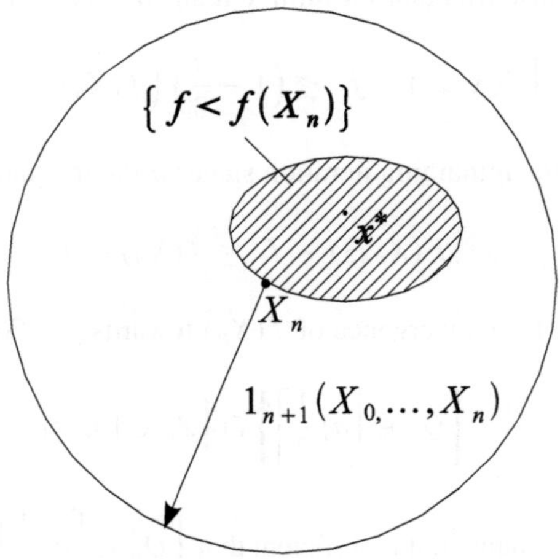

The mathematical formulation of this fact is:

$$\begin{aligned} P(Q_{n+1} &= \gamma_2 Q_n | \mathcal{A}) \\ &= 1 P\text{-a.s. on } \{Q_n > 0\}. \end{aligned}$$

This means that after entering the interval (c, ∞), the sequence $(Q_n : n \in \mathbb{N})$ will behave deterministically for some time, namely k times, where

$$k = \min\{i \in \mathbb{N} : \gamma_2^i Q_n < c\}$$

(here, n = time point at which the algorithm enters the interval (c, ∞)). Consequently, after each entry of $(Q_n : n \in \mathbb{N}_0)$ into the interval (c, ∞) the sequence will leave this interval after a finite number of iteration steps. Therefore, the event $\{Q_n < c\}$ will occur for infinite many time points $n \in \mathbb{N}$. Using probabilistic arguments, from property (i) we even obtain that for arbitrary $a > 0$ the event $\{Q_n < a\}$ occurs also for infinite many $n \in \mathbb{N}$ (cf. Lemma 16.8).

(iii) If $Q_n > \frac{1}{2}$, then in case of a failure in the $n+1$-th step we get $Q_{n+1} = \gamma_2 Q_n > \frac{1}{2}\gamma_2$.

In case of success we have

$$Q_{n+1} = \frac{\gamma_1 \ell_{n+1}(X_0, \ldots, X_n)}{|X_{n+1} - x^*|} = \gamma_1 Q_n \frac{|X_n - x^*|}{|X_{n+1} - x^*|} \geq b\gamma_1 Q_n > \frac{1}{2}\gamma_1 b$$

(here, b is according to assumption (16.1), cf. Remark 16.1, part (2).

If therefore $Q_n > \frac{1}{2}$, then in the next step Q_{n+1} cannot be arbitrarily close to zero. Hence, there exists $a_1 > 0$ such that the events

$$\left\{Q_n > \frac{1}{2} \cdot Q_{n+1} < a_1\right\} \quad (n \in \mathbb{N}_0)$$

are impossible, i.e. that their probability is zero (cf. Lemma 16.6).

If it were now also known that the event $\{Q_n \geq a_1\}$ occurs for infinite many $n \in \mathbb{N}$, then we would be finished. Then the sequence $(Q_n : n \in \mathbb{N})$ infinite many leave the interval $(0, a_1)$, and according to part (ii), infinite often enters this interval. However, then it would be necessary for $(Q_n : n \in \mathbb{N}_0)$ to be in the interval $\left[a_1, \frac{1}{2}\right]$ immediately before each entry. Since this process occurs infinite often, the event, too, must occur $\left\{Q_n \in \left[a_1, \frac{1}{2}\right]\right\}$ infinite often. This implies then, as already mentioned, the convergence of the algorithm. The proof of the existence of an $a_1 > 0$ such that the event $\{Q_n \geq a_1\}$ occurs for infinite many $n \in \mathbb{N}$ is given in Lemmas 16.2 to 16.5.

The basic idea for this is that that the event

$$\{Q_n < a_1, \ldots, Q_{n+K} < a_1\}_K$$

can occur only if the number of successes in the time interval $[n, n + K]$ does not exceed a certain number, which is essentially proportional to K. The corresponding probability can be estimated using the Binomial distribution.

To show a linear rate of convergence, it is necessary to evaluate in more detail how often the event

$$\left\{Q_n \in \left[a, \frac{1}{2}\right]\right\} \cap \left\{Z_n \in [0, \psi]\right\}$$

and therefore also the event

$$\left\{f(X_{n+1}) - f^* \leq \left(1 - \frac{a}{3}\right)\left(f(X_n) - f^*\right)\right\}$$

occurs.

For example, the following condition—unfortunately not known whether it holds—would be sufficient for the linear convergence:

$$P\left(\liminf_{n\to\infty}\frac{1}{2}\sum_{i=1}^{n}1_{\left\{Q_i\in[a,1/2]\right\}\cap\left\{Z_i\in[0,\psi]\right\}}\geq\alpha>0\right)=1.$$

For $\alpha = 1/3$ this would mean, for example, that the event

$$\left\{Q_n\in\left[a,\frac{1}{2}\right]\right\}\cap\left\{Z_n\in[0,\psi]\right\}$$

occurs "in the mean at every third step." Then it follows for large n that

$$f(X_{3n})-f^*\leq\left(1-\frac{a}{3}\right)^n\left(f(x_0)-f^*\right)$$

or

$$f(X_n)-f^*\leq\left(\left(1-\frac{a}{3}\right)^{1/3}\right)^n\left(f(x_0)-f^*\right),$$

i.e., linear convergence.

However, linear convergence is also obtained if

$$P\left(\liminf_{n\to\infty}\frac{1}{n}\sum_{i=1}^{n}\left(1_{\left\{Q_i\in[a,1/2],Z_i\in[0,\psi]\right\}}+1_{\{Q_i\geq c\}}\right)\geq\alpha>0\right)=1 \tag{16.4}$$

is satisfied for a sufficiently large $c = c(x_0)$.

The reason for this is that the event $\{Q_{n-1} < c, Q_n \geq c\}$ can only occur if there is a search success in the n-th step. It can be shown that this success must be greater, the greater c. Moreover, it can be shown that the following inequality is true for sufficiently large values of c (cf. Lemma 16.11):

$$f(X_{T_n})-f^*\leq\gamma_2^{M_n}\left(f(X_{T_{n-1}})-f^*\right)$$

(where T_n is the n-th time of entry of the sequence $(Q_n : n \in \mathbb{N}_0)$ into the interval $[c, \infty)$, and M_n denotes the n-th sojourn time of this sequence in the interval $[c, \infty)$). This means that the improvement of the function at the time of the n-th entry into the interval $[c, \infty)$ has within one iteration step the same effect (note the factor $\gamma_2^{M_n}$) as if the event

$$\left\{f(X_i)-f^*\leq\gamma_2\left(f(X_{i-1})-f^*\right)\right\}$$

occurred at any time $i = T_n, T_n + 1, \ldots, T_n + M_n - 1$.

From this consideration and from

$$\left\{Q_i \in \left[a, \frac{1}{2}\right], Z_i \in [0, \psi]\right\} \subset \left\{f(X_{i+1}) - f^* \leq \left(1 - \frac{a}{3}\right)\left(f(X_i) - f^*\right)\right\}$$

the following inclusion can be derived (cf. the proof of Theorem 16.1):

$$\left\{\sum_{i=1}^{n} (1_{\left\{Q_i \in [a,1/2], Z_i \in [0,\psi]\right\}}) = k\right\}$$
$$\subset \left\{f(X_{n+1}) - f^* \leq \gamma^k\left(f(x_0) - f^*\right)\right\}$$
$$\left(\text{with } \gamma := \max\left\{\gamma_2, 1 - \frac{a}{3}\right\}\right).$$

Then linear convergence follows, provided that (16.4) holds.

To prove (16.4), it is necessary to make a suitable estimate of the average sojourn times

$$\frac{1}{n}\sum_{i=1}^{n} M_i$$

of the sequence $(Q_n : n \in \mathbb{N}_0)$ in the intervals $(0, a)$ and $\left(\frac{1}{2}, c\right)$ (cf. Lemma 16.12). The basic tool for this estimate is Lemma C.1.

The following Lemmas 16.1 to 16.12 will serve exclusively the purpose to prepare the proof of Theorem 16.1.

Lemma 16.1 *For $n \in \mathbb{N}_0$, define*

$$Q_n := \begin{cases} \frac{1_{n+1}(X_0,\ldots,X_n)}{|X_n - x^*|} & \text{if } X_n \neq x^* \\ \infty & \text{else.} \end{cases}$$

Then, for all starting points $x_0 \in D\backslash D^*$ we have

(1)

$$P\Big(\bigcup_{n\in\mathbb{N}}\{Q_n < \infty\}\Big) = 1$$

(2)

$$P(Q_{n+1} < Q_n|\mathcal{A}_n) \geq P(Q_{n+1} = \gamma_2 Q_n|\mathcal{A}_n) \geq$$
$$\geq P\Big(f(X_{n+1}) \geq f(X_n)|\mathcal{A}_n\Big) \geq \frac{1}{2} \; P\text{-a.s.}$$

(3)

$$P(X_{n+1} = X_n|\mathcal{A}_n) = P(Q_{n+1} = \gamma_2 Q_n|\mathcal{A}_n) = 1$$
$$P\text{-a.s. on } \{Q_n \geq c\} \text{ for all } c \geq 2(1 + 1/b)/1_0$$

(with b according to assumption (16.1)*).*

For the notation, cf. Definition C.1 and Lemma C.2.

Proof

(1) For arbitrary $n \in \mathbb{N}$ and $x_0, \ldots, x_{n-1} \in D \backslash D^*$ we have

$$q_n\Big(x_0, \ldots, x_{n-1}, \{x^*\}\Big) = p_n\Big(x_0, \ldots, x_{n-1}, \{x^*\}\Big)$$
$$= \Big(\gamma \otimes \lambda_n(x_0, \ldots, x_{n-1}, \cdot)\Big)\Big(\Big\{(x, r) : x_{n-1} + rs = x^*\Big\}\Big)$$
$$\gamma\left(\left\{\frac{|x^* - x_{n-1}|}{|x^* - x_{n-1}|}\right\}\right) \cdot \lambda_n\Big(x_0, \ldots, x_{n-1}, \{|x^* - x_{n-1}|\}\Big) = 0.$$

This yields $q_n\Big(x_0, \ldots, x_{n-1}, D\backslash\{x^*\}\Big) = 1$ and therefore

$$P\left(\bigcap_{i=0}^{n}\{X_i \neq x^*\}\right) = \int\limits_{D\backslash\{x^*\}} \cdots \int\limits_{D\backslash\{x^*\}} q_n(x_0, \ldots, x_{n-1}, dx_n) \ldots q_1(x_0, dx_1) = 1.$$

This holds for arbitrary $n \in \mathbb{N}$.

Therefore $P\left(\bigcup\limits_{n\in\mathbb{N}_0} \{X_n = x^*\}\right) = 0.$

(2) Because of $\gamma_2 < 1$, the first inequality is obvious. Theorem 15.5 yields

$$q_n\left(x_0, \ldots, x_{n-1}, \Big\{f < f(x_{n-1})\Big\}\right)$$
$$= p_n\Big(x_0, \ldots, x_{n-1}, \Big\{f > f(x_{n-1})\Big\}\Big) \leq \frac{1}{2}$$
$$\text{for all } n \in \mathbb{N}, \quad x_0, \ldots, x_{n-1} \in D\backslash D^*$$

and therefore

$$q_n\left(x_0, \ldots, x_{n-1}, \Big\{f \geq f(x_{n-1})\Big\}\right) \geq \frac{1}{2}.$$

Hence,

$$P\Big(f(X_{n+1}) \geq f(X_n)|\mathcal{A}_n\Big) \geq \frac{1}{2} \ P\text{-a.s.}$$

The assertion follows now from

$$P(f(X_{n+1}) \geq f(X_n)|\mathcal{A}_n) = P(X_{n+1} = X_n|\mathcal{A}_n)$$
$$\leq P(Q_{n+1} = \gamma_2 Q_n|\mathcal{A}_n)\ P\text{-a.s.}$$

(3) Let $x_0 \in D\backslash D^*$ be an arbitrary starting point. Furthermore, let $b = b(x_0)$ be chosen according to assumption (16.1). Then, for $x, y \in D_{f(x_0)}$ with $f(y) \leq f(x)$, according to Remark 16.1, part (2) we have

$$|y - x| \leq \frac{1}{b}|x - x^*|,$$

and therefore

$$|y - x| \leq |y - x^*| + |x - x^*| \leq (1 + 1/b)|x - x^*|.$$

Hence, $\Big\{f \leq f(x)\Big\} \subset \overline{K\Big(x, (1 + 1/b)|x - x^*|\Big)}$.

Choose now an arbitrary $c \geq 2(1 + 1/b)/l_0$, and select

$$x_1, \ldots, x_n \in D_{f(x_0)} \text{ with } \frac{l_{n+1}(x_0, \ldots, x_n)}{|x_n - x^*|} \geq c.$$

Then,

$$P_{n+1}\Big(x_0, \ldots, x_n, \Big\{f < f(x_n)\Big\}\Big)$$
$$\leq P_{n+1}\left(x_0, \ldots, x_n, \overline{K\Big(x_n, (1 + 1/b)|x_n - x^*|\Big)}\right)$$
$$= \Big(\gamma \otimes \lambda_{n+1}(x_0, \ldots, x_n, \cdot)\Big)\left(\Big\{(s, r) : x_n + rs \in \overline{K\Big(x_n, (1 + 1/b)|x_n - x^*|\Big)}\Big\}\right)$$
$$= \gamma_{n+1}\Big(x_0, \ldots, x_n, \Big[0, (1 + 1/b)|x_n - x^*|\Big]\Big)$$
$$= T_{l_{n+1}(x_0,\ldots,x_n)}(\lambda_0)\Big(\Big[0, (1 + 1/b)|x_n - x^*|\Big]\Big)$$
$$= \lambda_0\left(\left[0, (1 + 1/b)\frac{|x_n - x^*|}{l_{n+1}(x_0, \ldots, x_n)}\right]\right) \leq \lambda_0\Big(\Big[0, (1 + 1/b)/c\Big]\Big)$$
$$\leq \lambda_0\left(\left[0, \frac{1}{2}l_0\right]\right) = 0, \text{ da } \lambda_0\Big([l_0, L_0]\Big) = 1.$$

Because of

$$P(X_n \neq X_{n+1}|X_n = x_n, \ldots, X_0 = x_0)$$
$$v = P_{n+1}\Big(x_0, \ldots, x_n, \Big\{f < f(x_n)\Big\}\Big) = 0$$

and $\{X_{n+1} = X_n\} \subset \{Q_{n+1} = \gamma_2 Q_n\}$, the assertion follows now.

□

Lemma 16.2 *Let Q_n be defined as in Lemma 16.2. Moreover, assume that assumption* (16.1) *of Theorem 16.1 holds, and let x_0 denote an arbitrary starting point from $D\backslash D^*$. Let b be chosen according to assumption* (16.1) *and Remark 16.1 such that for all $\alpha \in \Big(f^*, f(x_0)\Big)$ the inequality*

$$\inf\Big\{|y - x^*| : f(y) = \alpha\Big\} \geq b \cdot \sup\Big\{|y - x^*| : f(y) \leq \alpha\Big\}$$

is satisfied. Then, with $\rho_1 := \dfrac{\ln \gamma_2}{\ln \gamma_2 - \ln \gamma_1}$, for all $n, s \in \mathbb{N}$ and $\delta, \varepsilon > 0$, we have

$$\{Q_n \geq \delta, Q_{n+1} < \varepsilon, \ldots, Q_{n+s} < \varepsilon\} \subset \{Q_n \geq \delta, Q_{n+s} < \varepsilon\}$$

$$\subset \Big\{ \sum_{i=n+1}^{n+s} 1_{\{f(X_{i-1}) > f(X_i)\}} < \rho_1 s + \rho_2 \Big\} \ P\text{-a.s.}$$

$$\text{where } \rho_2 := \frac{\ln(\delta b) - \ln \varepsilon}{\ln \gamma_2 - \ln \gamma_1}.$$

Proof Let $\omega \in \{Q_n \geq \delta, Q_{n+s} < \varepsilon\}$.

Because of $P\Big(X_0 = x_0, f(X_0) \geq f(X_1) \geq f(X_2) \geq \ldots\Big) = 1$, without loss of generality we can assume that

$$X_0(\omega) \text{ und } f(x_0) \geq f\Big(X_1(\omega)\Big) \geq f\Big(X_2(\omega)\Big) \geq \ldots$$

holds.

Define $e(\omega) := \sum_{i=n+1}^{n+s} 1_{\{f(X_{i+1}) > f(X_i)\}}(\omega)$.

Obviously, $e(\omega) \leq s$. If, $e(\omega) \geq \rho_1 s + \rho_2$, then

$$e(\omega) \geq \frac{s \ln \gamma_2 + \ln(b\delta/\varepsilon)}{\ln \gamma_2 - \ln \gamma_1}$$

and therefore

$$\Big(s - e(\omega)\Big) \ln \gamma_2 + e(\omega) \ln \gamma_1 \geq \ln\left(\frac{\varepsilon}{\delta b}\right).$$

This yields $g\gamma_1^{e(\omega)} \gamma_2^{s-e(\omega)} \geq \frac{\varepsilon}{\delta}$.

Because of $f(X_{n+s}(\omega)) \leq f(X_n(\omega))$, we get

$$|X_{n+s}(\omega) - x^*| \leq \frac{1}{b}|X_n(\omega) - x^*|$$

and therefore (cf. Definition 16.1)

$$\begin{aligned} Q_{n+s}(\omega) &= \frac{1_{n+s+1}(X_0(\omega), \ldots, X_{n+s}(\omega))}{|X_{n+s}(\omega) - x^*|} \\ &\geq \frac{1_{n+s+1}(X_0(\omega), \ldots, X_{n+s}(\omega))}{|X_n(\omega) - x^*|} b \\ &= \frac{1_{n+1}(X_0(\omega), \ldots, X_n(\omega))}{|X_n(\omega) - x^*|} b\gamma_1^{e(\omega)} \gamma_2^{s-e(\omega)} \\ &= Q_n(\omega) b\gamma_1^{e(\omega)} \gamma_2^{s-e(\omega)} \geq Q_n(\omega)\frac{\varepsilon}{\delta} \geq \delta\frac{\varepsilon}{\delta} = \varepsilon \end{aligned}$$

in contradiction to $Q_{n+s}(\omega) < \epsilon$. □

Lemma 16.3 *Consider an arbitrary starting point $x_0 \in D \setminus D^*$. For $\varepsilon > 0$, let*

$$\begin{aligned} p_\varepsilon := \inf\{p_n(x_0, \ldots, X_{n-1}, \{f < f(x_{n-1})\}) : n \in \mathbb{N}, \\ x_1, \ldots, x_{n-1} \in D_{f(x_0)}, \frac{1_n(x_0, \ldots, x_{n-1})}{|x_{n-1} - x^*|} < \varepsilon\}. \end{aligned}$$

Then, for all $\delta, \varepsilon > 0$ and $n, s \in \mathbb{N}$ we have

$$P(\{Q_{n-1} \geq \delta\} \cap \bigcap_{i=1}^{n+s-1} \{Q_i < \varepsilon\} \leq \sum_{j=1}^{[p_1 s + p_2]} \binom{s}{j} p_\varepsilon^j (1 - p_\varepsilon)^{s-j}$$

with p_1, p_2 according to Lemma 16.2.

Proof According to the Definition of p_ε for all $n \in \mathbb{N}$ we have

$$P(f(X_n) < f(X_{n-1})|\mathcal{A}_{n-1}) \geq P_\varepsilon \ P\text{-a.s. on } \{Q_{n-1} < \varepsilon\}.$$

Lemma 16.2 yields:

$$\begin{aligned} &P(\{Q_{n-1} \geq \delta\} \cap \bigcap_{i=1}^{n+s-1} \{Q_i < \epsilon\}) \\ &\leq P\left(\left\{\sum_{i=n+1}^{n+s-1} 1_{\{f(X_{i-1}) > f(X_i)\}} < p_1 s + p_2\right\} \cap \bigcap_{i=1}^{n+s-1} \{Q_i < \varepsilon\}\right). \end{aligned}$$

Lemma C.4, part (2) applied to

$$Z_n = 1_{\{f(X_{i-1})>f(X_i)\}}(X_{n-1}, X_n), T = n, n = s$$

and $K = [p_1 s + p_2]$, yields the assertion. □

Lemma 16.4 *Consider an arbitrary starting point $x_0 \in D \setminus D^*$. Let p_ϵ be defined as in Lemma 16.3. Moreover, let the assumption* (16.2) *of Theorem 16.1 be fulfilled. Then there exists an $\varepsilon > 0$ with*

$$p_\varepsilon > p_1 = \frac{\ln \gamma_2}{\ln \gamma_2 - \ln \gamma_1}.$$

Proof Because of $\gamma_1^p \gamma_2^{1-p} > 1$, for continuity reasons there is $\tilde{\delta} > 0$ with

$$\gamma_1^{p-\tilde{\delta}} \gamma_2^{1-p+\tilde{\delta}} > 1.$$

This inequality is equivalent to

$$p_1 = \frac{\ln \gamma_2}{\ln \gamma_2 - \ln \gamma_1} < p - \tilde{\delta}.$$

According to assumption (16.2) (ii) there is an $\varepsilon > 0$ with

$$\gamma_{x,r}\{f < f(x)\} \geq p - \tilde{\delta} \text{ for all } x \epsilon D_{f(x_0)} \text{ with } \frac{r}{|x - x^*|} < L_0 \varepsilon.$$

Let now $n \in \mathbb{N}$, $x_1, \ldots, x_{n-1} \in D_{f(x_0)}$ be chosen arbitrarily with $\dfrac{l_n(x_0, \ldots, x_{n-1})}{|x_{n-1} - x^*|} < \varepsilon$.

Then we have

$$\begin{aligned}
&p_n(x_0, \ldots, x_{n-1}, \{f < f(x_{n-1})\}) \\
&= (\gamma \otimes \lambda_n(x_0, \ldots, x_{n-1}, \cdot))(\{(s, r) : x_{n-1} + rs \in \{f < f(x_{n-1})\} \\
&= \int_{\mathbb{R}_+} \gamma(\{s \in S : x_{n-1} + rs \in \{f < f(x_{n-1})\}\})\lambda_n(x_0, \ldots, x_{n-1}, dr) \\
&= \int_{\mathbb{R}_+} \gamma_{x_{n-1},r}\{f < f(x_{n-1})\}\lambda_n(x_0, \ldots, x_{n-1}, dr) \\
&= \int_{\mathbb{R}_+} \gamma_{x_{n-1},rl_n(x_0,\ldots,x_{n-1})}\{f < f(x_{n-1})\}\lambda_0(dr) \\
&= \int_{[l_0,L_0]} \gamma_{x_{n-1},rl_n(x_0,\ldots,x_{n-1})}\{f < f(x_{n-1})\}\lambda_0(dr).
\end{aligned}$$

For $r \in [l_0, L_0]$ we have $\dfrac{rl_n(x_0, \ldots, x_{n-1})}{|x_{n-1} - x^*|} < l_0\varepsilon$ and therefore

$\gamma_{x_{n-1}, rl_n(x_0,\ldots,x_{n-1})}\{f < f(x_{n-1})\} \geq p - \tilde{\delta}$.

From this it follows that

$$p_n(x_0, \ldots, x_{n-1}, \{f < f(x_{n-1})\}) \geq p - \tilde{\delta}$$

and therefore also $p_\varepsilon \geq p - \tilde{\delta} > p_1$. □

Lemma 16.5 *Assume that the conditions of Theorem 16.1 are satisfied. Then for all starting points $x_0 \in D \setminus D^*$ there exists an $a_0 = a_0(x_0) > 0$ such that*

$$P(Q_n \geq a_0 \textit{ for infinite many } n \in \mathbb{N}_0) = 1.$$

Proof Let $x_0 \in D \setminus D^*$ be an arbitrary starting point. According to Lemma 16.4 there is an $\varepsilon > 0$ with $p_\varepsilon > p_1$. Define $a_0 := \varepsilon$. From Lemma 16.3 it follows for all $n \in \mathbb{N}$, $s \in \mathbb{N}$ and $\delta > 0$ that

$$P(\{Q_{n-1} \geq \delta\}) \bigcap_{i=n}^{n+s-1} \{Q_i < a_0\} \leq \sum_{j=0}^{[p_1 s + p_2]} \binom{s}{j} p_{a_0}^j (1 - p_{a_0})^{s-j}.$$

Because of $P_{a_0} > p_1$, Lemma B.15 yields

$$\lim_{s\to\infty} P(\{Q_{n-1} \geq \delta\} \bigcap_{i=n}^{n+s-1} \{Q_i < a_0\}) = 0.$$

By definition we have for all $Q_n > 0$. Hence, for each $n \in \mathbb{N}$ we get

$$\begin{aligned}
P(\bigcap_{i\geq n}\{Q_i < a_0\} &= P(\{Q_{n-1} > 0\} \cap \bigcap_{i\geq n}\{Q_i < a_0\} \\
&= \lim_{\delta\to 0} P\left(\{Q_{n-1} \geq \delta\} \cap \bigcap_{i\geq n}\{Q_i < a_0\}\right) \\
&= \lim_{\delta\to 0}\left(\lim_{s\to\infty} P\left(\{Q_{n-1} \geq \delta\} \bigcap_{i=n}^{n+s-1} \{Q_i < a_0\}\right)\right) = 0.
\end{aligned}$$

for all $n \in \mathbb{N}$ we therefore have

$$P(\bigcap_{i\geq n}\{Q_i < a_0\}) = 0.$$

Hence,

$$P(\bigcup_{n\in\mathbb{N}}\bigcap_{i\geq n}\{Q_i < a_0\}) = 0.$$

Taking the complement, we get the assertion. □

Lemma 16.6 *Let the assumptions of Theorem 16.1 be satisfied. Then, for all starting points $x_0 \in D \setminus D^*$ there exists an $a_1 = a_1(x_0) \in (0, \frac{1}{2L_0})$ with*

$$P(Q_n < a_1, Q_{n+1} > \frac{1}{2L_0}|\mathcal{A}_k) = P(Q_n > \frac{1}{2L_0}Q_{n+1} < a_1|\mathcal{A}_k) = 0$$

P-a.s. for all $n \in \mathbb{N}$ and $k \leq n$.

Proof

(1) Let $k = n$, and assume that $a_1 := m$ in $\left\{\frac{\gamma_2}{2L_0}, \frac{\gamma_1^b}{2L_0}, \frac{1}{L_0 + 2L_0\gamma_1}\right\}$, where b is chosen, according to assumption (16.1), such that for all $\alpha \in (f^*, f(x_0)]$ the inequality

$$\inf\{|y - x^*| : f(y) = a\} \geq b \sup\{|y - x^* : f(y) = \alpha\}$$

holds.

Then for all

$$x_1, \ldots, x_n \in D_{f(x_0)} \text{ with } \frac{l_{n+1}(x_0, \ldots, x_n)}{|x_n - x^*|} < a_1$$

we have

$$\begin{aligned}
&q_{n+1}\left(x_0, \ldots, x_n, \left\{x_{n+1} \in \mathbb{R}^d : \frac{l_{n+2}(x_0, \ldots, x_n, x_{n+1})}{|x_{n+1} - x^*|} > \frac{1}{2L_0}\right\}\right) \\
&\leq q_{n+1}\left(x_0, \ldots, x_n, \left\{x_{n+1} : \frac{\gamma_1 l_{n+1}(x_0, \ldots, x_n)}{|x_{n+1} - x^*|} > \frac{1}{2L_0}\right\}\right) \\
&\leq q_{n+1}\left(x_0, \ldots, x_n, \left\{x_{n+1} : \frac{\gamma_1 a_1 |x_0, \ldots, x_n|}{|x_{n+1} - x^*|} > \frac{1}{2L_0}\right\}\right) \\
&= q_{n+1}(x_0, \ldots, x_n, \{x_{n+1} : |x_{n+1} - x^*| < 2L_0\gamma_1 a_1 |x_n - x^*|\}) \\
&\quad \times q_{n+1}(x_0, \ldots, x_n, \{x_{n+1} : |x_n - x^*| - |x_n - x_{n+1}| \\
&\quad < 2L_0 a_1 \gamma_1 |x_n - x^*|\}) \\
&= q_{n+1}(x_0, \ldots, x_n, \{x_{n+1} : |x_{n+1} - x_n| > (1 - 2L_0\gamma_1 a_1)|x_n - x^*|\} = 0
\end{aligned}$$

since on the one hand

$q_{n+1}(x_0, \dots, x_n, \{x_{n+1} : |x_{n+1} - x_n| \leq L_0 l_{n+1}(x_0, \dots, x_n)\}) = 0.$

However, because of

$$\frac{l_{n+1}(x_0, \dots, x_n)}{|x_n - x^*|} < a_1 \leq \frac{1}{L_O + 2\gamma_1 L_0}$$

the inequality

$$(1 - 2L_0\gamma_1 a_1)|x_n - x^*| > \frac{1 - 2L_0\gamma_1 a_1}{a_1} l_{n+1}(x_0, \dots, x_n)$$

$$\geq L_0 l_{n+1}(x_0, \dots, x_n)$$

is satisfied.

This yields

$$P(\{Q_{n+1} > \frac{1}{2L_0}\}|\mathcal{A}_n) = 0 P\text{-a.s. on } \{Q_n < a_1\}.$$

Because of $\{Q_n < a_1\} \in A_n$, for arbitrary $A_n \in \mathcal{A}_n$ we have

$$P(\{Q_n < a_1\} \cap \{Q_{n+1} > \frac{1}{2L_0}\} \cap \mathcal{A}_n) = \int_{A_n \cap \{Q_n < a_1\}} P(Q_{n+1} > \frac{1}{2L_0}|\mathcal{A}_n) dP = 0.$$

Lemma C.2 yields then

$$P(Q_n < a_1, Q_{n+1} > \frac{1}{2L_0}|\mathcal{A}_n) = 0 \; P\text{-a.s.}$$

For $k < n$ it follows that

$$P(Q_n < a_1, Q_{n+1} > \frac{1}{2L_0}|\mathcal{A}_n) = E(P(Q_n < a_1, Q_{n+1} > \frac{1}{2L_0}|\mathcal{A}_n)|\mathcal{A}_k) = 0$$

and thus the first part of the assertion.

(2) Select $x_1, \dots, x_n \in D_{f(x_0)}$ such that $\frac{l_{n+1}(x_0, \dots, x_n)}{|x_n - x^*|} > \frac{1}{2L_0}$.
Then,

$$\{x_{n+1} \in \mathbb{R}^d : \frac{l_{n+2}(x_0, \dots, x_n, x_{n+1})}{|x_{n+1} - x^*|} < a_1\}$$

$$= \{x_{n+1} : \frac{\gamma_2 l_{n+1}(x_0, \ldots, x_n)}{|x_n - x^*|} < a_1\} \cap \{x_{n+1} = x_n\} \cup$$

$$\{x_{n+1} : \frac{\gamma_1 l_{n+2}(x_0, \ldots, x_n)}{|x_{n+1} - x^*|} < a_1\} \cap \{x_{n+1} \neq x_n\}$$

$$= \{x_{n+1} : \frac{\gamma_1 l_{n+2}(x_0, \ldots, x_n)}{|x_{n+1} - x^*|} < a_1\} \cap \{x_{n+1} \neq x_n\}, \text{ since}$$

$$\frac{\gamma_2 l_{n+2}(x_0, \ldots, x_n)}{|x_n - x^*|} > \gamma_2 \frac{1}{2L_0} \geq a_1.$$

Consequently, we get

$$q_{n+1}\left(x_0, \ldots, x_n, \left\{x_{n+1} : \frac{\gamma_1 l_{n+1}(x_0, \ldots, x_n)}{|x_{n+1} - x^*|} < a_1\right\}\right)$$

$$= p_{n+1}\left(x_0, \ldots, x_n, \left\{x_{n+1} : \frac{l_{n+2}(x_0, \ldots, x_n, x_{n+1})}{|x_{n+1} - x^*|}\right.\right.$$

$$\left.\left. < a_1, f(x_{n+1}) < f(x_n)\right\}\right).$$

If $f(x_{n+1}) < f(x_n)$, then $|x_{n+1}| \leq \frac{1}{b}|x_n - x^*|$
(cf. Remark 16.1, part 2) and therefore

$$\left\{x_{n+1} : \frac{\gamma_1 l_{n+1}(x_0, \ldots, x_n)}{|x_{n+1} - x^*|} < a_1, f(x_{n+1}) < f(x_n)\right\}$$

$$\subset \left\{x_{n+1} : \frac{\gamma_1 l_{n+1}(x_0, \ldots, x_n)}{|x_{n+1} - x^*|} b < a_1\right\} = \emptyset, \text{ since}$$

$$\frac{\gamma_1 l_{n+1}(x_0, \ldots, x_n)}{|x_{n+1} - x^*|} b > \gamma_1 b \frac{1}{2L_0} \geq a_1.$$

The remainder is now done analogously to part 1.

□

Lemma 16.7 *Let the assumptions of Theorem 16.1 be satisfied.*
Define $B := [b_1, b_2]$ *with* $0 < b_1 < b_2$.
Let $K := \left\lceil \frac{\ln b_2 - \ln b_1}{-\ln \gamma_2} \right\rceil + 1$. *Then for all stopping times* T *with regard to* $(\mathcal{A}_n : n \in \mathbb{N}_0)$*(for the definition, cf. [1, p. 215]), for all* $i \in \mathbb{N}$ *and all starting points* $x_0 \in D \setminus D^*$ *we have*

$$P\left(\bigcap_{j=T}^{T+jK} \{Q_j \in B\} | \mathcal{A}_T\right) \leq (1 - 2^{-K})^i \ P\text{-a.s.}$$

Proof

(1) Let $n \in \mathbb{N}$ be arbitrary, but fixed. For $j \in \mathbb{N}_0$, define

$$A_j := \{Q_{n+Kj+1} = \gamma_2 Q_{n+Kj}\} \cap \cdots \cap \{Q_{n+K(j+1)} = \gamma_2 Q_{n+K(j+1)-1}\}.$$

A simple calculation yields

$$K = \inf\{n \in \mathbb{N} : b_2\gamma_2^n < b_1\}.$$

From this it follows (where $A_j^{\complement}$ denotes the complement of A_j)

$\bigcap_{j=0}^{i-1} A_j^{\complement} \supset \bigcap_{j=n}^{n+Ki} \{Q_j \in B\}$ for all $i \in \mathbb{N}$, since: Let $\omega \in \bigcap_{j=n}^{n+Ki} \{Q_j B\}$ with $\omega \notin \bigcap_{j=0}^{i=1} A_j^{\complement}$;

then there is a $j \in \{0, \ldots, i-1\}$ with $\omega \in A_j$.

Because of $Q_{n+Kj} \in B$, it follows that $\omega \in \{Q_{n+Kj} \in B\} \cap A_j$ and therefore $Q_{n+k(j+1)} < b_1$, hence, $Q_{n+k(j+1)}(\omega) \notin B$. However, because of $j+1 \leq i$, we find $Q_{n+k(j+1)}(\omega) \in B$.

(2) Because of $P(Q_{n+1} = \gamma_2 Q_2 | \mathcal{A}_j) \geq \frac{1}{2} P$-a.s. for all $j \in \mathbb{N}$ (cf. Lemma 16.1), we get

$$\begin{aligned} &P(\{Q_{j+1} = \gamma_2 Q_j\} \cap \{Q_{j+2} = \gamma_2 Q_{j+1}\} | \mathcal{A}_j) \\ &= E(E(l_{\{Q_{j+1}=\gamma_2 Q_j\}} l_{\{Q_{j+2}=\gamma_2 Q_{j+1}\}} | \mathcal{A}_{j+1}) | \mathcal{A}_j) \\ &= E(l_{\{Q_{j+1}=\gamma_2 Q_j\}} P(Q_{j+2} = \gamma_2 Q_{j+1} | \mathcal{A}_{j+1}) | \mathcal{A}_j) \\ &\geq \frac{1}{2} P(Q_{j+1} = \gamma_2 Q_j | \mathcal{A}_j) \text{ Continuing this way, one obtains} \end{aligned}$$

$$P(\bigcap_{m=j}^{j+L-l} \{Q_{m+1} \gamma_2 Q_m\} | \mathcal{A}_j) \geq (\frac{1}{2})^L \ P\text{-a.s.}$$

for all $L \in \mathbb{N}$ and $j \in \mathbb{N}$.

When considering the definition of A_j, it follows that

$$P(A_j | A_{n+Kj}) \geq (\frac{1}{2})^K$$

and therefore

$$P(A_j^{\complement} | A_{n+Kj}) \leq 1 - 2^{-K} \ P\text{-a.s. for all } j \in \mathbb{N}_0.$$

From this we get again

$$P(A_0^{\complement} \cap A_1^{\complement} | \mathcal{A}_n) = E(I_{A_0^{\complement}} \cdot P(A_1^{\complement} | A_{n+k}) | \mathcal{A}_n)$$
$$\leq (1 - 2^{-K}) P(A_0^{\complement} | \mathcal{A}_n) \leq (1 - 2^{-K})^2 \; P\text{-a.s.}$$

Proceeding this way, we finally get

$$P(\bigcap_{j=0}^{i-1} A_j^{\complement} | \mathcal{A}_n) \leq (1 - 2^{-K})^i \; P\text{-a.s.}$$

Now the assertion follows from Lemma C.3.

□

Lemma 16.8 *Let the assumptions of Theorem 16.1 be satisfied, and select an arbitrary starting point $x_0 \in D \setminus D^*$. Let $c \geq 2(1 + 1/b)/l_0$ (with b according to assumption* (16.1)*) be arbitrary. Denote by T_n, M_n, resp., the n-th time of entry, the sojourn time of the sequence $(Q_n : n \in \mathbb{N}_0)$ in the interval $[c, \infty)$ (for the definition, see Definition C.2). Then,*

(1)

$$M_n = 1 + \left[\frac{\ln Q_{T_n} - \ln c}{-\ln \gamma_2} \right] \; P\text{-a.s. on } \{T_n < \infty\}$$

(2)

$$P(Q_n \leq \alpha \text{ for infinite many } n \in \mathbb{N}_0) = 1 \text{ for all } \alpha > 0.$$

Proof

(1) (i) To prove the first assertion, by complete induction we prove next to the following assertion:

For all $k \in \mathbb{N}$ and any stopping time T with regard to $(\mathcal{A}_n : n \in \mathbb{N}_0)$ we have

$$P\left(\bigcap_{i=1}^{k} \{Q_{T+i} = \gamma_2 Q_{T+i-1} | \mathcal{A}_T\} \right) = 1 \; P\text{-a.s. for } \{\gamma_2^{k-1} Q_T \geq c\}.$$

$k = 1$:

From Lemma 16.1 it follows for all $n \in \mathbb{N}_0$ that

$$P(Q_{n+1} = \gamma_2 Q_n | \mathcal{A}_n) = 1 \; P\text{-a.s. on } \{Q_n \geq c\}.$$

Lemma C.3 yields then the basis of the induction.

$k \longrightarrow k+1$

Let now $A \in \mathcal{A}_T$ be arbitrary and define $E := \bigcap_{i=1}^{k}\{Q_{t+i} = \gamma_2 Q_{T+i-1}\}$.

Then,

$$P\left(\bigcap_{i=1}^{k+1}\{Q_{T+i} = \gamma_2 Q_{T+i-1}\} \cap A \cap \left\{\gamma_2^k Q_T \geq c\right\}\right)$$

$$= P(\{Q_{T+k+1} = \gamma_2 Q_{T+k}\} \cap E \cap A \cap \{\gamma_2^k Q_T \geq c\})$$

$$\geq P(\{Q_{T+k+1} = \gamma_2 Q_{T+k}\} \cap \{Q_{T+k} \geq c\} \cap E \cap A \cap \{\gamma_2^k Q_T \geq c\})$$

$$= \int P(\{Q_{T+k+1} = \gamma_2 Q_{T+k}|A_{T+k})dP$$

$$= \{Q_{T+k} \geq c\} \cap E \cap A \cap \{\gamma_2^k Q_T \geq c\})$$

$$= P(\{Q_{T+k} \geq c\} \cap E \cap A \cap \{\gamma_2^k Q_T \geq c\}) = (\text{ since } E \cap \{\gamma_2^k Q_T \geq c\}$$

$$\subset \{Q_{T+k} \geq c\})$$

$$P(E \cap A \cap \{\gamma_2^k Q_T \geq c\})$$

$$= \int P(E|\mathcal{A}_T)dP = P(A \cap \{\gamma_2^k Q_T \geq c\})$$

$$\times A \cap \{\gamma_2^k Q_T \geq c)$$

since $\{\gamma_2^k Q_T \geq c\} \subset \{\gamma_2^{k-1} Q_T \geq c\}$ and

$P(E|\mathcal{A}_T) = 1$ P-a.s. on $\{\gamma_2^{k-1} Q_T \geq c\}$.

With Lemma C.2 we obtain then

$$P(\bigcap_{i=1}^{k+1}\{Q_{T+i} = \gamma_2 Q_{T+i-1}\}|\mathcal{A}_T) = 1 \ P\text{-a.s. on}\{\gamma_2^k Q_T \geq c\}.$$

(ii) Let now T denote once again an arbitrary stopping time (with respect to $(\mathcal{A}_n : n \in \mathbb{N}_0)$), and let S be an $\mathcal{A}_T$-measurable random variable with values in $\mathbb{N}$.

Then,

$$P\left(\bigcap_{i=1}^{S}\{Q_{T+i} = \gamma_2 Q_{T+i-1}\}|\mathcal{A}_T\right) = P\text{-a.s. on } \left\{\gamma_2^{S-1} Q_T \geq c\right\}.$$

To prove this statement, let $A \in \mathcal{A}_T$ again be arbitrary.

Then,

$$
\begin{aligned}
&P\left(A\cap\bigcap_{i=1}^{S}\{Q_{T+i}=\gamma_2 Q_{T+i-1}\}\cap\left\{\gamma_2^{S-1}Q_T\geq c\right\}\right)\\
&=\sum_{k=1}^{\infty}P\left(\bigcap_{i=1}^{S}\{Q_{T+i}=\gamma_2 Q_{T+i-1}\}\cap A\cap\{\gamma_2^{S-1}Q_T\geq c\}\cap\{S=k\}\right)\\
&=\sum_{k=1}^{\infty}P\left(\bigcap_{i=1}^{k}\{Q_{T+i}=\gamma_2 Q_{T+i-1}\}\cap A\cap\{\gamma_2^{k-1}Q_T\geq c\}\cap\{S=k\}\right)\\
&=\sum_{k=1}^{\infty}\int_{A\cap\{\gamma_2^{k-1}Q_T\geq c\}\cap\{S=k\}} P\left(\bigcap_{i=1}^{k}\{Q_{T+i}=\gamma_2 Q_{T+i-1}\}|\mathcal{A}_T\right)dP\\
&=\sum_{k=1}^{\infty}P(A\cap\left\{\gamma_2^{k-1}Q_T\geq c\right\}\cap\{S=k\})\\
&=\sum_{k=1}^{\infty}P\Big(A\cap\left\{\gamma_2^{S-1}Q_T\geq c\right\}\cap\{S=k\}\Big)=P\left(A\cap\left\{\gamma_2^{S-1}Q_T\geq c\right\}\right).
\end{aligned}
$$

Now Lemma C.2 provides again the desired statement.

We will now apply this result to

$$
T=T_n \text{ and } S=1+\left[\frac{\ln Q_{T_n}-c}{-\ln\gamma_2}\right].
$$

From the definition of S we get the $\mathcal{A}_{T_n}$-measurability as well as

$$
S=\inf\{k\in\mathbb{N}:\gamma_2^k Q_{T_n}<c\}.
$$

Hence, $\{\gamma_2^{S-1}Q_{T_n}\geq c\}$ P-a.s. , and with the above statement we get

$$
P\left(\bigcap_{i=1}^{S}\{Q_{T_n+i}=\gamma_2 Q_{T_n+i-1}\}|\mathcal{A}_{T_n}\right)=1\ P\text{-a.s.}
$$

In particular we have

$$
P\left(\bigcap_{i=1}^{S}\left\{Q_{T_n+i}=\gamma_2 Q_{T_n+i-1}\right\}\right)=1.
$$

Because of

$$\bigcap_{i=1}^{S}\{Q_{T_n+i} = \gamma_2 Q_{T_n+i-1}\} = \bigcap_{i=1}^{S}\{Q_{T_n+i} = \gamma_2^i Q_{T_n}\}$$

$$\bigcap_{i=1}^{S}\{Q_{T_n+i} \geq c\}\{Q_{T_n+S} < c\}$$

we find

$$Q_{T_n}, \ldots, Q_{T_n+S-1} \geq c \text{ and } Q_{T_n+S} < c \ P\text{-a.s.}$$

This shows that S is the n-th time of sojourn in $[c, \infty)$.

Because of $S = M_n$, we get the first assertion.

(2) The definition of T_i and M_i yields

$$\{Q_n \geq c \text{ for finally all } n \in \mathbb{N}\} = \bigcup_{n\in\mathbb{N}} (\{T_n < \infty\} \cap \{M_n = \infty\}) .$$

From this and from Part (1) it follows that

$$\begin{aligned} &P\,(Q_n \geq c \text{ for finally any } n \in \mathbb{N}) \\ &\leq \sum_{n=1}^{\infty} P(\{T_n < \infty\} \cap \{M_n = \infty\}) \\ &= \sum_{n=1}^{\infty} \int_{\{T_n<\infty\}} P(M_n = \infty|\mathcal{A}_{T_n})dP = 0. \end{aligned}$$

Thus, the assertion is proven for $\alpha \geq c$.

Let now $0 < \alpha < c$ be arbitrary, but fixed.

Define

$$K := \inf\{n \in \mathbb{N} : \gamma_2^n c \leq \alpha\}.$$

Let $m \in \mathbb{N}$ be arbitrary, but fixed, and put $A_\alpha := \bigcap_{n\geq m} \{Q_n > \alpha\}$.

Now it must be shown that $P(A_\alpha) = 0$,
for then it follows that $P(\bigcup_{n\geq m} \{Q_n \leq \alpha\}) = 1$.

Since m was arbitrary, we have

$$P\left(\bigcap_{m\in\mathbb{N}}\bigcup_{n\geq m}\{Q_n \leq \alpha\}\right) = 1,$$

hence, the assertion.

Let

$$\begin{array}{llll} \tau_1 & := \inf\{k \geq m : Q_k < c\} & if & \{k \geq m : Q_k < c\} \neq 0 \\ & \infty & else & \\ \tau_2 & := \inf\{k \geq \tau_1 + K : Q_k < c\} & if & \{k \geq \tau_1 + K : Q_k < c\} \neq 0 \\ & \infty & else & \\ \vdots & \vdots & \vdots & \\ \tau_{n+1} & := \inf\{k \geq \tau_n + K : Q_k < c\} & if & \{k \geq \tau_n + K : Q_k < c\} \neq 0 \\ & \infty & else. & \end{array}$$

It is easy to see that for each $n \in \mathbb{N} \tau_n$ is a stopping time with respect to $(\mathcal{A}_n : n \in \mathbb{N})$ (cf. also Remark C.2).

Define

$$\Omega := \bigcap_{n \in \mathbb{N}} \{\tau_n < \infty\}.$$

Because of

$$P(\tilde{\Omega}) = P(Q_n < c \text{ for infinite many } n \in \mathbb{N})$$
$$1 - P(Q_n \geq c \text{ finally for all } n \in \mathbb{N}) = 1,$$

we have to show that $P(A_\alpha \cap \tilde{\Omega}) = 0$.

Let

$$A_n := \{Q_{T_n+1} = \gamma_2 Q_{T_n}\} \cap \cdots \cap \{Q_{T_n+K} = \gamma_2 Q_{T_n+K-1}\}.$$

Because of $Q_{T_n} < c$, for $\omega \in A_n \cap \tilde{\Omega}$ we have

$$Q_{T_n(\omega)+K}(\omega) = \gamma_2^k Q_{T_n(\omega)}(\omega) < \gamma_2^K c \leq \alpha.$$

This yields

$$\bigcap_{n \in \mathbb{N}} \{A_n \cap \tilde{\Omega}\} \supset \bigcap_{n \geq m} \{Q_n > \alpha\} \cap \tilde{\Omega} = A_\alpha \cap \tilde{\Omega}.$$

From Lemma 16.1 we get then

$$P(Q_{n+1} = \gamma_2 Q_n | \mathcal{A}_n) \geq \frac{1}{2} \; P\text{-a.s.}$$

Putting $p = \frac{1}{2}$, then, using the same iteration method as in the proof of Lemma 16.7, part 2, we have

$$P(A_n|\mathcal{A}_{\tau_n}) \leq p_k\ P\text{-a.s.}$$

Taking complements, we get

$$P(A_n^{\complement}|\mathcal{A}_{\tau_n}) \geq 1 - p_k\ P\text{-a.s.}$$

Hence,

$$\begin{aligned} P(A_1^{\complement} \cap A_2^{\complement}|\mathcal{A}_{\tau_1}) &= E(1_{A_1^{\complement}} * P(A_2^{\complement}|\mathcal{A}_{\tau_2})|\mathcal{A}_{\tau_1}) \\ &\leq (1 - p^K)P(A_1^{\complement}|\mathcal{A}_{\tau_1}) \leq (1 - p^K)^2\ P\text{-a.s.} \end{aligned}$$

With i-fold iteration we find

$$P(\bigcap_{n=1}^{i} A_n^{\complement}|\mathcal{A}_{\tau_1}) \leq (1 - p^K)^i\ P\text{-a.s.}$$

and therefore

$$P(\bigcap_{n=1}^{i} A_n^{\complement}) \leq (1 - p^K)^i\ P\text{-a.s.}$$

Consequently,

$$P\left(A_\alpha \cap \tilde{\Omega}\right) \leq P\left(\bigcap_{n\in\mathbb{N}} A_n^{\complement} \cap \tilde{\Omega}\right) = P\left(\bigcap_{n\in\mathbb{N}} A_n^{\complement}\right) = 0.$$

This yields the assertion.

□

Lemma 16.9 *Suppose that the assumptions of Theorem 16.1 hold. Then for all starting points $x_0 \in D \setminus D^*$ there exists an $a = a(x_0) > 0$ with*

$$P\left(Q_n \in \left[\frac{a}{l_0}, \frac{l}{2L_0}\right] \text{ for infinite many } n \in \mathbb{N}\right) = 1.$$

Proof Let $a(x_0) := \min\{l_0a_0(x_0), l_0a_1(x_0), \frac{l_0}{3L_0}\}$ with a_0, a_1, resp., according to Lemma 16.5, 16.6.

From Lemma 16.5, Lemma 16.8, resp., it follows that

$$P(Q_n < \frac{a}{l_0} \text{ finally for all } n \in \mathbb{N}) = 0,$$

$$P(Q_n < \frac{a}{2L_0} \text{ finally for all } n \in \mathbb{N}) = 0, \text{ respectively.}$$

Furthermore, from Lemma 16.6 for all $n \in \mathbb{N}$ we have

$$P\left(Q_n < \frac{a}{l_0}, Q_{n+1} > \frac{l}{2L_0}\right) = P\left(Q_n > \frac{l}{2L_0}, Q_{n+1} < \frac{a}{l_0}\right) = 0.$$

In addition, it is $\frac{a}{l_0} < \frac{1}{2L_0}$.

Put

$$A := \left(\bigcap_{n\in\mathbb{N}_0}\left\{Q_n \geq \frac{a}{l_0}\right\} \cup \left\{Q_{n+1} \leq \frac{1}{2L_0}\right\}\right)$$
$$\cap \left(\bigcap_{n\in\mathbb{N}_0}\left\{Q_n \leq \frac{1}{2L_0}\right\} \cup \left\{Q_{n+1} \geq \frac{a}{l_0}\right\}\right).$$

Because of

$$A = \left[\left(\left(\bigcup_{n\in\mathbb{N}_0}\left\{Q_n \geq \frac{a}{l_0}\right\} \cup \left\{Q_{n+1} \leq \frac{1}{2L_0}\right\}\right)\right.\right.$$
$$\left.\left.\cup \left(\bigcup_{n\in\mathbb{N}_0}\left\{Q_n \leq \frac{1}{2L_0}\right\} \cup \left\{Q_{n+1} \geq \frac{a}{l_0}\right\}\right)\right)\right],$$

we get $P(A) = 1$.

Based on

$$P(Q_n \in \left[\frac{a}{l_0}, \frac{1}{2L_0}\right] \text{ for infinite many } n \in \mathbb{N}) =$$

$$1 - P\left(A \cap \left\{Q_n \in (0, \frac{a}{l_0}) \cup (\frac{1}{2L_0}, \infty) \text{ finally for all } n \in \mathbb{N}\right\}\right),$$

the assertion follows, provided that

$$A \cap \left\{Q_n \in \left(0, \frac{a}{l_0}\right) \cup \left(\frac{1}{2L_0}, \infty\right) \text{ finally for all } n \in \mathbb{N}\right\} \subset$$

$$\left\{Q < \frac{a}{l_0} \text{ finally for all } n \in \mathbb{N}\right\} \cup \left\{Q_n > \frac{1}{2L_0} \text{ finally for all } n \in \mathbb{N}\right\}.$$

The proof of this inclusion is elementary and therefore left to the reader. □

Lemma 16.10 *Let the assumptions of Theorem 16.1 be satisfied.*

For $x_0 \in D \setminus D^*$ let $b = b(x_0)$ be chosen according to assumption (16.1).

In accordance with Lemma B.11, let $\phi = \phi(x_0)$ be chosen such that the inclusion

$$x + C(\phi, x - x^*, \frac{1}{2}|x - x^*|) \subset \{y \in D : f(y) \leq f(x)\}$$

is satisfied for all $x \in D_{f(x_0)} \setminus D^*$.

Let $a = a(x_0)$ be chosen according to Lemma 16.9.

Put $B = B(x_0) = [\frac{a}{l_0}, \frac{1}{2L_0}]$ and $C = C(x_0) =]0, \phi]$

Define $Z_n := \phi(X_{n+1} - X_n, x^* - X_n)$ for $n \in \mathbb{N}$ (for the Definition of $\phi(x, y)$, cf. Definition B.1).

Then for all starting points $x_0 \in D \setminus D^*$ there exists a $\beta = \beta(x_0) > 0$ such that

$$\lim_{n\to\infty} \inf \frac{\sum_{i=1}^{n} 1_{\{Q_i \in B, Z_i \in C\}}}{\sum_{i=1}^{n} 1_{\{Q_i \in B\}}} \geq \beta \ P\text{-a.s.}$$

Proof

(i) Let denote T_n, M_n, resp., the n-th time of entry, the n-th sojourn time of the sequence $(Q_n : n \in \mathbb{N})$ in B (cf. Definition C.2). According to Lemma 16.8, the sequence $(Q_n \in \mathbb{N})$ lies ∞-frequently in B and in $B^{\complement}$. Therefore, $T_n < \infty$, $M_n < \infty$ P-a.s. for all $n \in \mathbb{N}$.

Putting $K := 1 + [\frac{\ln l_0 - \ln 2L_0 a}{-\ln \gamma_2}]$, from Lemma 16.7 we get

$$\begin{aligned} P(M_n > Ki|\mathcal{A}_{T_n}) &= P(\{Q_{T_n} \in B, \ldots, Q_{T_n+Ki} \in B|\mathcal{A}_{T_n}\}) \\ &\leq (1 - 2^{-K})^i \ P\text{-a.s. for all } i \in \mathbb{N}_0. \end{aligned}$$

This implies (cf. [1, p. 47])

$$\begin{aligned} E({M_n}^2|\mathcal{A}_{T_n}) &= \sum_{k=0}^{\infty} P({M_n}^2 > k|\mathcal{A}_{T_n}) \\ &= \sum_{k=1}^{\infty} P(M_n > \sqrt{k-1}|\mathcal{A}_{T_n}) = \sum_{j=0}^{\infty} \sum_{k=j^2+1}^{(j+1)^2} P(M_n > \sqrt{k-1}|\mathcal{A}_{T_n}) \\ &\leq \sum_{j=0}^{\infty} ((j+1)_2 - j_2) P(M_n > j|\mathcal{A}_{T_n}) \leq 2 \sum_{j=1}^{\infty} j P(M_n > j-1|\mathcal{A}_{T_n}) \\ &= 2 \sum_{j=0}^{\infty} \sum_{k=jK+1}^{(j+1)K} k \cdot P(M_n > k-1|\mathcal{A}_{T_n}) \leq 2 \sum_{j=0}^{\infty} ((j+1)K)K \end{aligned}$$

$$\cdot P(M_n > jK|\mathcal{A}_{T_n})$$

$$\leq 2 + 2K^2 \sum_{j=0}^{\infty} jP(M_n > jK|\mathcal{A}_{T_n}) \leq 2 + 2K_2 2 \sum_{j=0}^{\infty} j(1-2^{-K})^j$$

$$= 2 + 2K^2 \frac{1-2^{-K}}{(1-(1-2^{-K}))^2} = 2 + 2K^2(1-2^{-K})2^{2K} =: q.$$

For all $n \in \mathbb{N}$ we therefore have $E({M_n}^2|\mathcal{A}_{T_n}) \leq q$ P-a.s.

According to Remark C.3, we have $\sigma(M_1, \ldots, M_n) \subset \mathcal{A}_{T_{n+1}}$.

By iteration of the conditional expectations we get

$$E({M_{n+1}}^2|M_n, \ldots, M_1) = E(E({M_{n+1}}^2|\mathcal{A}_{T_n})|M_n, \ldots, M_1) \leq q \ P\text{-a.s.}$$

In particular we have

$$E{M_n}^2 \leq q \text{ and } E(M_{n+1}|M_n, \ldots, M_1) \leq q \ P\text{-a.s.}$$

With Lemma C.1 the above yields

$$\limsup_{n\to\infty} \frac{1}{n} \sum_{i=1}^{\infty} M_i \leq q \ P\text{-a.s.}$$

(ii) Select $x_1, \ldots, x_n \in D_{f(x_0)} \setminus D^*$ such that $\frac{l_{n+1}(x_0,\ldots,x_n)}{|x_n - x^*|} \in B$.
Then,

$$p_{n+1}(x_0, \ldots, x_n, (x_n + C(\phi, x^* - x_n)) \cap \{f < f(x_n)\})$$

$$\geq p_{n+1}(x_0, \ldots, x_n, (x_n + C(\phi, x^* - x_n, \frac{1}{2}|x^* - x_n|)))$$

$$= \gamma(S \cap C(\phi, x^* - x_n)) \cdot \lambda_{n+1}(x_0, \ldots, x_n, [0, \frac{1}{2}|x^* - x_n|]).$$

Because of $\frac{l_{n+1}(x_0,\ldots,x_n)}{|x_n - x^*|} \leq \frac{1}{2L_0}$, we have

$$\lambda_{n+1}(x_0, \ldots, x_n, [0, \frac{1}{2}|x^* - x_n|])$$

$$= \lambda_0([0, \frac{1}{2}\frac{|x^* - x_n|}{l_{n+1}(x_0, \ldots, x_n)}]) \geq \lambda_0([0, L_0]) = 1.$$

With $p := \gamma(S \cap C(\phi, x^* - x_n))$ (this term depends only on ϕ and not on $x^* - x_n$, cf. Lemma A.3) it follows now that

$$\begin{aligned} &P(Z_n \in C | X_n = x_n, \dots X_0 = x_0) \\ &= q_{n+1}(x_0, \dots, x_n, x_n + C(\phi, x^* - x_n)) \\ &\geq p_{n+1}(x_0, \dots, x_n, (x_n + C(\phi, x^* - x_n)) \cap \{f < f(x_n)\}) \geq \quad p. \end{aligned}$$

Because of

$$P\left(\bigcup_{n \in \mathbb{N}_0} \{X_n \in D^*\}\right) = P\left(\bigcup_{n \in \mathbb{N}_0} \{X_n = x^*\}\right) = 0$$

(cf. Lemma 16.1), (1) we have therefore for all $n \in \mathbb{N}$

$$P(Z_n \in C | \mathcal{A}_n) \geq p \; P\text{-a.s. on } \{Q_n \in B\}.$$

Due to $P(Q_{T_n} \in B) = 1$, Lemma C.4 yields

$$P(Z_{T_n} \in C | \mathcal{A}_{T_n}) \geq p \; P\text{-a.s.}$$

From the definition of Z_n we get $\sigma(Z_n) \subset \mathcal{A}_{n+1}$ and therefore (cf. [1, p. 215] ff.)

$$\sigma(Z_{T_i}) \subset \mathcal{A}_{T_i+1} \subset \mathcal{A}_{T_i+1}.$$

Therefore we also have

$$\sigma(1_{\{Z_{T_1} \in C\}}, \dots, 1_{\{Z_{T_1} \in C\}} \subset \mathcal{A}_{T_i+1}.$$

By iteration of the conditional expectations we get again that

$$P(Z_{T_n} \in C | 1_{\{Z_{T_{n-1}} \in C\}}, \dots, 1_{\{Z_{T_1} \in C\}}) \geq \; P\text{-a.s.},$$

and Lemma C.1 yields

$$\liminf_{n \to \infty} \frac{1}{n} \sum_{i=1}^{n} 1_{\{Z_{T_i} \in C\}} \geq \; P\text{-a.s.}$$

(iii) For $K \in \mathbb{N}$ and $\omega \in \Omega$ let $n_k(\omega)$ denote the integer such that

$$T_{n_k(\omega)}(\omega) \leq k < kT_{n_k(\omega)+1}(\omega)$$

(i.e. $n_k = \sup\{n \in \mathbb{N} : T_n \leq k)$. Because of $T_n \leq n$, we have $T_n \to \infty P$-a.s. and therefore

$$T_{n_k} \overset{k\to\infty}{\longrightarrow} \infty \ P\text{-a.s.}$$

Define

$$R_n := \frac{\sum_{i=1}^{T_n} 1_{\{Q_i \in B, Z_i \in C\}}}{\sum_{i=1}^{T_{n+1}-1} 1_{\{Q_i \in B\}}} (n \in \mathbb{N}).$$

Because of

$$\sum_{i=1}^{T_{n+1}-1} 1_{\{Q_i \in B\}} = \sum_{i=1}^{n} M_i$$

and

$$\sum_{i=1}^{T_n} 1_{\{Q_i \in B, Z_i \in C\}} \geq \sum_{i=1}^{n} 1_{\{Q_{T_i} \in B, Z_{T_i} \in C\}} = \sum_{i=1}^{n} 1_{\{Z_{T_i} \in C\}},$$

we have

$$R_n \geq \frac{\frac{1}{n}\sum_{i=1}^{n} 1_{\{Z_{T_i} \in C\}}}{\frac{1}{n}\sum_{i=1}^{n} M_i}.$$

From part (i) and part (ii) and from Remark B.3 it follows that

$$\lim_{n\to\infty} R_n \geq \frac{\liminf_{n\to\infty} \frac{1}{n}\sum_{i=1}^{n} 1_{\{Z_{T_i} \in C\}}}{\liminf_{n\to\infty} \frac{1}{n}\sum_{i=1}^{n} M_i} \geq \frac{p}{q} \ P\text{-a.s.}$$

Because of $T_{n_k} \leq k$ and $T_{n_k+1} - 1 \geq k$ we find

$$\frac{\sum_{i=1}^{k} 1_{\{Q_i \in B, Z_i \in C\}}}{\sum_{i=1}^{k} 1_{\{Q_i \in B\}}} \geq R_{n_k}.$$

Now the assertion follows from

$$\liminf_{k\to\infty} R_{n_k} \geq \liminf_{k\to\infty} R_n.$$

□

Lemma 16.11 *Let the assumptions of Theorem 16.1 be satisfied.*

For $c > 0$, let denote T_n^c, M_n^c, resp., the n-th time of entry, the sojourn time of the sequence $(Q_n : n \in \mathbb{N}_0)$ in $[c, \infty)$.

Then for all starting points $x_0 \in D$ there exists a constant $c_0 = c_0(x_0) > 0$ such that for all $c \geq c_0$ and each $n \in \mathbb{N}$ we have

$$f(X_{T_n}c + M_n c) - f^* \leq f(X_{T_n}c) - f^* \leq \gamma_2^{M_n^c}(f(X_{T_n^c - 1}) - f^*)$$

$$P\text{-a.s. on } \{T_n^c < \infty\}.$$

Proof

(1) Select an arbitrary point $x_0 \in D \setminus D^*$.
Define $\tilde{c} := \frac{2}{l_0}(1 + \frac{1}{b})$ (with $b = b(x_0)$ according to assumption (16.1)).
Due to Lemma 16.1 we have

$$P(Q_{n+1} = \gamma_2 Q_n | \mathcal{A}_n) = 1 \ P\text{-a.s. on } \{Q_n \geq \tilde{c}\}.$$

Then, from Lemma 16.8 it follows for all $c = \tilde{c}$ that

$$M_n^c = 1 + \left[\frac{\ln QT_n^c - \ln c}{-\ln \gamma_2} \right] \ P\text{-a.s. on } \{T_n^c < \infty\}.$$

for all $n \in \mathbb{N}$ we have

$$\begin{aligned}
P(Q_{n-1} < \tilde{c}) &= 1 - P(Q_{n-1} \geq \tilde{c}) \\
&= 1 - \int\limits_{\{Q_{n-1} \geq \tilde{c}\}} P(Q_n = \gamma_2 Q_{n-1} | \mathcal{A}_{n-1}) dP \\
&= 1 - P(\{Q_{n-1} \geq \tilde{c}\} \cap \{Q_n = \gamma_2 Q_{n-1}\}) \\
&\geq 1 - P(\{Q_n = \gamma_2 Q_{n-1}) \geq P(Q_n \geq Q_{n-1})
\end{aligned}$$

Lemma C.4 shows that this inequality is also valid for stopping times. Therefore,

$$P(Q_{T_n}^{c-1} < \tilde{c} \geq P(Q_{T_n^c} \geq Q_{T_n^c - 1}).$$

From the definition of T_n^c we get

$$Q_{T_n^c} \geq c \text{ and } Q_{T_n^{c-1}} < c,$$

hence, in particular $Q_{T_n^c} \geq Q_{T_n^{c-1}}$.

Thus, for all $c > 0$ we have

$$P(Q_{T_n^{c-1}} > \tilde{c}) = 1.$$

For the same reason we have

$$Q_{T_n^c} = \frac{\gamma_1 1_{T_n^c}(X_0, \ldots, X_{T_n^{c-1}})}{|X_{T_n^c} - x^*|}$$

and therefore

$$Q_{T_n^c} = \gamma_1 Q_{T_n^{c-1}} \frac{|X_{T_n^{c-1}} - x^*|}{|X_{T_n^c} - x^*|}.$$

This yields

$$\frac{|X_{T_n^c} - x^*|}{|X_{T_n^{c-1}} - x^*|} = \frac{\gamma_1 Q_{T_n^{c-1}}}{Q_{T_n^c}} < \frac{1}{b} \frac{\gamma_1 \tilde{c}}{Q_{T_n^c}} \quad P\text{-a.s.}$$

With Remark 16.2 we obtain

$$\frac{f(X_{T_n^c}) - f^*}{f(X_{T_n^{c-1}}) - f^*} \leq \frac{1}{b} \frac{\gamma_1 \tilde{c}}{Q_{T_n^c}} \quad P\text{-a.s.}$$

Choosing now $c_0 \geq \frac{\gamma_1}{\gamma_2} \frac{1}{b} \tilde{c}$, then for arbitrary $c \geq c_0$ we get the inequality

$$1 + \frac{\ln Q_{T_n^c} - \ln c}{-\ln \gamma_2} \leq \frac{\ln Q_{T_n^c} - \ln \frac{\gamma_1 \tilde{c}}{b}}{-\ln \gamma_2}.$$

This yields

$$M_n^c \leq \frac{\ln Q_{T_n^c} - \ln \frac{\gamma_1 \tilde{c}}{b}}{-\ln \gamma_2}.$$

However, this inequality is equivalent to

$$\frac{\gamma_1 \tilde{c}}{b Q_{T_n^c}} \leq \gamma_2^{M_n^c}.$$

From this we get the assertion. □

Lemma 16.12 *Let the assumptions of Theorem 16.1 be satisfied. Select a point $x_0 \in D \setminus D^*$, and let $B = B(x_0)$, $C = C(x_0)$ and Z_n be defined as in Lemma 16.10. Suppose that $c \geq c(x_0)$ and $c > \frac{1}{2L_0}$ with $c_0(x_0)$ according to Lemma 16.11.*

Define $A = A(x_0) := (0, \frac{a}{2L_0}) \cup (\frac{1}{2L_0}, c)$ (with $a = a(x_0)$ according to Lemma 16.9). Then we have for all starting point $x_0 \in D$ a constant $\beta = \beta(x_0) > 0$ with

$$P\left(\liminf_{n\to\infty} \frac{1}{n} \sum_{i=1}^{n} \left(1_{\{Q_i \in B, Z_i \in C\}} + 1_{\{Q_i \geq c\}}\right) \geq \beta\right) = 1.$$

Proof

(1) Lemma 16.10 yields the existence of an $\alpha_2 = \alpha_2(x_0) > 0$ with

$$P\left(\liminf_{n\to\infty} \frac{\sum_{i=1}^{n} 1_{\{Q_i \in B, Z_i \in C\}}}{\sum_{i=1}^{n} 1_{\{Q_i \in B\}}} \geq \alpha_2\right) = 1.$$

The assertion follows then from Lemma C.6, provided that an $\alpha_1 = \alpha_1(x_0) > 0$ exists such that

$$P\left(\liminf_{n\to\infty} \frac{\sum_{i=1}^{n} 1_{\{Q_i \in B\}}}{\sum_{i=1}^{n} 1_{\{Q_i < c\}}} \geq \alpha_1\right) = 1.$$

In the following the existence of such an $\alpha_1 > 0$ will be shown.

(2) Let T_n be the n-th time of entry of the sequence $(Q_n : n \in \mathbb{N})$ into A.

Let S_n denote the n-th time of entry of the sequence $(Q_n : n \in \mathbb{N})$ into $A^{\complement}$.

Let M_n be the n-th time of sojourn of the sequence $(Q_n : n \in \mathbb{N})$ in A. According to Lemma 16.8 and Lemma 16.9 we have $Q_n \in A$ and $Q_n \in A^{\complement}$ for infinite many $n \in \mathbb{N}$.

Hence,

$$S_n > \infty \text{ and } T_n > \infty P\text{-a.s. for all } n \in \mathbb{N}.$$

For $k \in \mathbb{N}$ and $\omega \in \Omega$ let denote $n_k(\omega)$ the integer such that

$$T_{n_k(\omega)}(\omega) \leq k < T_{n_k(\omega)+1}(\omega)$$

holds.

Then,

$$\frac{\sum_{i=1}^{k} 1_{\{Q_i \in B\}}}{\sum_{i=1}^{k} 1_{\{Q_i < c\}}} = \frac{\sum_{i=1}^{k} 1_{\{Q_i \in B\}} - \sum_{i=1}^{k} 1_{\{Q_i < c, Q_i \in B\}}}{\sum_{i=1}^{k} 1_{\{Q_i < c\}}}$$

$$1 - \frac{\sum_{i=1}^{k} 1_{\{Q_i \in A\}}}{\sum_{i=1}^{k} 1_{\{Q_i < c\}}} \leq 1 - \frac{\sum_{i=1}^{T_{n_k+1}-1} 1_{\{Q_i \in A\}}}{\sum_{i=1}^{T_{n_k}} 1_{\{Q_i < c\}}} \text{ (since } k \leq T_{n_k+1} - 1) =$$

$$1 - \frac{\sum_{i=1}^{n_k} M_i}{\sum_{i=1}^{T_{n_k}} 1_{\{Q_i \in A\}} + \sum_{i=1}^{T_{n_k}} 1_{\{Q_i \in B\}}} \text{ (note the Definition of } T_n M_n) \geq$$

$$1 - \frac{\sum_{i=1}^{n_k} M_i}{\sum_{i=1}^{n_k} M_i + \sum_{i=1}^{S_{n_k-1}} 1_{\{Q_i \in B\}}} \text{(since } T_i > S_{j-1}) \geq$$

$$1 - \frac{\sum_{i=1}^{n_k} M_i + M_{n_k}}{\sum_{i=1}^{n_k-1} M_i + \sum_{i=1}^{n_k-1} 1_{\{Q_{S_i} \in B\}}} = \frac{\sum_{i=1}^{n_k-1} 1_{\{Q_{S_i} \in B\}} + M_{n_k}}{\sum_{i=1}^{n_k-1} M_i + \sum_{i=1}^{n_k-1} 1_{\{Q_{S_i} \in B\}}}$$

$$\frac{1 - \frac{M_{n_k}}{\sum_{i=1}^{n_k-1} 1_{\{Q_{S_i} \in B\}}}}{1 - \frac{\sum_{i=1}^{n_k-1} M_i}{\sum_{i=1}^{n_k-1} 1_{\{Q_{S_i} \in B\}}}} = \frac{1 - \frac{\frac{1}{n_k-1} M_{n_k}}{\frac{1}{n_k-1}\sum_{i=1}^{n_k-1} 1_{\{Q_{S_i} \in B\}}}}{1 - \frac{\frac{1}{n_k-1}\sum_{i=1}^{n_k-1} M_i}{\frac{1}{n_k-1}\sum_{i=1}^{n_k-1} 1_{\{Q_{S_i} \in B\}}}} \leq$$

$$\leq \frac{1 + \frac{1}{n_k - 1} M_{n_k}}{1 + \frac{\frac{1}{n_k-1}\sum_{i=1}^{n_k-1} M_i}{\frac{1}{n_k-1}\sum_{i=1}^{n_k-1} 1_{\{Q_{S_i}\in B\}}}}.$$

Therefore, according to Remark B.3 we get

$$\liminf_{k\to\infty} \frac{\sum_{i=1}^{k} 1_{\{Q_i \in B\}}}{\sum_{i=1}^{k} 1_{\{Q_i < c\}}} \leq$$

$$\leq \frac{1 + \liminf_{k\to\infty} \frac{1}{n_k-1} M_{n_k}}{1 + \frac{\limsup_{k\to\infty} \frac{1}{n_k-1}\sum_{i=1}^{n_k-1} M_i}{\limsup_{k\to\infty} \frac{1}{n_k-1}\sum_{i=1}^{n_k-1} 1_{\{Q_{S_i}\in B\}}}}.$$

Hence, the assertion follows, provided that:

(a) $\lim_{n\to\infty} \frac{1}{n} M_n = 0$ P-a.s.;

(b) There exists $d_1 < \infty$ with $\limsup_{n\to\infty} \frac{1}{n} \sum_{i=1}^{n} M_i \leq d_1$ P-a.s. ;

(c) There exists $d_2 > 0$ such that $\limsup_{n\to\infty} \frac{1}{n} \sum_{i=1}^{n} 1_{\{Q_{S_i}\in B\}} \geq d_2$ P-a.s.

Indeed, in this case we have

$$\liminf_{n\to\infty} \frac{\sum_{i=1}^{k} 1_{\{Q_i \in B\}}}{\sum_{i=1}^{k} 1_{\{Q_i < c\}}} \geq \frac{1}{1 + \frac{d_1}{d_2}} \quad P\text{-a.s.}$$

As already mentioned, the assertion follows now from Lemma C.6.

(3) Let us now prove (i) and (ii): For $s \in \mathbb{N}$ we have

$$P(M_n > s - 1 | \mathcal{A}_{T_n}) = P \bigcap_{i=T_n}^{T_n s-1} \{Q_i \in A\} | \mathcal{A}_{T_n}) =$$

$$P(\{Q_{T_n} \in (0, \frac{a}{l_0})\} \cap \bigcap_{i=T_n}^{T_n+s-1} \{Q_i \in A\} | \mathcal{A}_{T_n}) +$$

$$P(\{Q_{T_n} \in (\frac{1}{2L_0}, c)\} \cap \bigcap_{i=T_n}^{T_n+s-1} \{Q_i \in A\} | \mathcal{A}_{T_n}) =$$

$$P\left(\bigcap_{i=T_n}^{T_n+s-1} \left\{Q_i \in (0, \frac{a}{l_0})\right\} | \mathcal{A}_{T_n}\right) + P\left(\bigcap_{i=T_n}^{T_n+s-1} \left\{Q_i \in (\frac{1}{2L_0}, c)\right\} | \mathcal{A}_{T_n}\right).$$

The last equation follows by means of induction with respect to s. Obviously, the assertion holds for $s = 1$. The induction step is performed then by using the equality:

$$P\left(Q_n < \frac{a}{l_0}, Q_{n+1} > \frac{1}{2L_0} | \mathcal{A}_k\right) = P\left(Q_n > \frac{1}{2L_0}, Q_{n+1} < \frac{a}{l_0} | \mathcal{A}_k\right) = 0 \ P\text{-a.s.}$$

(cf. Lemma 16.6 and the Definition of a in the proof of Lemma 16.9).

Because of the $\mathcal{A}_{T_n}$—measurability of Q_{T_n}, with $\delta := \min\{\frac{l}{x_0 - x^*}, \frac{a}{l_0}\}$ it follows that

$$P(Q_{T_n - 1} \geq \delta | \mathcal{A}_{T_n}) = 1_{Q\{T_n \geq \delta\}} = 1 \ P\text{-a.s.}$$

This yields the following estimate of the first summand:

$$P(\bigcap_{i=T_n}^{T_n+s-1} \{Q_i < \frac{a}{l_0})\} | \mathcal{A}_{T_n})$$

$$\leq \sum_{j=0}^{[\rho_1 s + \rho_2]} \binom{s-1}{j} (p_{a/l_0})^j (1 - p_{a/l_0})^{s-1-j} \ P\text{-a.s.}$$

defining $\rho_1 = \frac{\ln \gamma_2}{\ln \gamma_2 - \ln \gamma_1}$ and $\rho_2 = \frac{\ln(b l_0) - \ln a}{\ln \gamma_2 - \ln \gamma_1}$
as well as with $p_{\frac{a}{l_0}}$ defined according to Lemma 16.3.

For the proof of the above estimate, let $E \in \mathcal{A}_{T_n}$ be arbitrary. Then,

$$P\left(E \cap \bigcap_{i=T_n}^{T_n+s-1} \{Q_i < \frac{a}{l_0}\}\right) = P\left(E \cap \{Q_{T_n - 1} \geq \rho\} \cap \bigcap_{i=T_n}^{T_n+s-1} \{Q_i < \frac{a}{l_0}\}\right)$$

$$\leq P(E \cap \{\sum_{i=T_n}^{T_n+s-1} 1_{\{f(X_{i-1}) > f(X_i)\}} < \rho_1 s + \rho_2\} \cap \bigcap_{i=T_n}^{T_n+s-1} \{Q_i < \frac{a}{l_0}\})$$

$$= \int_{E\cap\{Q_{T_n}<\frac{a}{l_0}\}} P(\{\sum_{i=T_n+1}^{T_n+s-1} 1_{\{f(X_{i-1})>f(X_i)\}}$$

$$< \rho_1 s + \rho_2\} \cap \bigcap_{i=T_n+1}^{T_n+s-1} \{Q_i < \frac{a}{l_0}\} | \mathcal{A}_{T_n}) dP$$

$$\leq P(E) \sum_{j=0}^{[\rho_1 s+\rho_2]} \binom{s-1}{j} (p_{a/l_0})^j (1-p_{a/l_0})^{s-1-j}.$$

The first inequality follows from Lemma 16.2. The last inequality follows from Lemma C.5, applied to $T = T_n, n = s-1$ and $K = [\rho_1 s + \rho_2]$.

For the second term, from Lemma 16.7 with $b_1 = \frac{1}{2L_0}$, $b_2 = c$ and $K = \left[\frac{\ln 2cL_0}{-\ln \gamma_2}\right] + 1$ we get

$$P\left(\bigcap_{i=T_n}^{T_n+Ks} \{Q_i \in (\frac{1}{2L_0}, 0)\} | \mathcal{A}_{T_n}\right) \leq (1-2^{-K})^s \ P\text{-a.s.}$$

Hence,

$$P(M_n^2 | \mathcal{A}_{T_n}) = \sum_{s=1}^{\infty} P(M_n^2 > s-1 | \mathcal{A}_{T_n}) =$$

$$= \sum_{s=1}^{\infty} P(M_n > \sqrt{s-1} | \mathcal{A}_{T_n}) = \sum_{m=0}^{\infty} \sum_{s=m^2+1}^{(m+1)^2} P(M_n > \sqrt{s-1} | \mathcal{A}_{T_n})$$

$$\leq \sum_{m=0}^{\infty} ((m+1)^2 - m^2) P(M_n > m | \mathcal{A}_{T_n}) \leq 2 \sum_{s=1}^{\infty} s P(M_n > s-1 | \mathcal{A}_{T_n})$$

$$\leq 2 \sum_{s=1}^{\infty} (s \sum_{j=0}^{[\rho_1 s+\rho_2]} \binom{s-1}{j} (p_{a/l_0})^j (1-p_{a/l_0})^{s-1-j}$$

$$+ 2 \sum_{s=1}^{\infty} s P(\bigcap_{i=T_n}^{T_n+s-1} \{Q_i \in (\frac{1}{2L_0}, c)\} | \mathcal{A}_{T_n}).$$

Because of $\frac{a}{l_0}$ (cf. the proof of Lemma 16.9) and the fact that p_ϵ is strictly decreasing in ϵ, which can be seen immediately from the definition of p_ϵ, it follows that (cf. Lemma 16.4 and the proof of Lemma 16.5)

$$p_{a/l_0} \geq p_{a_0} > \rho_1.$$

The first sum converges according to Lemma B.15.

For the second sum we have the following estimate:

$$\sum_{s=1}^{\infty} s P(\bigcap_{i=T_n}^{T_n+s-1} \{Q_i(\frac{1}{2L_0}, c)\}|\mathcal{A}_{T_n}) =$$

$$= \sum_{s=0}^{\infty} \sum_{k=Ks+1}^{K(s+1)} k P(\bigcap_{i=T_n}^{T_n+k-1} \{Q_i(\frac{1}{2L_0}, c)\}|\mathcal{A}_{T_n})$$

$$\le \sum_{s=0}^{\infty} K(s+1) K P(\bigcap_{i=T_n}^{T_n+Ks} \{Q_i(\frac{1}{2L_0}, c)\}|\mathcal{A}_{T_n})$$

$$\le K^2 \cdot \sum_{s=0}^{\infty} (s+1)(1-2^{-K})^s \ P\text{-a.s.}$$

Thus, also the second sum converges.

Consequently, there exists a $d_1 < \infty$ with $E(M_n^2|\mathcal{A}_{T_n}) \le d_1$ P-a.s. for all $n \in \mathbb{N}$.

Because of $\sigma(M_1, \ldots, M_{n-1}) \subset \mathcal{A}_{T_n}$ (cf. Remark C.3) we have $E(M_n^2|M_{n-1}, \ldots, M_1) \le d_1$ P-a.s. , in particular therefore $E(M_n^2) \le d_1$ and $E(M_n|M_{n-1}, \ldots, M_1) \le d_1$ P-a.s. .

Now (ii) follows from Lemma C.1.

The above estimates also yield the existence of constants a_k with

$$\sum_{k=1}^{\infty} a_k < \infty \text{ and } P(M_n^2 \ge k|\mathcal{A}_{T_n}) \le a_k \ P\text{-a.s. for all } k \in \mathbb{N}.$$

In particular, we therefore have $P(M_n \ge k) \le a_k$.

Now (i) follows with Lemma C.6.

(4) Let us now prove (iii): Let $i \in \mathbb{N}$ be arbitrary. Then

$$P(Q_{S_{i+1}} \in B|\mathcal{A}_{S_i}) = 1 - P(Q_{S_{i+1}} \notin B|\mathcal{A}_{S_i}) =$$

$$1 - P(Q_{S_{i+1}} \ge c|\mathcal{A}_{S_i}) \quad \text{(according to the definition of } S_{i+1}) =$$

$$1 - P(Q_{S_{i+1}} \ge c, Q_{S_{i+1}-1} < \frac{a}{l_0}|\mathcal{A}_{S_i}) - P(Q_{S_{i+1}} \ge c, Q_{S_{i+1}-1} \in \left(\frac{1}{2L_0}, c\right)|\mathcal{A}_{S_i})$$

since $P(Q_{S_{i+1}-1} \in A) = 1$. For arbitrary $A \in \mathcal{A}_{S_i}$ we now have

$$P(\{Q_{S_{i+1}} \ge c\} \cap \{Q_{S_{i+1}-1} \in (0, \frac{a}{l_0})\} \cap A) =$$

$$\sum_{k=1}^{\infty} P(\{Q_{S_{i+1}} \ge c\} \cap \{Q_{k-1} < \frac{a}{l_0}\} \cap \{S_{i+1} = k\}) = 0, \text{ since}$$

$$P(Q_k \geq c, Q_{k-1} < \frac{a}{l_0}) \leq P(Q_k > \frac{1}{2L_0}, Q_{k-1} < \frac{a}{l_0}) = 0$$

according to Lemma 16.3 (cf. the definition of a in the proof of Lemma 16.9). Furthermore, for arbitrary $i \in \mathbb{N}$, according to Lemma 16.1 we have

$$P(Q_{i+1} \geq Q_i | \mathcal{A}_i) \leq \frac{1}{2} \ P\text{-a.s.}$$

From this it follows that (cf. Lemma C.4)

$$P(Q_{S_{i+1}} > Q_{S_{i+1}-1} | \mathcal{A}_{S_i}) \leq \frac{1}{2} \ P\text{-a.s.}$$

and therefore

$$P(\{Q_{S_{i+1}} \geq c\} \cap \{Q_{S_{i+1}-1} \in (\frac{1}{2L_0}, c)\} | \mathcal{A}_{S_i}) \leq \frac{1}{2} \ P\text{-a.s.}$$

Thus,

$$P(Q_{S_{i+1}} \in B | \mathcal{A}_{S_i}) \geq \frac{1}{2} \ P\text{-a.s.}$$

Because of $\sigma(Q_{S_1}, \ldots, Q_{S_i}) \subset \mathcal{A}_{S_i}$, by means of iteration of the conditional expectations we get

$$E(1_{\{Q_{S_{i+1}}} \in B\} | 1_{\{Q_{S_i}} \in B\}, \ldots, 1_{\{Q_{S_1}} \in B\}) \geq \frac{1}{2} \ P\text{-a.s.}$$

Now Lemma C.1 yields (iii) with $d_2 = \frac{1}{2}$.

Therefore, (i), (ii), and (iii) are satisfied, which yields the assertion.

□

Proof Proof of Theorem 16.1

(1) Suppose that the assumptions of Theorem 16.1 hold. Let $x_0 \neq x^*$ denote an arbitrary starting point.

Define $B = B(x_0)$, $C = C(x_0)$ and Z_n as in Lemma 16.10. Because of

$$\begin{aligned}
&P(\{X_{n+1} = X_n\} \cup \{l_0 l_{n+1}(X_0, \ldots, X_n) \leq |X_{n+1} - X_n| \\
&\quad \leq L_0 l_{n+1}(X_0, \ldots, X_n)\} | X_n = x_n, \ldots, X_0 = x_0) \\
&\quad = q_{n+1}(x_0, \ldots, x_n, \{x_n\} \cup \{x_{n+1} : l_0 \leq \frac{|x_{n+1} - x_n|}{l_{n+1}(x_0, \ldots, x_n)} \leq L_0\}) \\
&\quad = p_{n+1}(x_0, \ldots, x_n, \{y : l_0 \leq \frac{|y - x_n|}{l_{n+1}(x_0, \ldots, x_n)} \leq L_0\} \cap \{f < f(x_n)\})
\end{aligned}$$

$$
\begin{aligned}
&+ p_{n+1}(x_0, \ldots, x_n, \{f \geq f(x_n)\}) \\
&\geq p_{n+1}(x_0, \ldots, x_n, \{y : l_0 \leq \frac{|y - x_n|}{l_{n+1}(x_0, \ldots, x_n)} \leq L_0\}) \\
&= \lambda_{n+1}(x_0, \ldots, x_n, [l_0 L_{n+1}(x_0, \ldots, x_n), L_0 l_{n+1}(x_0, \ldots, x_n)]) \\
&\quad \lambda_0([l_0, L_0]) = 1,
\end{aligned}
$$

it follows that

$$
\begin{aligned}
P(\{X_{n+1} = X_n\} \cup \{&l_0 l_{n+1}(X_0, \ldots, X_n) \\
&\leq |X_{n+1} - X_n| \leq L_0 l_{n+1}(X_0, \ldots, X_n)\}) = 1.
\end{aligned}
$$

From

$$
\{Z_n \in C\} \subset \{X_{n+1} \neq X_n\}
$$

and

$$
\{Q_n \in B\} = \{\frac{a}{l_0} \leq \frac{l_{n+1}(X_0, \ldots, X_n)}{|X_n - x^*|} \leq \frac{1}{2L_0}\}
$$

we get

$$
\begin{aligned}
\{Q_n \in B\} \cap \{Z_n \in C\} \subset \{&a \leq \frac{|X_{n+1} - X_n|}{|x^* - X_n|} \\
&\leq \frac{1}{2}\} \cap \{\varphi(X_{n+1} - X_n, x^* - X_n) \leq \phi\} \ P\text{-a.s.}
\end{aligned}
$$

Now it follows from Lemma B.9 that

$$
\begin{aligned}
\{Q_n \in B\} \cap \{Z_n \in C\} &\subset \{f(X_n) - f(X_{n+1}) \\
&\geq \frac{|X_{n+1} - X_n|}{3|x^* - X_n|}(f(X_n) - f^*)\} \cap \{\frac{|X_{n+1} - X_n|}{|x^* - X_n|} \geq a\} \\
&\subset \{f(X_n) - f(X_{n+1}) \geq \frac{a}{3}(f(X_n) - f^*)\} \\
&= \{f(X_{n+1}) - f^* \leq (1 - a/3)(f(X_n) - f^*)\} \ P\text{-a.s.}
\end{aligned}
$$

(2) Let $c_0 = c_0(x_0)$ be chosen according to Lemma 16.11.
Select $c \geq c_0$ with $c > \frac{1}{|x_0 - x^*|}$.
Let $n \in \mathbb{N}$ be arbitrary, but fixed. Consider $k \in \mathbb{N}_0$ with

$$\sum_{i=1}^{k}(1_{\{Q_i \in B, Z_i \in C\}} + 1_{\{Q_i \geq c\}}) = k.$$

Denote by T_i the time of the i-th entry of the sequence $(Q_n : n \in \mathbb{N}_0)$ into the interval $[c, \infty)$.

Let $j_n := \max\{i : T_i \leq n\}$ if $\{i : T_i \leq n\} \neq 0$ and $j_n := 0$ else. $T_1, \ldots, T_{j_n}$ are then the entry times of the finite sequence $(Q_i : 0 \leq i \leq n)$ into the interval $[c, \infty)$.

The case $j_n = 0$ means that $Q_i < c$ for $0 \leq i \leq n$.

Let then $M_1, \ldots, M_{j_n}$ denote the associated sojourn times in the interval $[c, \infty)$.

Then a subset

$$S_n = S_n(\omega) \subset \{1, 2, \ldots, n\}$$

exists with

$$Q_n \in B, Z_n \in C \text{ for all } i \in S_n,$$

$$S_n \cap (\{T_1, T_1 + 1, \ldots, T_1 + M_1 - 1\} \cup \cdots \cup \{T_{j_n}, \ldots, T_{j_n} + M_{j_n} - 1\}) \neq 0$$

and

$$|S_n| + M_1 + \cdots + M_j = k.$$

According to the definition, we have for $i = 1, \ldots, j_n$

$$Q_{T_i} \geq c, \ldots, Q_{T_i+M_i-1} \geq c, Q_{T_i+M_i} < c.$$

For $i = 2, \ldots, j_n$, because of $T_i - 1 \geq 0$, by definition we also have $Q_{T_i-1} < c$.

However, this is valid also for $i = 1$.

Indeed, because of $Q_0 = \frac{1}{|x_0 - x^*|}$ P-a.s., we have $Q_0 < c$ P-a.s.

Thus, $T_i \geq 1$, i.e. $T_1 - 1 \geq 0$ P-a.s.

Now it follows from Lemma 16.11 that

$$f(X_{T_i+M_i}) - f^* \leq \gamma_2^{M_i}(f(X_{T_i-1}) - f^*)P\text{-a.s. for all } i = 1, \ldots, j_n.$$

For $i \in S_n$ we have $Q_i \in B, Z_i \in C$ and thus according to part 1:

$$F(X_{i+1}) - f^* \leq (1 - \frac{a}{3})(f(X_i) - f^*) \ P\text{-a.s.}$$

Now because of

$$F(X_{i+1}) - f^* \leq f(X_j) - f^* \ P\text{-a.s. for all } j \geq 0$$

it follows that

$$\begin{aligned} f(X_{n+1}) - f^* &\le (1 - \frac{a}{3})^{|S_n|} \gamma_2^{M_1 + \cdots + M_{j_N}} (f(x_0) - f^*) \\ &\le \gamma^k (f(x_0) - f^*) \; P\text{-a.s. where } \gamma := \max \gamma_2, 1 - a/3. \end{aligned}$$

Therefore for $k \in \mathbb{N}_0$ we have

$$\begin{aligned} &\left\{ \sum_{i=1}^{n} (1_{\{Q_i \in B, Z_i \in C\}} + 1_{\{Q_i \ge c\}}) = k \right\} \\ &\subset \{ f(X_{n+1}) - f^* \le \gamma^k (f(x_0) - f^*) \} \; P\text{-a.s.} \end{aligned}$$

Consequently, for all $n \in \mathbb{N}$ with the definition

$$R_n = \sum_{i=1}^{n} 1_{\{Q_i \in B, Z_i \in C\}} + 1_{\{Q_i \ge c\}}$$

we get

$$\begin{aligned} &P(f(X_{n+1}) - f^* \le (f(x_0) - f^*)\gamma^{R_n}) \\ &= \sum_{k=0}^{n} P(\{f(X_{n+1}) - f^* \le \gamma^k (f(x_0) - f^*)\} \cap \{R_n = k\}) \\ &= \sum_{k=0}^{n} P(R_n = k) = 1. \end{aligned}$$

(3) According to Lemma 16.12, there exists a constant $\beta = \beta(x_0) > 0$ with

$$P(R_n \ge (\beta - \epsilon)n \text{ finally for all } n \in \mathbb{N}) = 1.$$

From part 2 it follows that

$$P(f(X_{n+1}) - f^* \le (f(x_0) - f^*)\gamma^{R_n} \text{ finally for all } n \in \mathbb{N}) = 1$$

and therefore

$$P(f(X_{n+1}) - f^* \le (f(x_0) - f^*)\gamma^{(\beta - \epsilon)n} \text{ finally for all } n \in \mathbb{N}) = 1.$$

For arbitrary $\alpha > 0, \varepsilon > 0$ we therefore have

$$P(\alpha^n (f(X_{n+1}) - f^*) \le (\alpha \gamma^{\beta - \epsilon})^n (f(x_0) - f^*) \text{ finally for all } n \in \mathbb{N}) = 1.$$

Because of $\gamma < 1$, we have $\alpha > 1$.

This yields the assertion for arbitrary starting points $x_0 \neq x^*$.

For $x_0 = x^*$, the assertion is trivial.

□

References

1. Gänssler, P., Stute, W.: Wahrscheinlichkeitstheorie. Springer, Berlin (1977)
2. Tarasenko, G.: Über die Konvergenzgeschwindigkeit der adaptiven zufälligen Suche. (Russisch), Problemy slucajnogo poiska (1980)

Chapter 17
Hybrid Methods

Depending on the type of the random search method, it may happen that the procedure finds rather fast a local extremum x_{loc}^* of the objective function f under consideration, but get then stuck in this point. On the other hand, there are procedures having features to omit this behavior. For example, a special property of simulated annealing methods [1–3] is that also non-improving steps are possible with decreasing probability (cooling). Thus, steps out of the neighborhood of a local extremum are possible with a certain probability.

In the following, a whole class of random search methods, called *hybrid methods*, is presented now which is based on a mixture or combination of two or more different search algorithms among them there is at least one method allowing also larger steps. We start now with a simple, but nevertheless instructive example.

Example 17.1 Define $D = \mathbb{R}$, $f(x) = \frac{1}{4}x^4 - x^3 + x^2$, and let $p(x, \cdot)$ denote the normal distribution with mean x and standard deviation $\sigma = \sigma(x)$. In $x^* = 0$ the function f takes its absolute minimum $f^* = 0$.

Furthermore, the function has a relative maximum at $x = 1$ and a relative minimum at $x = 2$.

Choosing

$$\sigma(x) = |\bigtriangledown f(x)| = |f'| = |x^3 - 3x^2 + 2x| = |x||x-1||x-2|,$$

$$\text{then } \frac{3}{4} \le \frac{\sigma(x)}{|x - x^*|} \le \frac{15}{4} \quad \text{for all } x \in \left[-\frac{1}{2}, \frac{1}{2}\right].$$

Because of $f''(0) = 2 > 0$, the assumptions (15.1) and (15.2) are satisfied (cf. also Remark 15.1).

From Theorem 15.3 or also from Theorem 15.2 for all starting points $x_0 \in [-\frac{1}{2}, \frac{1}{2}]$ we obtain linear convergence in the mean and P-a.s.

K. Marti, *Optimization Under Stochastic Uncertainty*, International Series in Operations Research & Management Science 296,
https://doi.org/10.1007/978-3-030-55662-4_17

On the other hand, for the starting points $x_0 = 1$ or $x_0 = 2$ not even the convergence of associated procedure can be obtained. Indeed, in both cases we have $\sigma(x_0) = 0$, hence, $p(x_0, \cdot) = \epsilon_{x_0}$.

Consequently, the procedure will get stuck already at the starting point. Choosing now $\sigma(x) \equiv \sigma > 0$, for all starting points $x_0 \in \mathbb{R}$ Theorem 12.1 and Remark 12.3 (note that D_a is compact for all $a > f^*$) yield the convergence of the associated algorithm. However, this is achieved at the expense of the convergence rate (cf. Chap. 14).

Now the question arises, whether by a suitable combination of the two methods it is not possible to transfer the positive characteristics of the two procedures—in the first case the linear rate of convergence and in the second case the convergence for each starting point—to **one** algorithm? For example, we could select $\sigma(x) \equiv \sigma > 0$ each second time and the "variance control" $\sigma(x) = |x||x-1||x-2|$ at the remaining times.

The idea behind this is simply that an algorithm designed this way will initially have the property to converge for all starting points. This means, however, that after a finite number of steps the interval $[-\frac{1}{2}, \frac{1}{2}]$ will be reached, which will not be left again. But once the algorithm is within this interval, the "control" $\sigma(x) = |x||x-1||x-2|$ applied at each second step must have an accelerating effect on the rate of convergence.

In the following we show that these considerations are not mere speculations, but can be put in precise mathematical terms.

Definition 17.1 (Model) Select a subsequence $(n_i : i \in \mathbb{N})$ of integers such that $n_1 < n_2 < \ldots$.

Let denote $N := \{n_1, n_2, \ldots\}$ the set of time points at which a control is applied For $n \in \mathbb{N}$ define $h_n := |\{1, 2, \ldots, n\} \cap N|$.

A random search procedure $(X_n : n \in \mathbb{N}_0)$ having the mutation sequence $(p_n : n \in \mathbb{N})$ is called a **hybrid** method, provided that there exist two transition probabilities $p^{(1)}, p^{(2)}$ from $\mathbb{R}^d$ to $\mathbb{R}^d$ such that

$$p_n(x_0, \ldots, x_{n-1}, \cdot) = p^{(1)}(x_{n-1}, \cdot) \text{ for } n \in N \text{ and}$$

$$p_n(x_0, \ldots, x_{n-1}, \cdot) = p^{(2)}(x_{n-1}, \cdot) \text{ for } n \in \mathbb{N}\backslash N \text{ holds.}$$

Obviously, a hybrid method is in any case a Markovian procedure.

Of course, it may also be stationary, e.g. if $p^{(1)} = p^{(2)}$ or $N = \mathbb{N}$.

Lemma 17.1 *Select an $a > f^*$. Assume that there exists $c \in (0, 1)$ with*

$$\int_{\{y \in D: f(y) < f(x)\}} (f(x) - f(y))p^{(1)}(x, dy) \geq c(f(x) - f^*) \textit{ for all } x \in D_a.$$

Furthermore, suppose that for all $x_0 \in D$ there is a $\delta \in (0, 1)$ with

$$p^{(2)}(x, D_a) \geq \delta \textit{ for all } x \in D_{f(x_0)} \backslash D_a.$$

Then for all starting points $x_0 \in D$ *we have a* $b \in (0, 1)$ *such that*

$$E(f(X_n) - f^*) \leq (a - f^*)(1 - c)^{h_n} \sum_{k=1}^{n} b^{k-1}((1-c)b)^{-h_k} + (f(x_0) - f^*)b^{n-h_n}$$

for all $n \in \mathbb{N}$.

Proof

(1) Let $x_0 \in D$ denote an arbitrary starting point.
According to the assumption, there is a $\delta \in (0, 1)$ with

$$p^{(2)}(x, D_a) \geq \delta \text{ for all } x \in D_{f(x_0)} \backslash D_a.$$

For $i \in \mathbb{N} \setminus N$ and $x \in D_{f(x_0)} \backslash D_a$ we have

$$\int\limits_{D_{f(x_0)} \backslash D_a} q_i(x, dz) = 1 - q_i(x, D_a) = 1 - p_i(x, D_a)$$

$$= 1 - p^{(2)}(x, D_a) \leq 1 - \delta.$$

For $i \in N$ this integral can be estimated from above by the constant 1. By n-fold iteration it follows therefore that

$$P(X_n \notin D_a) = \int\limits_{D_{f(x_0)} \backslash D_a} \cdots \int\limits_{D_{f(x_0)} \backslash D_a} q_n(x_{n-1}, dx_n) \ldots q_1(x_0, dx_1)$$

$$\leq (1 - \delta)^{n - |\{1, \ldots, n\} \cap N|} = (1 - \delta)^{n - h_n}.$$

(2) Let

$$T := \begin{cases} \min\{n \in \mathbb{N} : X_n \in D_a\} & \text{if } \{n \in \mathbb{N} : X_n \in D_a\} \neq 0 \\ \infty & \text{else.} \end{cases}$$

Then we have

$$E(f(X_n) - f^*) = \sum_{k=1}^{\infty} E(f(X_n) - f^* | T = k) P(T = k)$$

$$\leq \sum_{k=1}^{n} E(f(X_n) - f^* | T = k) P(T = k)(f(x_0) - f^*) \sum_{k>n} P(T = k)$$

$$\leq E(f(X_n) - f^*|T = 1) + \sum_{k=2}^{n} E(f(X_n) - f^*|T = k)P(X_{k-1} \notin D_a)$$

$$+ (f(x_0) - f^*)P(T > n)$$

$$\leq \sum_{k=1}^{n} E(f(X_n) - f^*|T = k)(1 - \delta)^{k-1-h_{k-1}} + (f(x_0) - f^*)(1 - \delta)^{n-h_n},$$

where $h_0 := 0$.

(3) For $k = 1, 2, \dots, n-1$ we have

$$\int_{\{T=k\}} (f(X_n) - f^*)dP = \int_{\{T=k\}} E(f(X_n) - f^*|X_k, \dots, X_0)dP$$

$$= \int_{\{T=k\}} (f(X_n) - f^*|X_k)dP$$

$$= \int_{\{T=k\}} \int \dots \int (f(x_n) - f^*)q_n(x_{n-1}, dx_n) \dots$$

$$q_{k+2}(x_{k+1}, dx_{k+2})q_{k+1}(X_k, dx_{k+1}).$$

If $i \in \{k+1, \dots, n\} \cap N$ and $x_{i-1} \in D_a$, then

$$\int (f(x) - f^*)q_i(x_{i-1}, dx)$$

$$= f(x_{i-1}) - f^* - \int (f(x_{i-1}) - f(x))q_i(x_{i-1}, dx)$$

$$= f(x_{i-1}) - f^* - \int_{\{f<f(x_{i-1})\}} (f(x_{i-1}) - f(y))p_i(x_{i-1}, dy)$$

$$= f(x_{i-1}) - f^* - \int_{\{f<f(x_{i-1})\}} (f(x_{i-1}) - f(y))p^{(1)}(x_{i-1}, dy)$$

$$\leq (1 - c)(f(x_{i-1}) - f^*).$$

For $i \in \mathbb{N} \setminus N$ we have

$$\int (f(x) - f^*)q_i(x_{i-1}, dx) \leq f(x_{i-1}) - f^*.$$

Because $\{t = k\} \subset \{X_k \in D_a\} \subset \dots \subset \{X_n \in D_a\}$ P-a.s., iteration yields

$$\int\limits_{\{T=k\}} (f(X_n) - f^*)dP \leq (1-c)^{|N \cap \{k+1,\ldots,n\}|} \int\limits_{\{T=k\}} (f(X_n) - f^*)dP$$
$$\leq (1-c)^{h_n - h_k}(a - f^*)P(t = k).$$

This inequality also holds for $k = n$.

Thus,

$$E(f(X_n) - f^*|T = k) \leq (1-c)^{h_n - h_k}(a - f^*)P\text{-a.s. for } k = 1, \ldots, n.$$

From part (2) we get now that

$$E(f(X_n) - f^*) \leq (a - f^*)(1-c)^{h_n} \sum_{k=1}^{n} (1-\delta)^{k-1-h_{k-1}}(1-c)^{-h_k}$$
$$+ (f(x_0) - f^*)(1-\delta)^{n-h_n}.$$

Because of $h_{k-1} \leq h_k$, the assertion follows now with $b := 1 - \delta$. □

Lemma 17.2 *Assume that the assumptions of Lemma 17.1 are satisfied.*

Furthermore, suppose that $\limsup_{n\to\infty} \frac{h_n}{n} < 1$.

Then for all starting points $x_0 \in D$ *there exists a constant* $\beta > 1$ *such that*

$$\beta^{h_n} E(f(X_n) - f^*) \to 0.$$

Proof Because of $\limsup_{n\to\infty} \frac{h_n}{n} < 1$, there is a $K \in (0, 1)$ with $h_n \leq Kn$ finally for all $n \in \mathbb{N}$.

According to the assumption, for a $c \in (0, 1)$ we have

$$\int\limits_{\{z \in D: f(y) < f(x)\}} (f(x) - f(y))p^{(1)}(x, dy) \geq c(f(x) - f^*) \text{ for all } x \in D_a.$$

Obviously, this inequality holds true if c is replaced by a $\tilde{c} \leq c$.

Without loss of generality we may therefore assume that $c < 1 - b^{1/K-1}$ (with b according to Lemma 17.1).

However, then we have $\frac{b^{1-K}}{(1-c)^K} < 1$.

For sufficiently large $i \in \mathbb{N}$ this yields the estimate

$$\frac{b^{i-1}}{((1-c)b)^{h_i}} \leq \frac{1}{b} \frac{b^{i-Ki}}{(1-c)^{Ki}} = \frac{1}{b}(b^{1-K}(1-c)^{-K})^i.$$

However, from this we obtain the convergence of the series $\sum_{k=1}^{\infty} b^{k-1}((1-c)b)^{-h_k}$.

With a suitable constant c', Lemma 17.1 yields now

$$\begin{aligned} E(f(X_n) - f^*) &\leq c'(1-c)^{h_n} + c'bn - h_n \\ &\leq c'((1-c)^{h_n} + b\left(\frac{1}{k} - 1\right) h_n) \leq 2c'(1-c)^{h_n} \text{ finally for all } n \in \mathbb{N}. \end{aligned}$$

Choosing $\beta \in (1, (1-c)^{-1})$, because of $h_n \to \infty$ we get the assertion. □

Lemma 17.3 *Suppose that* $\lim_{n\to\infty} (n - h_n) = \infty$ *(i.e.* $|\mathbb{N}\backslash N| = \infty$*), and assume that there exist quantities* $a > f^*$, $c_1 \in [0, 1)$ *and* $c_2 > 0$ *such that*

$$p^{(1)}(x, \{y \in D : f(y) - f^* \leq c_1(f(x) - f^*)\}) \geq c_2 \text{ for all } x \in D_a.$$

Moreover, assume that for all $x_0 \in D$ *a* $\delta \in (0, 1)$ *exists with*

$$p^{(2)}(x, D_a) \geq \delta \text{ for all } x \in D_{f(x_0)} \backslash D_a.$$

Then for all starting points $x_0 \in D$ *we have a constant* $\beta > 1$ *such that*

$$\beta^{h_n}(f(X_n) - f^*) \to 0 \text{ } P\text{-a.s.}$$

Proof

(1) Select arbitrary, but fixed $k \in \mathbb{N}$ and $x \in D_a$.

Because of the theorem of Ionescu-Tulcea (cf. [4]), without loss of generality we can assume the existence of a probability measure $p^{x,k}$ on $(\Omega, \mathcal{A})$ and the existence of a sequence of random variables $(\tilde{X}_n : n \in \mathbb{N}_0)$ having the following properties:

(i) $p^{x,k}(\hat{X}_0 = x) = 1$.
(ii) $p^{x,k}(\hat{X}_{n+1}, \hat{X}_{n+2}, \ldots) \in B|(\hat{X}_n, \ldots, \hat{X}_0) = p^{x,k}(\hat{X}_{n+1}, \hat{X}_{n+2}, \ldots) \in B|\hat{X}_n) = P((X_{k+n+1}, X_{k+n+2}, \ldots) \in B|X_{k+n})$ for all $n \in \mathbb{N}$, $B \subset (\mathbb{R}^d)^{\mathbb{N}}$ measurable.

In particular, for all measurable $C \subset \mathbb{R}^d$ and index $n \in \mathbb{N}_0$ we have

$$(*) \begin{cases} p^{x,k}(\hat{X}_{n+1} \in C|\hat{X}_n = z) = P(X_{k+n+1} \in C|X_{k+n} = z) = \\ q_{k+n+1}(z, C) = q^{(1)}(z, C) \underline{\text{ if }} k + n + 1 \in \mathbb{N} \end{cases}$$

($q^{(1)}$ is the selection transition probability associated with the probability measure $p^{(1)}$, cf. (9.5)).

Let $E^{x,k}$ denote the mean value operator related to the probability measure $p^{x,k}$.

(2) Define $m := \min\{j \in \mathbb{N} : n_j \geq k\}$. From the definition of h_n follows that $m = h_{k-1} + 1 (h_{-1} = h_0 := 0)$.
For $i \in \mathbb{N}$ put

$$\hat{Z}_i := 1_{\{f(\hat{X}_{n_{m+i}-k}) - f^* \leq c_1(f(\hat{X}_{n_{m+i}-k-1}) - f^*)\}}.$$

Because of $n_m \geq k$, we have $n_{m+i} - k \geq 1$ for all $i \in \mathbb{N}$.
Due to $p^{x,k}(\hat{X}_0 = x) = 1$ and $c \in D_a$, we get $p^{x,k}(\hat{X}_i \in D_a) = 1$ for all $i \in \mathbb{N}_0$. □
Hence,

$$\begin{aligned}
&E^{x,k}(\hat{Z}_{i+1}|\hat{Z}_i, \ldots, \hat{Z}_1)\\
&= E^{x,k}(E^{x,k}(\hat{Z}_{i+1}|\hat{X}_{n_{m+i+1}-k}, \ldots, \hat{X}_0|\hat{Z}_i, \ldots, \hat{Z}_1))\\
&= E^{x,k}(p^{x,k}(f(\hat{X}_{n_{m+i+1}-k}) - f^* \leq c_1(f(\hat{X}_{n_{m+i}-k-1}) - f^*)|\hat{X}_{n_{m+i}-k-1})|\\
&\quad |\hat{Z}_i, \ldots, \hat{Z}_1) =\\
&= E^{x,k}(q_{n_{m+i+1}}(\hat{X}_{n_{m+i+1}-1}), \{y \in D : f(y) - f^* \leq c_1(f(\hat{X}_{n_{m+i+1}})\\
&\quad - f^*)\})|(\hat{Z}_i, \ldots, \hat{Z}_1)\\
&= E^{x,k}(p^{(1)}(\hat{X}_{n_{m+i+1}-1}), \{f - f^* \leq c_1(f(\hat{X}_{n_{m+i+1}-1}) - f^*)\})|\\
&\quad |(\hat{Z}_i, \ldots, \hat{Z}_1) \geq c_2
\end{aligned}$$

according to the assumption, since $p^{x,k}(\hat{X}_{n_{m+i+1}-1} \in D_a) = 1$.
From Lemma C.1 and property (i) for each $\tilde{c}_2 < c_2$ we get

$$p^{x,k}(\frac{1}{i}\sum_{j=1}^{i} \hat{Z}_j \geq \tilde{c}_2 \text{ finally for all } i \in \mathbb{N}|\hat{X}_0 = x) = 1.$$

(3) For $i \in \mathbb{N}$, put $N_{k,k+i} := \{k, \ldots, k+i\} \cap N$.
Choosing $i \in \mathbb{N}$ as large that $h_i \geq 2k$ is satisfied, we obtain

$$\frac{h_i}{|N_{k,k+i-1}|} = \frac{h_i}{h_{k+i-1} - h_{k-1}} \leq \frac{h_i}{h_i - k} \leq 2.$$

For all $c > 0$ it follows that

$$\left\{\frac{h_i}{|N_{k,k+i-1}|} \sum_{j=1}^{|N_{k,k+i-1}|} \hat{Z}_j \geq 2c\right\}$$

$$\subset \left\{ \frac{h_i}{|N_{k,k+i-1}|} \sum_{j=1}^{|N_{k,k+i-1}|} \hat{Z}_j \geq \frac{h_i}{|N_{k,k+i-1}|} c \right\}$$

$$= \left\{ \frac{1}{h_i} \sum_{j=1}^{|N_{k,k+i-1}|} 1_{\left\{f(\hat{X}_{n_{m+j}-k}) - f^* \leq c_1 (f(\hat{X}_{n_{m+j}-k-1}) - f^*)\right\}} \geq c \right\}$$

$$\subset \left\{ \frac{1}{h_i} \sum_{j=1}^{k+i-1} 1_{\left\{f(\hat{X}_{j+1-k}) - f^* \leq c_1 (f(\hat{X}_{j-1}) - f^*)\right\}} \geq c \right\}.$$

This yields the inclusion

$$\{\frac{h_i}{|N_{k,k+i-1}|} \sum_{j=1}^{|N_{k,k+i-1}|} \hat{Z}_j \geq \tilde{c}_2 \quad \text{finally for all } i \in \mathbb{N}\}$$

$$\subset \{\frac{1}{h_i} \sum_{j=1}^{i} 1_{\{f(\hat{X}_j) - f^* \leq c_1 (f(\hat{X}_{j-1}) - f^*)\}} \geq \frac{\tilde{c}_2}{c_2} \quad \text{finally for all } i \in \mathbb{N}\}$$

$$\subset \{f(\hat{X}_i) - f^* \leq c_1^{\tilde{c}_2 h_i/2} (f(\hat{X}_0) - f^*) \quad \text{finally for all } i \in \mathbb{N}\} p^{x,k} - a.s.$$

From part (2) it follows therefore that

$$p^{x,k}(f(\hat{X}_i) - f^* \leq c_1^{\tilde{c}_2 h_i/2} (f(x) - f^*) \text{ finally for all } i \in \mathbb{N} | \hat{X}_0 = x) = 1.$$

Together with property (ii) this yields

$$P(f(X_{k+i}) - f^* \leq c_1^{\tilde{c}_2 h_i/2} (f(x) - f^*) \text{ finally for all } i \in \mathbb{N} | X_k = x) = 1.$$

Since x lies in D_a, we get $f(x) - f^* \leq a - f^*$.
For arbitrary $x \in D_a$ and $\beta \in (1, c_1^{-\tilde{c}_2/2})$ we therefore have

$$P(\lim_{i \to \infty} \beta^{h_i} (f(X_{k+i}) - f^*) = 0 | X_k = x) = 1.$$

Because of $h_{k+i} - h_i \leq$ we get $\beta^{h_{k+i} - h_i} \leq \beta^k$.
Thus,

$$P(\lim_{n \to \infty} \beta^{h_n} (f(X_n) - f^*) = 0 | X_k = x) = 1$$

$$\text{for all } k \in \mathbb{N}_0, x \in D_a \text{and } \beta \in (1, c_1^{-\tilde{c}_2/c_2}).$$

(4) Let now $x_0 \in D$ denote an arbitrary starting point.
Then we have $P(X_n \notin D_a) \leq (1 - \delta)^{n - h_n}$ (cf. proof of Lemma 17.1).

Because of $\lim_{n\to\infty}(n - h_n) = \infty$, this yields $P(\bigcup_{n\in\mathbb{N}}\{X_n \in D_a\}) = 1$.
For $T := \min\{n \in \mathbb{N} : X_n \in D_a\}$ we therefore have $P(T < \infty) = 1$.
Put $A := \{\lim_{n\to\infty} \beta^{h_n}(f(X_n) - f^*) = 0\}$.
Due to $\{T = k\} \subset \{X_k \in D_a\}$, according to part (2) we have

$$\int_{\{T=k\}} P(A|X_k)dP = P(T = k).$$

This implies finally that

$$\begin{aligned} P(A) &= \sum_{k=0}^{\infty} P(A \cap \{T = k\}) \\ &= \sum_{k=0}^{\infty}\left(\int_{\{T=k\}} P(A|X_k, \dots, X_0)dP\right) \\ &= \sum_{k=0}^{\infty}\left(\int_{\{T=k\}} P(A|X_k)dP\right) = \sum_{k=0}^{\infty} P(T = k) = 1. \end{aligned}$$

□

Theorem 17.1 *Suppose that the assumptions* (15.1) *and* (15.2) *hold. Moreover, assume that $p^{(2)}$ satisfies the assumption* (14.1), *i.e.:*

for all $x_0 \in D\backslash D^*$ there exists an $a_0 > 0$ such that $p^{(2)}(x, B) \geq a_0\lambda^d(B)$.
for all $x \in D_{f(x_0)}\backslash D^*$ and measurable $B \subset D_{f(x_0)}$.

In addition, assume that for $p_n(x_0, \dots, x_{n-1}, \cdot) := p^{(1)}(x_{n-1}, \cdot)$ one of the following assumptions is fulfilled:

(i) $p^{(1)}$ is of the random direction type with the step length distribution $\lambda^{(1)}$, where $\lambda_n(x_0, \dots, x_{n-1}, \cdot) := \lambda^{(1)}(x_{n-1}, \cdot)$ is defined according to Theorem 15.2. Also assumption 15.6 is satisfied for an $a > f^*$.
(ii) $p^{(1)}$ is absolutely continuous with the density $g^{(1)}$, where $g_n(x_0, \dots, x_{n-1}, \cdot) := g^{(1)}(x_{n-1}, \cdot)$ is defined according to Theorem 15.3. Moreover, assumption (15.7) is satisfied for an $a > f^*$.
(iii) $p^{(1)}$ is defined according to Theorem 15.4. Also assumption (15.9) *holds for an $a > f^*$.*

If in addition the condition $\lim_{n\to\infty}\sup \frac{h_n}{n} < 1$ holds, then to each starting point $x_0 \in D$ there exists $\beta > 1$ such that

$$\beta^{h_n} E(f(X_n) - f^*) \to 0 \textit{ and}$$

$$\beta^{h_n}(f(X_n) - f^*) \to 0 \ P\textit{-a.s.}$$

Proof Let one of the assumptions (i), (ii), and (iii) be satisfied.

According to assumption (15.1), we can assume without loss of generality that f is continuous on D_a.

Because of $a > f^*$, we have $\lambda^d(D_a) > 0$ (cf. Remark 12.2).

Defining $\delta := \lambda^d(D_a)$, according to assumption (14.1) we have the inequality

$$p^{(2)}(x, D_a) \geq a_0\delta \text{ for all } x_0 \in D \text{ and } x \in D_{f(x_0)} \backslash D_a.$$

Considering the proofs of the Theorems 15.2,15.3, and 15.4, one observes that each of the assumptions (i), (ii), and (iii) implies that all assumptions of Theorem 15.1and thus also of Lemma 15.1 and Lemma 15.2 are satisfied.

Taking into account that due to

$$n - h_n = n(1 - \frac{h_n}{n})$$

the inequality $\limsup\limits_{n\to\infty} \frac{h_n}{n} < 1$ implies that

$$\lim_{n\to\infty} (n - h_n) = \infty,$$

it can be seen that all assumptions of Lemmas 17.1, 17.2, and 17.3 are fulfilled, which yields the assertion. Now we can answer the question posed subsequent to Example 17.1.

Choosing in the case of Example 17.1

$$N = \{1, 3, 5, \dots\} \text{ and } \sigma_n(x) = |f'(x)| \text{ for } n \in \mathbb{N} \quad \text{ and } \sigma_n(x) \equiv \sigma \text{ for } n \in \mathbb{N}\backslash N,$$

then Remark 14.4 shows that $p^{(2)}$ satisfies assumption (14.1). Obviously, $p^{*1)}$ fulfills condition (ii) of Theorem 17.1 (cf. also part (2) of Example 15.2).

$$\text{Because of } h_n \sim \frac{n}{2} \text{ and } \beta^{n/2} = (\sqrt{\beta})^n,$$

Theorem 17.1 yields now that for all starting points the procedure according to the above construction converges linearly in the mean and P-a.s. □

References

1. Aarts, E., Korst, J.: Simulated Annealing and Boltzmann Machines: A Stochastic Approach to Combinatorial Optimization and Neural Computing. Wiley, Chichester (1989)
2. Kirkpatrick, S., Gelatt, C.D., Vecchi, M.P.: Optimization by simulated annealing. Science **220**(4598), 671–680 (1983)

3. Metropolis, N., et al.: Equations of state calculations by fast computing machines. J. Chem. Phys. **21**, 1087–1092 (1953)
4. Neveu, J.: Mathematische Grundlagen der Wahrscheinlichkeitstheorie. Oldenburg Verlag, Munich (1969)

Part IV
Optimization Under Stochastic Uncertainty by Random Search Methods (RSM)

Chapter 18
Solving Optimization Problems Under Stochastic Uncertainty by Random Search Methods (RSM)

18.1 Introduction

In many optimization problems arising in practice

$$\min F_0(x) \quad \text{s.t. } x \in D \tag{18.1}$$

only statistical information is available about the objective function F_0. Hence, we have then only access to a realization of a random function $f = f(\omega, x)$ on a probability space $(\Omega, \mathcal{A}, \mathcal{P})$ such that $F_0(x)$ is the expectation of $f(\cdot, x)$:

$$F_0(x) = Ef(\omega, x), \; x \in D, \tag{18.2a}$$

see, e.g., [1, 2]. A special case is given by

$$f(\omega, x) = F_0(x) + n(\omega, x), \tag{18.2b}$$

where $n(\omega, x)$ is an additional zero-mean random noise term. Having a certain number m of independent sample functions

$$f_k(x) = f(\omega_k, x), \; k \in M, \tag{18.2c}$$

of the random function in (18.2a), we may use the estimated objective function

$$\hat{F}(x) := \frac{1}{m} \sum_{k \in M} f_k(x). \tag{18.3}$$

K. Marti, *Optimization Under Stochastic Uncertainty*, International Series in Operations Research & Management Science 296,
https://doi.org/10.1007/978-3-030-55662-4_18

The basic random search procedure as considered in Part II and Part III has then to be replaced by

$$X_{t+1} = \left\{ \begin{array}{ll} Z_{b+1}, \text{ if } Z_{t+1} \in D \text{ and } \hat{F}_t(Z_{t+1}) < \hat{F}_t(X_t) \\ X_t, \text{ if } Z_{t+1} \notin D \text{ or } \hat{F}_t(Z_{t+1}) \geq \hat{F}_t(X_t). \end{array} \right\} \tag{18.4}$$

Depending on the type of sampling, the estimate $\hat{F}_t$ is defined, cf. (18.3), by

$$\hat{F}_t(x) = \hat{F}_{t,m_t}(x) = \frac{1}{m_t} \sum_{k \in M_t} f(\omega_k, x), \tag{18.5a}$$

where for the definition of the index set $M = M_t$ we may consider the following cases:

$$\begin{aligned} &(I) \qquad M_t = \left\{k_{t1}, k_{t2}, \ldots, k_{m_t}\right\} \\ &(II) \qquad M_t = \{1, 2, \ldots, m_t\} \qquad\qquad t = 0, 1, \ldots, \end{aligned} \tag{18.5b}$$

where in the first case we assume that $M_t \cap M_s = \emptyset$ for $s \neq t$.

The problem is then to find conditions for the convergence of the resulting search process (X_t) to a solution x^* of the original problem (18.1). Note that in the first case (X_t) is a Markov process.

18.2 Convergence of the Search Process (X_t)

The convergence is studied by means of the probabilities $P(X_t \in B_\epsilon)$, where ϵ is a (small) positive number and

$$B_\epsilon := \left\{y \in D : F_0(y) \leq F_o^* + \epsilon\right\}, \tag{18.6a}$$

$$F_0^* = \inf_{y \in D} F_0(y). \tag{18.6b}$$

We have

$$\begin{aligned} P(X_{t+1} \notin B_\epsilon) &= 1 - P(X_{t+1} \in B_\epsilon) \\ &= P(X_t \notin B_\epsilon) + P(X_t \in B_\epsilon) - P(X_{t+1} \in B_\epsilon) \\ &= P(X_t \notin B_\epsilon) + P(X_t \in B_\epsilon) - \int_{x_t \notin B_\epsilon} P(X_{t+1} \in B_\epsilon | X_t = x_t)\, P_{X_t}(dx_t) \\ &\quad - \int_{x_t \in B_\epsilon} P(X_{t+1} \in B_\epsilon | X_t = x_t)\, P_{X_t}(dx_t) \end{aligned}$$

$$= P(X_t \notin B_\epsilon) + \int\limits_{x_t \in B_\epsilon} (1 - P(X_{t+1} \in B_\epsilon | X_t = x_t))\, P_{X_t}(dx_t)$$

$$- \int\limits_{x_t \notin B_\epsilon} P(X_{t+1} \in B_\epsilon | X_t = x_t)\, P_{X_t}(dx_t)$$

$$= P(X_t \notin B_\epsilon) + \int\limits_{x_t \in B_\epsilon} P(X_{t+1} \notin B_\epsilon | X_t = x_t) P_{X_t}(dx_t)$$

$$- \int\limits_{x_t \notin B_\epsilon} P(X_{t+1} \in B_\epsilon | X_t = x_t) P_{X_t}(dx_t). \tag{18.7a}$$

Estimating the probability in the second, third term from above, below, resp., we get

$$P(X_{t+1} \notin B_\epsilon) \le P(X_t \notin B_\epsilon) + \sup_{x_t \in B_\epsilon} P(X_{t+1} \notin B_\epsilon | X_t = x_t) P(X_t \in B_\epsilon)$$

$$- \inf_{x_t \notin B_\epsilon} P(X_{t+1} \in B_\epsilon | X_t = x_t) P(X_t \notin B_\epsilon). \tag{18.7b}$$

Defining the coefficients

$$\alpha_t(\epsilon) := \inf_{x_t \notin B_\epsilon} P(X_{t+1} \in B_\epsilon | X_t = x_t), \tag{18.8a}$$

$$r_t(\epsilon) := \sup_{x_t \in B_\epsilon} P(X_{t+1} \notin B_\epsilon | X_t = x_t), \tag{18.8b}$$

from (18.7a)–(18.7b) we get the recursive inequalities

$$P(X_{t+1} \notin B_\epsilon) \le P(X_t \notin B_\epsilon) - \alpha_t(\epsilon) P(X_t \notin B_\epsilon) + r_t(\epsilon) \left(1 - P(X_t \notin B_\epsilon)\right),$$

hence,

$$P(X_{t+1} \notin B_\epsilon) \le (1 - \alpha_t(\epsilon) - r_t(\epsilon)) P(X_t \notin B_\epsilon) + r_t(\epsilon) \quad t = 0, 1, \ldots. \tag{18.9}$$

Given an arbitrary $\epsilon > 0$, for the consideration of the behavior of the recurrence inequality (18.9) for $t \to \infty$ we put

$$p_t := P(X_t \notin B_\epsilon), \quad t = 0, 1, 2, \ldots. \tag{18.10a}$$

Hence, with the coefficients

$$a_t := \alpha_t + r_t, \qquad \delta_t := \frac{r_t}{\alpha_t + r_t} = \frac{r_t}{a_t} \tag{18.10b}$$

Equation (18.9) can be represented by

$$p_{t+1} \leq (1 - a_t) p_t + a_t \delta_t, \quad t = 0, 1, \ldots, \tag{18.10c}$$

where $0 \leq \delta_t \leq 1$. Defining for $m, n \in \mathbb{N}_0,\ m \leq n$, the sequence

$$b_{m,n} := \begin{cases} \prod_{j=m}^{n-1} (1 - a_j) & , m < n \\ 1 & , m = n, \end{cases} \tag{18.11}$$

by means of complete induction, cf. Lemma A.4, Appendix A.1 in [3], for the probability not reaching B_ϵ we have the following result:

Lemma 18.1 *If* $0 < a_t < 1,\ t = 0, 1, 2, \ldots$, *then*

$$p_n \leq b_{0,n} p_0 + \sum_{n=0}^{n-1} b_{m,n} a_m \delta_m, \ n = 1, 2, \ldots. \tag{18.12a}$$

Moreover, if $\sum_{n=0}^{\infty} a_n = +\infty$, *then*

(*i*) $b_{m,n} \to 0,\ n \to \infty$, *for each* $m = 0, 1, 2, \ldots$

and

$$(ii) \ \lim_{n\to\infty} \inf \delta_n \leq \lim_{n\to\infty} \inf p_n \leq \lim_{n\to\infty} \sup p_n \leq \lim_{n\to\infty} \sup \delta_n. \tag{18.12b}$$

Consequently, for the convergence of (p_t) we find this result:

Theorem 18.1 *For given* $\epsilon > 0$, *suppose that*

$$\sum_{t=0}^{\infty} a_t = +\infty, \ 0 < a_t \leq 1, \quad t = 0, 1, 2, \ldots, \ \text{and} \ \lim_{t\to\infty} \delta_t = 0.$$

Then,

$$P(X_t \notin B_\epsilon) \to 0, \ t \to \infty, \qquad \text{and} \qquad \lim_{t\to\infty} P(X_t \in B_\epsilon) = 1.$$

For the consideration of the properties of recursions of the type (18.10c) there are similar conditions as given, e.g., in [4, p. 45], Lemma 3:

Let denote $(u_t)_{t\in\mathbb{N}}$ a sequence of nonnegative numbers such that

$$p_{t+1} \leq q_t p_t + r_t, \quad t = 0, 1, 2, \ldots, \tag{18.13a}$$

with coefficients $0 \leq q_t < 1,\ r_t \geq 0$. If

$$\sum_{t=0}^{\infty}(1 - q_t) = +\infty, \qquad \lim_{t\to\infty} \frac{r_t}{1 - q_t} = 0, \tag{18.13b}$$

then $p_t \to 0,\ t \to \infty$.

18.3 Estimation of the Minimum, Maximum Entry, Leaving Probability, Resp., α_t, r_t

In the following we derive bounds for probability entry, leaving, resp., the set B_ϵ of ϵ-optimal points from a point $X_t = x_t$ lying outside, inside, resp., of B_ϵ. According to (18.4) we have for $X_t = x_t$

$$X_{t+1} = T_{\hat{F}_t,x_t}(Z_{t+1}), \tag{18.14a}$$

where

$$T_{\hat{F}_t,x_t}(z) = \begin{cases} z, & \text{if } z \in G_{\hat{F}_t(x_t)} \\ x_t, & \text{else} \end{cases} \tag{18.14b}$$

$$G_{\hat{F}_t}(x) = \left\{z \in D : \hat{F}_t(z) < \hat{F}_t(x)\right\}. \tag{18.14c}$$

Given a point $x_t \notin B_\epsilon$, we have

$$\begin{aligned} P(X_{t+1} \in B_\epsilon | X_t = x_t) &= P(T_{\hat{F}_t(\omega,\cdot),x_t}(Z_{t+1}) \in B_\epsilon | X_t = x_t) \\ &= P(Z_{t+1} \in G_{\hat{F}_t(\omega,\cdot)}(x_t) \cap B_\epsilon | X_t = x_t). \end{aligned} \tag{18.15a}$$

Assuming now Case I in (18.5b), hence, the independence of the function estimator $\hat{F}_t(\omega, \cdot) = \hat{F}_{t,m_t}(\omega, \cdot)$ and the search variable Z_{t+1}, we get

$$P(X_{t+1} \in B_\epsilon | X_t = x_t) = E_\omega \pi_t(x_t, G_{\hat{F}_t(\omega,\cdot)}(x_t) \cap B_\epsilon). \tag{18.15b}$$

In order to derive lower bounds of (18.15b), for $x_t \notin B_\epsilon$, hence $F_0(x_t) > F_0^* + \epsilon$, see (18.8a), we consider $y \in B_{\frac{\epsilon}{2}}$. Then, $y \in D,\ \ F_0(y) \leq F_0^* + \frac{\epsilon}{2} < F_0(x_t)$ and therefore

$$F_0(y) - F_0(x_t) = F_0(y) - F_0^* - (F_0(x_t) - F_0^*) < \frac{\epsilon}{2} - \epsilon < 0. \tag{18.16a}$$

For the function estimator $\hat{F}_t$, this yields now

$$\begin{aligned}\hat{F}_t(y) - \hat{F}_t(x_t) &= (\hat{F}_t(y) - F_0(y)) + (F_0(y) - \hat{F}_t(x_t)) \\ &= \hat{F}_t(y) - F_0(y) + (F_0(y) - F_0(x_t)) + F_0(x_t) - \hat{F}_t(x_t) \\ &< \Delta\hat{F}_t(y) - \frac{\epsilon}{2} - \Delta\hat{F}_t(x_t) = -\frac{\epsilon}{2} + \Delta\hat{F}_t(y) - \Delta\hat{F}_t(x_t).\end{aligned} \tag{18.16b}$$

Supposing uniform convergence of the function estimator $\hat{F}_t$ w.p.1, hence

$$\hat{F}_t = \hat{F}_{t,m_t}(\omega, x) \to F_0(x), \quad t \to \infty, \text{ w.p.1 uniformly for } x \in D, \tag{18.17}$$

then there exists $t_{\frac{\epsilon}{2}}$ such that for all $y \in D$ we have

$$\hat{F}_t(\omega, y) < \hat{F}(\omega, x_t), \text{ w.p.1} \quad t > t_{\frac{\epsilon}{2}} \text{ for } x_t, y \in D.$$

This means that $y \in G_{\hat{F}_t(\omega,\cdot)}(x_t)$ w.p.1, if $t > t_{\frac{\epsilon}{2}}$, and therefore

$$G_{\hat{F}_t(\omega,\cdot)}(x_t) \supset B_{\frac{\epsilon}{2}} \text{ and } G_{\hat{F}_t(\omega,\cdot)}(x_t) \cap B_\epsilon \supset B_{\frac{\epsilon}{2}} \ w.p.1, \ t > t_{\frac{\epsilon}{2}}. \tag{18.18a}$$

For the minimum probability of accessibility of B_ϵ from a point x outside B_ϵ we now have this result (Fig. 18.1):

Lemma 18.2 *Under condition* (18.17) *and assuming that* Z_{t+1} *and* $\hat{F}_{t,m_t}$ *are stochastic independent, given* $X_t = x_t$*, we have*

$$\alpha_t(\epsilon) \geq \inf_{x_t \notin B_\epsilon} \pi_t(x_t, B_{\frac{\epsilon}{2}}), \quad t > t_{\frac{\epsilon}{2}}. \tag{18.18b}$$

Consider now the maximum probability of leaving B_ϵ, hence

$$r_t(\epsilon) = \sup_{x_t \in B_\epsilon} P(X_{t+1} \notin B_\epsilon | X_t = x_t).$$

Using again (18.14a)–(18.14c), for $x_t \in B_\epsilon$ we have

$$\begin{aligned}P(X_{t+1} \notin B_\epsilon | X_t = x_t) &= P(T_{\hat{F}_t, x_t}(Z_{t+1}) \notin B_\epsilon | X_t = x_t) \\ &= P(Z_{t+1} \in G_{\hat{F}_t}(x_t), Z_{t+1} \notin B_\epsilon | X_t = x_t) \\ &= P(Z_{t+1} \in G_{\hat{F}_t}(x_t) \setminus B_\epsilon | X_t = x_t) \\ &= E_\omega \pi_t(x_t, G_{\hat{F}_t(\omega,\cdot)}(x_t) \setminus B_\epsilon | X_t = x_t).\end{aligned} \tag{18.19a}$$

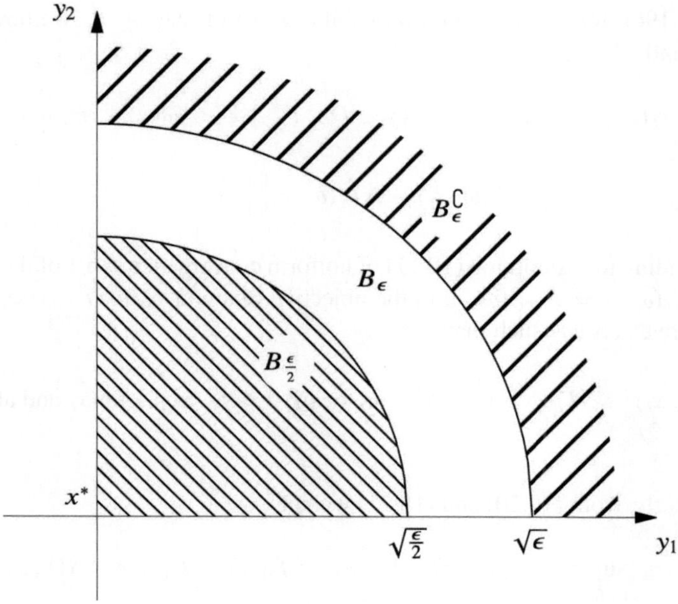

Fig. 18.1 Sets $B_{\frac{\epsilon}{2}}$, B_ϵ, and complement $B_\epsilon^\complement$ for $D = \mathbb{R}^2_+$, $F_0(y) := y_1^2 + y_2^2$

Corresponding to (18.16b) we get

$$\begin{aligned} G_{\hat{F}_t(\omega,\cdot)}(x_t) &= \left\{ y \in D : \hat{F}_t(\omega, y) < \hat{F}_t(\omega, x_t) \right\} \\ &= \left\{ y \in D : F_0(y) < F_0(x_t) + \Delta\hat{F}_t(\omega, x_t) - \Delta\hat{F}_t(\omega, y) \right\}. \end{aligned} \tag{18.19b}$$

Since $Z_{t+1} = y \in G_{\hat{F}_t(\omega,\cdot)}(x_t)$, $y \notin B_\epsilon$ yields $F_0(y) > F_0^* + \epsilon$, from (18.19b) and because of $x_t \in B_\epsilon$ we obtain

$$\begin{aligned} & G_{\hat{F}_t(\omega,\cdot)}(x_t) \setminus B_\epsilon \\ &= \left\{ y \in D : F_0^* + \epsilon < F_0(y) < F_0(x_t) + \Delta\hat{F}_t(\omega, x_t) - \Delta\hat{F}_t(\omega, y) \right\} \\ &\subset \left\{ y \in D : F_0^* + \epsilon < F_0(y) < F_0^* + \epsilon + \Delta\hat{F}_t(\omega, x_t) - \Delta\hat{F}_t(\omega, y) \right\} \\ &\subset \left\{ y \in D : F_0^* + \epsilon < F_0(y) < F_0^* + \epsilon + \left\| \Delta\hat{F}_t(\omega, x_t) - \Delta\hat{F}_t(\omega, y) \right\| \right\}. \end{aligned} \tag{18.19c}$$

From (18.19c), for the maximum probability $r_t(\epsilon)$ of leaving B_ϵ we now find the upper bound

$$r_t(\epsilon) \le E_\omega \sup_{x_t \in B_\epsilon} \pi_t \Big(x_t, \Big\{y \in D: \ F_0^* + \epsilon < F_0(y) < F_0^* + \epsilon + \Big\|\Delta \hat{F}_t(\omega, x_t) - \Delta \hat{F}_t(\omega, y)\Big\|\Big\}.\Big) \tag{18.20}$$

Corresponding to assumption (18.17) of uniform convergence w.p.1 of the estimator (18.5a), $\hat{F}_t(\omega,\cdot) = \hat{F}_{t,m_t}(\omega,\cdot)$, to the objective function F_0 for $t \to \infty$, for given $\delta > 0$ there exists a t_δ such that

$$\Big\|\Delta \hat{F}_t(\omega, x_t) - \Delta \hat{F}_t(\omega, y)\Big\| < \delta, \quad m_t \ge \overline{m_t},\ t > t_\delta,\ \text{w.p.1 for } x_t \text{ and all } y \in D. \tag{18.21}$$

Consequently, from (18.20) and (18.21) we get

$$r_t(\epsilon) \le \sup_{x \in B_\epsilon} \pi_t \Big(x, \Big\{y \in D: \ F_0^* + \epsilon < F_0(y) < F_0^* + \epsilon + \delta\Big\}\Big), t > t_\delta. \tag{18.22a}$$

For $r_t(\epsilon)$ we now have the following result (Fig. 18.2):

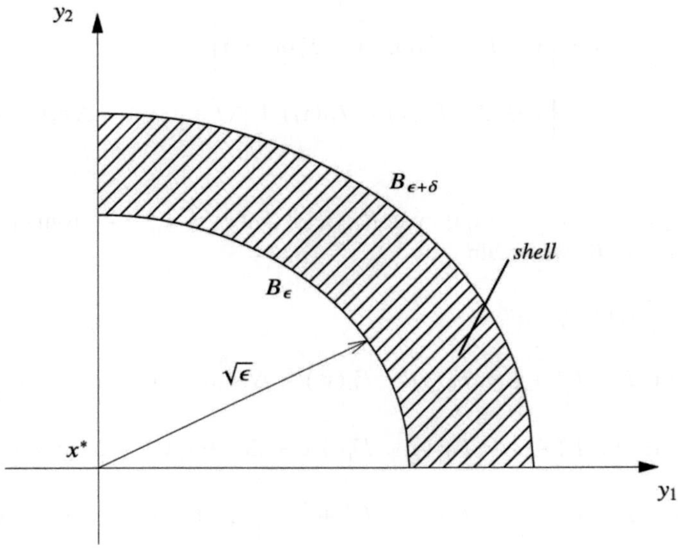

Fig. 18.2 Shell for $D = \mathbb{R}^2_+,\ F_0(y) := y_1^2 + y_2^2$

Lemma 18.3 *With condition* (18.21) *and assuming stochastic independence of* Z_{t+1} *and* $\hat{F}_t$*, given* $X_t = x_t$*, it holds*

$$r_t(\epsilon) \leq \sup_{x \in B_\epsilon} \pi_t(x, B_{\epsilon+\delta} \setminus B_\epsilon), \quad t > t_\delta. \tag{18.22b}$$

Note 18.1 The set $B_{\epsilon+\delta} \setminus B_\epsilon$ is a shell of the level set $B_{\epsilon+\delta} = \{y \in D : F_0(y) \leq \epsilon + \delta\}$:

References

1. Devroye, L., Krzyzak, A.: Random Search Under Additive Noise, pp. 383–417. Kluwer Academic Publishers, Boston (2002)
2. Marti, K.: Minimizing noisy objective function by random search methods. ZAMM **62**, 377–380 (1982)
3. Marti, K.: Stochastic Optimization Methods, 2nd edn. Springer, Berlin (2008). https://doi.org/10.1007/978-3-540-79458-5
4. Polyak, B.: Introduction to optimization. Translation series in mathematics and engineering. Optimization Software, New York (1987)

Appendix A
Properties of the Uniform Distribution on the Unit Sphere

Let denote $S := \{x \in \mathbb{R}^d : |x| = 1\}$ the unit sphere in the space $\mathbb{R}^d (d \in \mathbb{N})$.

Moreover, for $d \geq 2$ let

$T : \mathbb{R}_+ \times [o, \pi]^{d-2} \times [o, 2\varphi] \to \mathbb{R}^d$ be the d-dimensional polar coordinate representation, i.e.,

$$T(r, \varphi_1, \ldots, \varphi_{d-1}) := \begin{pmatrix} r \cos \varphi_1 \\ r \sin \varphi_1 \cos \varphi_2 \\ \vdots \\ r \sin \varphi_1 \sin \varphi_2 \ \ldots \sin \varphi_{d-2} \cos \varphi_{d-1} \\ r \sin \varphi_1 \sin \varphi_2 \ \ldots \sin \varphi_{d-2} \cos \varphi_{d-1} \end{pmatrix}$$

and define

$T_1 : [o, \pi]^{d-2} \times [o, 2\pi] \to \mathbb{R}^d$,

$T_1(\varphi_1, \ldots, \varphi_{d-1}) := T(1, \varphi_1, \ldots, \varphi_{d-1})$.

We know that $T_1([o, \pi]^{d-2} \times [o, 2\pi]) = S$.

Definition A.1 The probability measure γ on $(S, \mathcal{B}|S)$, defined by

(i) $\gamma(B) := \frac{1}{2}\epsilon_{-1}(B) + \frac{1}{2}\epsilon_1(B) \quad$ for $d = 1$

(ii) $\gamma(B) := \frac{\Gamma(d/2)}{2\pi^{d/2}} \int\limits_{T_1^{-1}(B)} \sin^{d-2} \varphi_1 \sin^{d-3} \varphi_2 \ldots \sin \varphi_{d-2} d(\varphi_1, \ldots, \varphi_{d-1})$

for $d \geq 2$

is called the normalized uniform distribution on S.

Obviously, this is a measure. The fact that this measure is normalized follows from

K. Marti, *Optimization Under Stochastic Uncertainty*, International Series in Operations Research & Management Science 296,
https://doi.org/10.1007/978-3-030-55662-4

$$\int_0^{\pi} \dots \int_0^{\pi} \int_0^{2\pi} \sin \varphi_{d-2} d\varphi_{d-1} \dots d\varphi_1 = \frac{2\pi^{d/2}}{\Gamma(d/2)}$$

Furthermore, from the definition of γ, for each nonnegative function $g : S \to \mathbb{R}$ we have

$$\int_S g(s)\gamma(ds) = \frac{\Gamma(d/2)}{2\pi^{d/2}} \int_0^{\pi} \dots \int_0^{\pi} \int_0^{2\pi} g(T_1(\varphi_1, \dots, \varphi_{d-1}))|\frac{dT_1}{d\varphi}|d\varphi_{d-1} \dots d\varphi_1,$$

where $|\frac{dT_1}{d\varphi}| := \sin^{d-2} \varphi_1 \sin^{d-3} \varphi_2 \dots \sin \varphi_{d-2}$.

Lemma A.1 *Given a nonnegative, measurable function $h : \mathbb{R}^d \to \mathbb{R}$, we have*

$$\begin{aligned}\int_{\mathbb{R}^d} h(x)\lambda^d(dx) &= \frac{2\pi^{d/2}}{\Gamma(d/2)} \int_{\mathbb{R}_+} \int_S r^{d-1} h(rs)\gamma(ds)\lambda^1(dr) \\ &= \frac{2\pi^{d/2}}{\Gamma(d/2)} \int_S \int_{\mathbb{R}_+} r^{d-1} h(rs)\lambda^1(dr)\gamma(ds).\end{aligned}$$

Proof Using the d-dimensional substitution rule, we get

$$\int_{\mathbb{R}^d} h(x)\lambda^d(dx) = \int_{T^{-1}(\mathbb{R}^d)} h(T(r, \varphi_1, \dots, \varphi_{d-1}))|\frac{dT}{d(r, \varphi)}|\lambda^d(d(r, \varphi)).$$

Because of $|\frac{dT}{d(r,\varphi)}| = r^{d-1} \sin^{d-2} \varphi_1 \sin^{d-3} \varphi_2 \dots \sin \varphi_{d-2} = r^{d-1}|\frac{dT_1}{d\varphi}|$ and due to $T(r, \varphi_1, \dots, \varphi_{d-1}) = rT_1(\varphi_1, \dots, \varphi_{d-1})$, we find

$$\begin{aligned}\int_{\mathbb{R}^d} h(x)\lambda^d(dx) &= \int_0^{\infty} \int_0^{\pi} \dots \int_0^{\pi} \int_0^{2\pi} r^{d-1} h(rT_1(\varphi_1, \dots, \varphi_{d-1}))|\frac{dT_1}{d\varphi}|d\varphi_{d-1} \dots d\varphi_1 dr \\ &= \frac{2\pi^{d/2}}{\Gamma(d/2)} \int_0^{\infty} \int_S r^{d-1} h(rs)\gamma(ds)dr.\end{aligned}$$

Thus, the first equality is shown, and the second one follows from Fubini's theorem. □

Corollary A.1 *Let X be a d-dimensional random vector such its distribution P_X has a Lebesgue density g.*

Defining $Y := \frac{X}{|X|}$ *and* $R := |X|$, *we get*

(1)

$$P((Y, R) \in C) = \frac{2\pi^{d/2}}{\Gamma(d/2)} \int_C r^{d-1} g(rs), \gamma \otimes \lambda^1(d(s, r))$$

$$C \in (\mathcal{B}^1|\mathbb{R}_+).$$

(2) If $g(x) = h(|x|)$ *(with a measurable function* $h : \mathbb{R} \to \mathbb{R}$*), then* Y *and* R *are independent. Moreover,* $P_Y = \gamma$.

Proof It is sufficient to show the first assertion for product sets $C = A \times B (A \in \mathcal{B}^d|S, B \in \mathcal{B}^1|\mathbb{R})$, since those form an intersection-stable generator of the product$-\sigma-$algebra $(\mathcal{B}^d|S) \otimes (\mathcal{B}^1|\mathbb{R}_+)$.

Hence, let $C = A \times B$.

Furthermore, define $D := \{x \in \mathbb{R}^d \setminus \{0\} : \frac{x}{|x|} \in A, |x| \in B\}$.

Because of $\{(Y, R) \in C\} = \{X \in D\}$, we get

$$\begin{aligned}
P((Y, R) \in C) &= \int_D g(x) \lambda^d(dx) \\
&= \frac{2\pi^{d/2}}{\Gamma(d/2)} \int_S \int_{R_+} r^{d-1} 1_D(rs) g(rs) dr \gamma(ds) \\
&= \frac{2\pi^{d/2}}{\Gamma(d/2)} \int_S \int_{R_+} r^{d-1} 1_A(s) 1_B(r) g(rs) dr \gamma(ds) \\
&= \frac{2\pi^{d/2}}{\Gamma(d/2)} \int_A \int_{R_+} r^{d-1} g(rs) dr \gamma(ds) \\
&= \frac{2\pi^{d/2}}{\Gamma(d/2)} \int_{A \times B} r^{d-1} g(rs) \gamma \otimes \lambda^1(d(s, r)).
\end{aligned}$$

Assertion (2) follows now directly from the first one, since

$$g(rs) = h(|rs|) = h(r)$$

and

$$\int_{\mathbb{R}_+} r^{d-1} h(r) dr = \frac{\Gamma(d/2)}{2\pi^{d/2}} \quad \text{according to A.2.}$$

□

Lemma A.2 *If $T : \mathbb{R}^d \to \mathbb{R}$ is an orthogonal mapping , then $T(\gamma) = \gamma$.*

Proof Because of $|T(x)| = |x|$ for arbitrary $x \in \mathbb{R}^d$, we have $T(S) \subset S$. Since T is bijective and the mapping T^{-1} is also orthogonal, it follows that $T(S) = S$.

Let X denote a d-dimensional standardized normal distributed random vector. Then P_X has the density

$$f_X(x) = (2\pi)^{-d/2} \exp\left(-\frac{|x|^2}{2}\right).$$

According to A.1, (2), we have $P_Y = \gamma$, where $Y = \frac{X}{|X|}$.

It is known that the random vector $T(X)$ has the density

$$f_{T(X)}(y) = \left|\frac{dT^{-1}}{dy}\right| f_X(T^{-1}(y)) = f_X(y);$$

indeed, because of the orthogonality of T we have

$$|\frac{dT^{-1}}{dy}| = 1 \text{ and } |T^{-1}| = |y|.$$

Thus, $T(X)$ is also a d-dimensional standardized normal distributed random vector. Because of

$$T(\frac{X}{|X|}) = \frac{1}{|X|} T(X) = \frac{T(X)}{|T(X)|},$$

we get

$$T(\gamma) = T(P_\gamma) = P_{T(Y)} = P_{T(X/|X|)} = P_{\frac{T(X)}{|T(X)|}} = \gamma,$$

which yields the assertion. □

Lemma A.3 *Consider a cone $C(\alpha, x)$ with the axis of symmetry x and the apex angle 2α (cf. Definition B.1).*

Then for all $x, y \neq 0$ and $\alpha \in [0, \pi]$ we have

$$\gamma(S \cap C(\alpha, x)) = \gamma(S \cap C(\alpha, y)).$$

Proof Consider $x, y \neq 0$ and $\alpha \in [0, \pi]$.

Without loss of generality we may assume that $|x| = |y| = 1$. Let T denote the reflection at the hyperbola $\mathcal{H}$ through the point $\frac{x+y}{2}$ and with the normal vector $x - y$, i.e.

$$\mathcal{H} = \{z \in \mathbb{R}^d : (x-y)^T \left(z - \frac{x+y}{2}\right) = 0\}.$$

We know that T is orthogonal, hence, $T = T^{-1}$, moreover, $T(x) = y$. Because of

$$x^T z = (T(x))^T T(z) = y^T T(z),$$

we have $z \in S \cap C(\alpha, x)$ and therefore $x^T z \leq \cos\alpha$ if and only if $T(z) \in S \cap C(\alpha, y)$.

This yields $T(S \cap C(\alpha, x)) = S \cap C(\alpha, y)$.

Now the assertion follows with Lemma (A.2) . □

Definition A.2 For given $x \in \mathbb{R}^d$, $r > 0$ and $B \in \mathcal{B}^d$ define

$$\gamma_{x,r}(B) := \gamma(\{s \in S : x + rs \in B\}).$$

The probability measure $\gamma_{x,r}$ is called the uniform distribution on the sphere around x and with radius r.

In particular we have $\gamma_{0,1} = \gamma$.

Appendix B
Analytical Tools

At first we will compile some characteristics of convex functions and convex sets.

Lemma B.1 *Let $f : \mathbb{R}^d \to \mathbb{R}$. Then the following statements are equivalent:*

(i) *f is convex*
(ii) $\frac{f(x)-f(y)}{|x-y|} \geq \frac{f(x)-f(z)}{|x-z|}$ *for all $x, z \in \mathbb{R}^d$, $x \neq z$ and for all y between x and z (i.e. $y = (1-t)x + tz$ with $t \in (0,1)$).*
(iii) $\frac{f(x)-f(y)}{|x-y|} \geq \frac{f(x)-f(z)}{|x-z|}$ *for all $x, z \in \mathbb{R}^d$, $x \neq z$ and any y between x and z.*

Proof To prove the first equivalence, let $x, z \in \mathbb{R}^d$, $x \neq z$ and $y = (1-t)x + tz$ with $t \in (0,1)$.

From the continuity of f it follows that $f(y) \leq (1-t)f(x) + tf(z)$, therefore $f(x) - f(y) \geq t(f(x) - f(z))$ and thus

$$\frac{f(x)-f(y)}{|x-y|} = \frac{f(x)-f(y)}{|x-z|} \geq \frac{f(x)-f(z)}{|x-z|}.$$

If, conversely, (ii) is satisfied, it follows that

$$\begin{aligned} f((1-t)x+tz) = f(y) &\leq f(x) - |z-y| \frac{f(x)-f(z)}{|x-z|} \\ &= f(x) - t(f(x)-f(z)) = (1-t)f(x) + tf(z), \end{aligned}$$

i.e. f is convex.

The equivalence of (i) and (iii) is proven analogically. □

Corollary B.1 *Let $f : \mathbb{R}^d \to \mathbb{R}$ be convex and differentiable.*

K. Marti, *Optimization Under Stochastic Uncertainty*, International Series in Operations Research & Management Science 296,
https://doi.org/10.1007/978-3-030-55662-4

Then we have for all $x, z \in \mathbb{R}^d$

$$(\nabla f(z))^T (x - z) \leq f(x) - f(z) \leq (\nabla f(x))^T (x - z)$$

Proof Let first be $d = 1$ and let $x, z \in \mathbb{R}^d$.

Without loss of generality let $z > x$.

For all $y \in (x, z)$ Lemma B.1, (ii) yields the inequality $\frac{f(x)-f(z)}{|z-x|} \leq \frac{f(x)-f(y)}{|y-x|}$ and thus

$$f(x) - f(z) \leq (z - x) \lim_{y \downarrow x} \frac{f(x) - f(y)}{|y - x|} = f'(x)(x - z).$$

The inequality $f(x) - f(z) \geq (f'(z))(x - z)$ can be shown analogically.

Let now $d \in \mathbb{N}$ be arbitrary.

Let $x, z \in \mathbb{R}^d$ be arbitrary, but fixed.

Let $g(t) := f((1 - t)x + tz)$.

From the differentiability of f follows the differentiability of g. If f is convex, g is also convex.

Therefore

$$-g'(1) \leq g(0) - g(1) \leq -g(0)', \text{ i.e.}$$

$$(\nabla f(z))^T (x - z) \leq f(x) - f(z) \leq (\nabla f(x))^T (x - z).$$

□

Lemma B.2 *Let* $D \subset \mathbb{R}^d$ *be open and convex, and let* $f : D \to \mathbb{R}$ *be convex.*

Then f *is continuous.*

Lemma B.3 *Let* $D \subset \mathbb{R}^d$ *be open and convex, and let* $f : D \to \mathbb{R}$ *be twice continuously differentiable. Then* f *is convex if and only if the Hessian Matrix of* f *is positive semidefinite definite in all points* $x \in D$.

Lemma B.4 *Let* $D \subset \mathbb{R}^d$ *be convex and let* $f : D \to \mathbb{R}$ *be continuous and convex.*

Let there be an $\hat{a} \in \mathbb{R}$ *such that the set* $\{x \in D : f(x) \leq \hat{a}\}$ *is not empty and bounded.*

Then the sets $\{x \in D : f(x) \leq a\}$ *are bounded and convex for any* $a \geq \inf\{f(x) : x \in D\}$.

Lemma B.5 *Let* $A \subset \mathbb{R}^d$ *be convex and let* $x \in rd A$.

Then A *has supporting hyperplane at* x, *i.e. there is a* $z \in \mathbb{R}^d \backslash \{0\}$ *with*

$$A \subset x + \{y \in \mathbb{R}^d : y^T z \leq 0\}.$$

Lemma B.6 *Let* $D \subset \mathbb{R}^d$ *be convex. Let* $f : D \to \mathbb{R}$ *be convex.*

Then we have for all $x \in D$ *with* $f(x) > \inf\{f(z) : z \in D\}$:

$$\{y \in D : f(y) = f(x)\} \subset rd\{y \in D : f(y) \leq f(x)\}$$

Proof Let $x \in D$ with $f(x) > \inf\{f(z) : z \in D\}$ and let $r > 0$ be arbitrary. Then it is trivial that we have for $y \in D$ with $f(y) = f(x)$:

$$K(y,r) \cap \{u \in D : f(u) \leq f(x)\} \neq \emptyset.$$

But we also have $K(y,r) \subset (\mathbb{R}^d \backslash \{u \in D : f(u) \leq f(x)\} \neq \emptyset$.

For if this intersection were empty, then we would have

$$K(y,r) \subset \{u \in D : f(u) \leq f(x)\}.$$

Now Lemma B.1 yields a contradiction to the convexity of f. For since we have $f(x) > \inf\{f(u) : u \in D\}$, there is a $z \in D$ with $f(x) > f(z)$.

If we now choose $y_1 := y + \frac{r}{2|y-z|}(y-z)$, then

$$y_1 \in K(y,r), \text{ therefore } f(y_1) \leq f(x) = f(y).$$

Since y lies between y_1 and z (for we have $y = y_1 + \frac{r}{2|y-z|}(z - y_1)$), it follows from Lemma B.1

$$\frac{f(y_1) - f(y)}{|y_1 - y|} \geq \frac{f(y) - f(z)}{|y - z|}, \text{ therefore}$$

$$f(y_1) \leq f(y) + \frac{r}{2}\frac{f(y) - f(z)}{|y - z|} > f(y).$$

□

Definition B.1

(1) For $x, y \in \mathbb{R}^d \backslash \{0\}$ let $\varphi(x,y) := \arccos \frac{x^T y}{|x||y|}$ (angle between x and y)
(2) For $\varphi \in [0, \pi]$ and $x \in \mathbb{R}^d \backslash \{0\}$ let $C(\varphi, x) := \{y \in \mathbb{R}^d : \varphi(x,y) \leq \varphi\}$ (a cone with the "axis of symmetry" x and the "apex angle" 2φ)
(3) For $\varphi \in [0, \pi]$, $x \in \mathbb{R}^d \backslash \{0\}$ and $r > 0$ let
$C(\varphi, x, r) := \{y \in \mathbb{R}^d : \varphi(x,y) \leq \varphi, |y| \leq r\}$
(4) For $\varphi \in [0, \pi]$, $x \in \mathbb{R}^d \backslash \{0\}$, and $r_1, r_2 > 0$ let
$C(\varphi, x, r_1, r_2) := \{y \in \mathbb{R}^d : \varphi(x,y) \leq \varphi, r_1 \leq |y| \leq r_2\}$

Lemma B.7 *Let $(A_t : t \in K)$ be a family of invertible $d \times d$ matrices with a compact index set K. Let the representation $t \to A_t$ be continuous.*

Then:

(1) $\forall \phi > 0 \, \exists \, \varphi > 0 \, \forall \, x \in \mathbb{R}^d \setminus \{0\}, t \in K : C(\phi, A_t^{-1}x) \subset A_t^{-1}(C(\phi, x))$
(2) $\forall \, x \in \mathbb{R}^d \setminus \{0\}, t \in K : \sqrt{\lambda_t^{\min}} \, |x| \leq |A_t x| \leq \sqrt{\lambda_t^{\max}} \, |x|$, *where* $\lambda_t^{\min} := \min\{\lambda : \lambda$ *eigenvalue of* $A_t^T A_t\}$
and $\lambda_t^{\max} := \max\{\lambda : \lambda$ *eigenvalue of* $A_t^T A_t\}$

(3) $\forall \phi > 0 \, \exists \varphi > 0 \, \forall x \in \mathbb{R}^d \setminus \{0\}, t \in K:$
$C(\varphi, A_t^{-1}x, \frac{r_1}{\sqrt{\lambda_t^{\min}}}, \frac{r_2}{\sqrt{\lambda_t^{\max}}}) \subset A_t^{-1}(C(\phi, x, r_1, r_2))$
with $\lambda_t^{\min}$ *and* $\lambda_t^{\max}$ *according to (2).*

Proof

(i)

$$\forall \phi > 0 \, \exists \varphi > 0 \, \forall x, y \neq 0, t \in K : [\varphi(A_t^{-1}x, y) \le \varphi \Rightarrow \varphi(x, A_t y) \le \phi]$$
$$\Longleftrightarrow \forall \phi > 0 \, \exists \varphi > 0 \, \forall x, y \neq 0, t \in K : [\varphi(x, y) \le \varphi \Rightarrow \varphi(A_t x, A_t y) \le \phi]$$
$$\Longleftrightarrow \forall \phi > 0 \, \exists \varphi > 0 \, \forall x, y \text{ with } |x| = |y| = 1, t \in K :$$
$$[x^T y \ge 1 - \varphi \Rightarrow \arccos\left(\frac{(A_t x)^T A_t y}{|A_t x||A_t y|}\right) \le \phi]$$
$$\Longleftrightarrow \forall \epsilon > 0 \, \exists \delta > 0 \, \forall x, y \text{ with } |x| = |y| = 1, t \in K :$$
$$[x^T y \ge 1 - \delta \Rightarrow \frac{(A_t x)^T A_t y}{|A_t x||A_t y|} \le \phi]$$

For an arbitrary $d \times d$ matrix A and for arbitrary $x, y \in \mathbb{R}^d$ it follows that

$$\begin{aligned} 1 - \frac{(Ax)^T Ay}{|Ax||Ay|} &\le |1 - \frac{(A_t x)^T A_t y}{|A_t x||A_t y|}| \\ &= \frac{1}{|Ay|}||Ay| - |Ax| - (\frac{Ax}{|Ax|})^T A(y - x)| \le \\ &\le \frac{1}{|Ay|}(||Ay| - |Ax|| + |(\frac{Ax}{|Ax|})^T A(y - x)|) \le \\ &\le \frac{1}{|Ay|}(|A(y - x)| + |A(y - x)|) = \frac{2|A(y - x)|}{|Ay|} \le \\ &\le \frac{2|A||y - x|}{|Ay|}, \end{aligned}$$

where $|A|$ stands for the Euclidean norm of the matrix A.

For $t \in K$ and $y \in \mathbb{R}^d$ let $g(t, y) := |A_t y|$.

From the continuity of the representation $t \to A_t$ follows the continuity of the representation g. From the compactness of $K \times \{y \in \mathbb{R}^d : |y| = 1\} = 1$ and the invertibility of the matrices A_t it follows that there is a $c > 0$ with

$$\inf\{|A_t y| : t \in K, |y| = 1\} \ge c.$$

From the continuity of the representation $t \to A_t$ it follows also that the representation $t \to |A_t|$ is continuous and thus that there is an $R < \infty$ with

$$\sup\{|A_t| : t \in K\} \le R.$$

For arbitrary $x, y \in \mathbb{R}^d$ with $|x| = |y| = 1$ and arbitrary $t \in K$ and because of

$$|x - y|^2 = |x|^2 + |y|^2 - 2x^T y = 2(1 - x^T y)$$

this now yields the inequality

$$1 - \frac{(A_t x)^T A_t y}{|A_t x||A_t y|} \le \frac{2R}{c}\sqrt{2(1 - x^T y)}.$$

Thus the assertion is proven.

(ii) From the invertibility of A_t it follows that $A_t^T A_t$ is a positive definite symmetric matrix.

There is therefore a diagonal matrix Λ with positive diagonal elements and an orthogonal matrix S with

$$A_t^T A_t = S^T \Lambda S.$$

Hence

$$|A_t x|^2 = x^T (A_t^T A_t) x = (Sx)^T \Lambda (Sx).$$

Since the diagonal elements of Λ are exactly the eigenvalues of $A_t^T A_t$, it follows that

$$\lambda_t^{\min} |Sx|^2 \le |A_t x|^2 \le \lambda_t^{\max} |Sx|^2$$

Because of the orthogonality of S we have $|Sx| = |x|$.

This yields the assertion.

(iii) If $\frac{r_1}{\sqrt{\lambda_t^{\min}}} \le |y| \le \frac{r_2}{\sqrt{\lambda_t^{\max}}}$, the inequality

$$r_1 \le |A_t y| \le r_2$$

follows from (2). Together with (1) this yields the assertion.

□

Lemma B.8 *Let $x, z \in \mathbb{R}^d$ and $r > 0$ with $r < |x - z|$.*
Then with $\varphi := \arcsin \frac{r}{|x-z|}$:

(i) for all $y \in x + C(\varphi, z - x)$ there is a $t > 0$ with

$$|x + t(y - z) - z| \le r.$$

(ii)

$$x + C(\varphi, z - x, |z - x| - r) \subset conv\ (\{x\} \cup \overline{K(z, r)})$$

$(conv\ A = convex\ closure\ of\ A)$

Proof

(i) Since y lies in $x+C(\varphi, z-x)$, it follows that $\varphi(y-z, z-x) \leq \arcsin \frac{r}{|x-z|} < \frac{\pi}{2}$ and hence $(y - x)^T(x - z) > 0$.

Let $t := \frac{(y-x)^T(x-z)}{|y-x|^2}$ and let $z' := x + t(y - x)$ and $u := x + \frac{1}{t}(z - x)$ (z' is the projection of z on the straight line through x and y and y the projection of u on this straight line).

Then $(y - x)^T(u - y) = (z' - x)^T(z - z') = 0$, which yields

$$|u - y|^2 + |y - x|^2 = |u - x|^2$$

hence

$$\frac{|u - y|^2}{|u - x|^2} + \frac{|y - x|^2}{|u - x|^2} = 1$$

and

$$\frac{|z - z'|^2}{|z - x|^2} + \frac{|z' - x|^2}{|z - x|^2} = 1.$$

According to the definition of z' and u, we have

$$\frac{|y - x|}{|u - x|} + \frac{|z' - x|}{t \cdot \frac{1}{t}|z - x|} = \frac{|z' - x|}{|z - x|}$$

From this and from the above equality it follows that

$$\frac{|u - y|}{|u - x|} = \frac{|z - z'|}{|z - x|}$$

and thus $|z - z'| = |z - x|\frac{|u-y|}{|u-x|}$.

Because of

$$\sqrt{1-\frac{|u-y|^2}{|u-x|^2}} = \frac{|y-x|}{|u-x|} = \frac{(y-x)^T(y-x)}{|u-x|\cdot|y-x|}$$
$$= \frac{(u-x)^T(y-x)+(y-u)^T(y-x)}{|u-x|\cdot|y-x|} = \frac{(u-x)^T(y-x)}{|u-x|\cdot|y-x|}$$
$$= \cos\varphi(u-x, y-x) = \cos\varphi(z-x, y-x)$$

it follows that

$$\sin\varphi(z-x, y-x) = \frac{|u-y|}{|u-x|}$$

and thus

$$\frac{|u-y|}{|u-x|} \le \sin\varphi = \frac{r}{|x-z|}$$

hence $|z-z'| \le r$.

(ii) Let $y \in x + C(\varphi, z-x, |z-x|-r)$ be arbitrary.

According to part (1), there is a $t > 0$ with $|x+t(y-x)-z| \le r$.

Let $z' := x + t(y-x)$.

Because of $|z'-z| \le r$ we have $|z'-x| \ge |z-x| - |z'-z| \ge |z-x| - r$ and thus $\frac{|y-x|}{|z'-x|} \le \frac{|z-x|-r}{|z-x|-r} = 1$.

With $\alpha := \frac{|y-x|}{|z'-x|}$ we have

$$y = x + (y-x) = x + \frac{|y-x|}{|z'-x|}(z'-x) = \alpha z' + (1-\alpha)x.$$

Since $\alpha \in [0,1]$ and $z' \in \overline{K(z,r)}$, we have $y \in \text{conv}\,(\{x\} \cup \overline{K(z,r)}) = 1$.

□

Lemma B.9 *Let $D \subset \mathbb{R}^d$ be convex. Let $f : D \to \mathbb{R}$ be convex and continuous.*
Let $f^* := \inf\{f(x) : x \in D\} > -\infty$.
Let the minimum set $D^* := \{x \in D : f(x) = f^*\}$ be bounded.
Furthermore, let there exist an $x^* \in D^* \cap \mathring{D}$.
Then:

(i) for all $x \in D \backslash D^*$ there is a $\phi > 0$ with

$$x + C(\phi, x^*-x, \frac{1}{2}|x^*-x|) \subset \{y \in D : f(x) - f(y) \ge \frac{|y-x|}{3|x^*-x|}(f(x) - f^*)\}$$

(ii) If $\tilde{a} > f^*$ and $\tilde{b} > 0$ exist in such a way that the condition

$$\inf\{|y-x^*| : y \in D, f(y) = f(x)\}$$

$$\geq \tilde{b} \cdot \sup\{|y - x^*| : y \in D,\ f(y) = f(x)\}$$

is satisfied for all $x \in \{y \in D : f(y) \leq \tilde{a}\}\backslash D^*$, *then we have furthermore:*
for all $a > f^*$ *there is a* $\phi \in (0, \frac{\pi}{2}]$ *with*

$$x+C(\phi, x^*-x, \frac{1}{2}|x^*-x|)\subset \left\{y \in D : f(x) - f(y) \geq \frac{|y - x|}{3|x^* - x|}(f(x) - f^*)\right\}$$

for all $x \in \{y \in D : f(x) \leq a\}\backslash D^*$

Proof Because of $x^* \in D^0$ there is a $\rho > 0$ with $\overline{K(x^*, \rho)} \subset D$.
For $x \in D$ let $r(x) := \inf\{|y - x^*| : y \in D,\ f(y) = f(x)\}$.
From the continuity of f it follows that $r(x) > 0$ for all $x \in D\backslash D^*$.
For $x \in D\backslash D^*$ let

$$\rho(x) := \min\{\rho, r(x)\}$$

and

$$\phi(x) := \arcsin \frac{\rho(x)}{2|x - x^*|}.$$

Let now $x \in D\backslash D^*$ be arbitrary.
Let $z \in \overline{K(x^*, \frac{1}{2}\rho(x))}$, $z \neq x^*$.
Let $\tilde{x} := x^* + \frac{\rho(x)}{|z-x^*|}(z - x^*)$.
Because of $|\tilde{x} - x^*| = \rho(x) > 0$, we have $f(\tilde{x}) \leq f(x)$.
For if $f(\tilde{x}) > f(x)$, then there would be a $t \in [0, 1)$ with $f(x^* + t(\tilde{x} - x^*)) = f(x)$ and thus

$$\rho(x) \leq r(x) \leq |x^* + t(\tilde{x} - x^*) - x^*| = t|\tilde{x} - x^*| = t\rho(x),$$

i.e. $\rho(x) = 0$, which is in contradiction to $\rho(x) > 0$.
Then it follows from Lemma B.1 that

$$\frac{f(z) - f^*}{|z - x^*|} \leq \frac{f(\tilde{x}) - f^*}{|\tilde{x} - x^*|} \leq \frac{f(x) - f^*}{\rho(x)}$$

and thus

$$f(z) - f^* \leq \frac{1}{2}(f(x) - f^*).$$

Let now $y \in x + C(\varphi(x), x^* - x, \frac{1}{2}|x^* - x|)$ be arbitrary.
According to Lemma B.8, we now have because of the convexity of D

$$x + C(\varphi(x), x^* - x, \frac{1}{2}|x^* - x|) \subset$$

$$\overline{\operatorname{conv}(\{x\} \cup K(x^*, \frac{\rho(x)}{2}))} \subset D.$$

Likewise according to Lemma B.8, there is a $t > 0$ with

$$|x + t(y - x) - x^*| \le \frac{\rho(x)}{2}.$$

Because of $|y - x| \le \frac{1}{2}|x^* - x|$, we even have $t > 1$, because for $t < 1$ it follows that

$$|x + t(y - x) - x^*| \ge |x - x^*| - t|y - x| >$$

$$|x - x^*| - |y - x| \ge |x - x^*| - \frac{1}{2}|y - x|$$

$$\ge \frac{r(x)}{2} \ge \frac{\rho(x)}{2},$$

which is in contradiction to

$$|x + t(y - x) - x^*| \le \frac{\rho(x)}{2}.$$

From Lemma B.1 it follows furthermore with $z := x + t(y - x)$ that

$$\frac{f(x) - f(y)}{|x - y|} \ge \frac{f(x) - f(z)}{|x - z|}.$$

Because of $|z - x^*| \le \frac{1}{2}\rho(x)$ is $f(z) - f^* \le \frac{1}{2}(f(x) - f^*)$ and because of

$$|x - z| \le |x - x^*| + |z - x^*| \le$$

$$\le |x - x^*| + \frac{1}{2}\rho(x) \le$$

$$\le |x - x^*| + \frac{1}{2}r(x)| \le \frac{3}{2}|x - x^*|$$

it finally follows that

$$f(x) - f(y) \ge \frac{|x - y|}{|x - z|}(f(x) - f^* - (f(z) - f^*))$$

$$\ge \frac{|x - y|}{|x - z|}\frac{f(x) - f^*}{2} \ge \frac{1}{3}\frac{|y - x|}{|x^* - x|}(f(x) - f^*).$$

Thus the first assertion is proven.

The second assertion follows obviously from the first, if for all $a > f^*$ there is a $\phi > 0$ such that we have $\phi(x) \geq \phi$ for all $x \in D$ with $f^* < f(x) \leq a$.

Now we will show that such a ϕ exists.

Let $a \geq \tilde{a}$ be arbitrary.

Let $R := \sup\{|x - x^*| : x \in D, f(x) \leq a\}$. According to Lemma B.4, we have $R < \infty$.

For $x \in \{y \in D : \tilde{a} \leq f(y) \leq a\}$ we have

$$\frac{\inf\{|y - x^*| : y \in D, f(y) = f(x)\}}{\sup\{|y - x^*| : y \in D, f(y) = f(x)\}} \geq \frac{\inf\{|y - x^*| : y \in D, f(y) \geq \tilde{a}\}}{R} := \hat{b}$$

Because of $\tilde{a} > f^*$ and since f is continuous, we have $\hat{b} > 0$.

According to the condition, it follows now with $b := \min\{\hat{b}, \tilde{b}\}$ that

$$\inf\{|y - x^*| : y \in D, f(y) = f(x)\} \geq b \cdot \sup\{|y - x^*| : y \in D, f(y) = f(x)\}$$

for all $x \in D$ with $f^* < f(x) \leq a$.

Let now $x \in D$ with $f^* < f(x) \leq a$.

If $\rho \leq r(x)$ it follows in accordance with the definition that $\rho(x) = \rho$ and thus

$$\frac{\rho(x)}{2|x - x^*|} \geq \frac{\rho}{2R}.$$

If $\rho > r(x)$, it follows in accordance with the definition that $\rho(x) = r(x)$ and thus

$$\frac{\rho(x)}{2|x - x^*|} = \frac{1}{2}\frac{r(x)}{2|x - x^*|} \geq \frac{b}{2}.$$

If we now choose $\phi := \min\{\arcsin \frac{b}{2}, \arcsin \frac{\rho}{2R}\}\phi(x) \geq \phi$, and thus the assertion, follows for all $x \in D$ with $f^* < f(x) \leq a$.

In the following it is a matter of formulating conditions that are sufficient so that the conditions (14.2) and (14.3) and the conditions (15.1) and (15.2) are satisfied.

□

Lemma B.10 *Let $D \subset \mathbb{R}^d$ be not empty and let $f : D \to \mathbb{R}$ be a map with the following properties:*

(1) There is an $x^* \in D^0$ with $f^* := f(x^*) < f(x)$ for $x \neq x^*$.
(2) f is twice continuously differentiable in a neighborhood of x^*.
(3) The Hessian matrix of f is positive definite at x^*.
(4) An $r_0 > 0$ exists such that there is a $c > 0$ with $\{y \in D : f(y) - f^* < c\} \subset K(x^*, r)$ for all $r \in (0, r_0]$

Then $c_0, a_1, a_2 > 0$ exist with

$$K(x^*, a_1\sqrt{c}) \subset \{y \in D : f(x) - f^* < c\} \subset K(x^*, a_2\sqrt{c}) \qquad \textit{for all } c \in (0, c_0]$$

Proof

(1) Let H be the Hessian Matrix of f at x^*.

According to (2), f is twice continuously differentiable on a sphere $K(x^*, r)$. From the multidimensional Taylor formula it follows then for $h = (h_1, \ldots, h_d)^T \in D(0, r)$ that

$$f(x^*+h)+f(x^*)+(\nabla f(x^*))^T+\frac{1}{2}\sum_{i,j=1}^{d}\frac{\partial^2 f(x^*+\vartheta h)}{\partial x_i \partial x_j}h_i h_j \quad \text{with } \vartheta \in [0,1]$$

Because of (1) and (2), we have $\nabla f(x^*) = 0$.
Therefore $f(x^* + h) + f(x^*) + \frac{1}{2}h^T Hh + R(h)$, with

$$R(h) = \frac{1}{2}\sum_{i,j=1}^{d}\left(\frac{\partial^2 f(x^*+\vartheta h)}{\partial x_i \partial x_j} - \frac{\partial^2 f(x^*)}{\partial x_i \partial x_j}\right)h_i h_j$$

Let

$$\rho(h) := \frac{1}{2}\sum_{i,j=1}^{d}\left|\frac{\partial^2 f(x^*+\vartheta h)}{\partial x_i \partial x_j} - \frac{\partial^2 f(x^*)}{\partial x_i \partial x_j})\right|$$

For $i, j \in \{1, \ldots, d\}$ we have $|h_i h_j| \le |h|^2$ and therefore

$$|R(h)| \le \rho(h)|h|^2.$$

Furthermore, it follows from (2) that

$$\lim_{h\to 0}\rho(h) = 0.$$

For $y \in K(x^*, r)$ we have therefore

$$f(y) = f^* + \frac{1}{2}(y - x^*)^T H(y - x^*) + R(y - x^*),$$

where $|R(h)| \le |h|^2\rho(h)$ and

$$\lim_{h\to 0}\rho(h) = 0.$$

Let λ_0 and λ_1 be the smallest and the largest eigenvalue of H, respectively. Then

$$\frac{1}{2}\lambda_0|y-x^*|^2 \le \frac{1}{2}|y-x^*|^T H(y-x^*) \le \frac{1}{2}\lambda_1|y-x^*|^2$$

(cf., for example, the proof of Lemma B.7, Part 2)).

Hence

$$\frac{1}{2}\lambda_0|y-x^*|^2 - |R(y-x^*)| \le f(y) - f^* \le \frac{1}{2}\lambda_1|y-x^*|^2 + |R(y-x^*)|$$

and because of $|R(y-x^*)| \le |y-x^*|^2\rho(y-x^*)$ we have

$$\left(\frac{1}{2}\lambda_0 - \rho(y-x^*)\right)|y-x^*|^2 \le f(y) - f^* \le \left(\frac{1}{2}\lambda_1 - \rho(y-x^*)\right)|y-x^*|^2$$

This inequality is valid for all $y \in K(x^*, r)$.

(2) Let now $r_1 \le r$ be chosen so that $\rho(y-x^*) < \frac{1}{4}\lambda_0$ is satisfied for all $y \in K(x^*, r_1)$. Then the inequality

$$\frac{1}{2}\lambda_0|y-x^*|^2 \le f(y) - f^* \le \lambda_1|y-x^*|^2$$

follows for $y \in K(x^*, r_1)$. According to Condition (4) $r_0 \le r_1$ and $c_1 > 0$ exist with

$$\{y \in D : f(y) - f^* < c_1\} \subset K(x^*, r_0).$$

Let now

$$c_0 := \min\left\{c_1, r_0^2\lambda_0\frac{1}{4}\right\} \text{ and let } c \in (0, c_0].$$

Because of $c \le c_1$ we have $\{y \in D : f(y) - f^* < c\} \subset K(x^*, r_0)$.

Furthermore, because of $c \le r_0^2\lambda_0 \cdot \frac{1}{4}$ it follows that

$$K(x^*, \frac{2\sqrt{c}}{\sqrt{\lambda_0}})$$

and of course also

$$K\left(x^*, \frac{\sqrt{c}}{\sqrt{\lambda_1}}\right) \subset K(x^*, r_0).$$

Therefore

$$K\left(x^*, \frac{2\sqrt{c}}{\sqrt{\lambda_0}}\right) = \left\{y \in \mathbb{R}^d : \frac{1}{4}\lambda_0|y-x^*|^2 < c\right\} \cap K(x^*, r_0) \supset$$

$$\supset \left\{y \in D : f(y) - f^* < c\right\} \cap K(x^*, r_0) =$$
$$= \left\{y \in D : f(y) - f^* < c\right\} \supset$$
$$\supset \left\{y \in \mathbb{R}^d : \lambda_1 |y - x^*|^2 < c\right\} \cap K(x^*, r_0) = K\left(x^*, \frac{\sqrt{c}}{\sqrt{\lambda_1}}\right)$$

□

Lemma B.11 *Property (4) of Lemma B.10 follows from the properties (1), (2), (3) and one of the additional properties (4′) D is compact, and f is continuous or (4″) D is convex and f is convex.*

Proof Let property (4') be satisfied.

From (2) and (3) it follows that the Hessian matrix is positively definite—and therefore, according to Lemma B.3, that f is convex—on a closed neighborhood $\overline{K(x^*, r_1)}$.

Since D is compact, also $D \backslash K(x^*, r_1)$ is compact.

Let $a := \min\{f(y) : y \in D \backslash K(x^*, r_1)\}$.

Because of (1) we have $a > f^*$.

Because f is continuous and x^* lies in the interior of D, there is an $r_0 \in (0, r_1)$ with

$$K(x^*, r_1) \subset \{x \in D : f(x) < a\}.$$

Let $r \in (0, r_0)$ be arbitrary and $c := \min\{f(y) : |y - x^*| = r\}$.

To prove property (4) it is necessary to prove the inclusion

$$\{y \in D : f(y) < c\} \subset K(x^*, r).$$

Let for this purpose $x \in D \backslash K(x^*, r)$ be arbitrary.

If $|x - x^*| \geq r_1$, then we have by definition $f(x) \geq a$.

But now the inequality $c \leq a$ is satisfied because of $r \leq r_0$.

For if $c > a$, a $y \in \mathbb{R}^d$ would exist with

$$|y - x^*| = r \text{ and } f(y) = c > a.$$

For $z := x^* + \frac{r_0}{r}(y - x^*)$ it follows then that $|z - x^*| = r_0$.

Since y lies between x^* and z and because of the convexity of f on $K(x^*, r_1)$, Lemma B.1 yields the inequality

$$\frac{f(z) - f^*}{r_0} \geq \frac{f(y) - f^*}{r}, \text{ therefore}$$

$$f(z) \geq f^* + \frac{r_0}{r}(f(y) - f^*).$$

Because of $\frac{r_0}{r} \geq 1$, this yields the inequality

$$f(z) \geq f(y) > a,$$

which is in contradiction to $\overline{K(x^*, r_0)} \subset \{x \in D : f(x) < a\}$.

therefore we have $c \leq a$ and thus $f(x) \geq c$.

If $r \leq |x - x^*| < r_1$, we have with $z := x^* + \frac{r}{|x-x^*|}(x - x^*)$ at first $|z - z^*| = r$ and thus $f(z) \geq c$.

From the convexity of f on $K(x^*, r_1)$ it follows again with Lemma B.1 that

$$\frac{f(z) - f^*}{r} \leq \frac{f(x) - f^*}{|x - x^*|}.$$

From this it follows that $f(z) \leq f(x)$ and from this in turn that $f(x) \geq c$.

In both cases we therefore have $f(x) \geq c$.

Thus $D \backslash K(x^*, r) \subset \{y \in D : f(y) \geq c\}$ and therefore the Property (4) is satisfied.

If D and f are convex, the inclusion

$$\{y \in D : f(y) < c_r\} \subset K(x^*, r),$$

and thus the assertion, follows analogously for all $r > 0$ with

$$c_r := \inf\{f(y) : y \in D, |y - x^*| = r\}.$$

□

Corollary B.2 *Let the conditions of Lemma B.10 be satisfied.*

Then $r, a, b > 0$ exist with

$$a|x - x^*|^2 \leq f(x) - f^* \leq b|x - x^*|^2 \text{ for all } x \in K(x^*, r)$$

Proof The proof follows from the first inequality in part (2) of the proof of Lemma B.11. □

Corollary B.3 *Let the conditions of Lemma B.10 be satisfied.*

Then for all $x_0 \in D \backslash D^*$ constants $c \in (0, 1]$ and $d > 0$ exist with

$$\left\{y \in D : f(y) - f^* \leq \epsilon(f(x) - f^*)\right\} \subset K(x^*, d\sqrt{\epsilon}|x - x^*|)$$
$$\text{for all } x \in \{y \in D : f(y) \leq f(x_0)\} \text{ and any } \epsilon \in (0, c].$$

Proof Let $x_0 \in D \backslash D^*$.

According to Lemma B.10 there is a $c_0 > 0$ and an $a > 0$ with

$$\{y \in D : f(y) - f^* \leq \rho\} \subset K(x^*, a\sqrt{\rho}) \text{ for all } \rho \in (0, c_0].$$

If we choose $c = \frac{c_0}{f(x_0)-f^*}$, then we have for arbitrary $x \in \{y \in D : f(x) \leq f(x_0)\}$ and for arbitrary $\epsilon \in (0, c]$

$$\{y \in D : f(y) - f^* \leq \epsilon(f(x) - f^*)\} \subset K(x^*, a\sqrt{\epsilon}\sqrt{f(x) - f^*}).$$

According to Corollary B.2, $r > 0$ and $b > 0$ exist with

$$f(x) - f^* \leq b|x - x^*|^2 \text{ for all } x \in K(x^*, r).$$

If therefore $x \in K(x^*, r)$, we have

$$K(x^*, a\sqrt{\epsilon}\sqrt{f(x) - f^*}) \subset K(x^*, a\sqrt{b\epsilon}|x - x^*|)$$

If $|x - x^*| \geq r$, we have

$$f(x) - f^* \leq f(x_0) - f^* \leq \frac{f(x_0) - f^*}{|x - x^*|^2}(|x - x^*|^2) \quad \leq \frac{f(x_0) - f^*}{r^2}|x - x^*|^2$$

and thus

$$K(x^*, a\sqrt{\epsilon}\sqrt{f(x) - f^*}) \subset K(x^*, \frac{a}{r}\sqrt{f(x_0) - f^*\epsilon}|x - x^*|).$$

Now the assertion follows with $d := a \cdot \max\{\sqrt{b}, \frac{1}{r}\sqrt{f(x_0) - f^*}\}$. □

Corollary B.4 *Let $D \subset \mathbb{R}^d$ be convex. Let $f : D \to \mathbb{R}$ be convex and continuous.*

Assume there is $x^* \in D^0$ with $f(x) > f^*$ for all $x \in D\backslash\{x^*\}$. Let f be twice continuously differentiable in a neighborhood of x^* with a Hessian matrix that is positively definite at x^*.

Then for all $x_0 \in D$ constants $m > 0$, $M < \infty$ exist with

$$m|x - x^*|^2 \leq f(x) - f^* \leq M|x - x^*|^2$$
$$\textit{for all } x \in \{y \in D : f(y) \leq f(x_0)\}.$$

Proof Since D and f are convex, it follows from Lemma B.11 that all conditions of Lemma B.10 are satisfied. According to Corollary B.2, constants $\tilde{m}, r > 0$ and $\tilde{M} < \infty$ exist with

$$\tilde{m}|x - x^*|^2 \leq f(x) - f^* = \tilde{M}|x - x^*|^2 \text{ for all } x \in K(x^*, r).$$

Let now $x_0 \in D$ be arbitrary.

According to Lemma B.4 there is an $R \geq r$ with

$$\{y \in D : f(y) \leq f(x_0)\} \subset K(x^*, R).$$

Let $m := \frac{r}{R}\tilde{m}$ and $M := \max\{\frac{f(x_0)-f^*}{r^2}, \tilde{M}\}$.

Let $x \in \{y \in D : f(y) \leq f(x_0)\}$ be arbitrary.

If $x \in K(x^*, r)$, the assertion is trivial.

If $r \leq |x - x^*|$, we initially have

$$\frac{f(x) - f^*}{|x - x^*|^2} \leq \frac{f(x_0) - f^*}{r^2} \leq M, \text{ therefore } f(x) - f^* \leq M|x - x^*|^2.$$

To prove the second inequality, let $y := x^* + \frac{r}{|x-x^*|}(x - x^*)$.

Because of $|y - x^*| = r$ and according to Lemma B.1 it follows from the convexity of f that

$$\begin{aligned}\frac{f(x) - f^*}{|x - x^*|^2} &= \frac{1}{|x - x^*|}\frac{f(x) - f^*}{|x - x^*|} \leq \frac{1}{|x - x^*|}\frac{f(y) - f^*}{|y - x^*|} \\ &= \frac{|y - x^*|}{|x - x^*|}\frac{f(y) - f^*}{|y - x^*|^2} \leq \frac{r}{R}\tilde{m} = m.\end{aligned}$$

□

Remark B.1 The conditions of Lemma B.10 imply the conditions (14.2), (14.3), (15.1), and (15.2).

Proof Because of

$$\begin{aligned}\{y \in D : f(y) - f^* \leq \frac{c}{2}\} &\subset \{y \in D : f(y) - f^* < c\} \\ &\subset \{y \in D : f(y) - f^* \leq c\}\end{aligned}$$

($c > 0$ arbitrary), the conditions (14.2) and (14.3) follow directly from Lemma B.10.

Let $c_0, a_1, a_2 > 0$ be chosen according to Lemma B.10.

Then we have (cf. the proof of Corollary B.2) for all $x \in \{y \in D : f(y) - f^* \leq \frac{c_0}{2}\}$ the inequality

$$\frac{1}{2a_0^2}|x - x^*|^2 \leq f(x) - f^* \leq \frac{1}{a_1^2}|x - x^*|^2.$$

If we choose $\tilde{a} := f^* + \frac{c_0}{2}$, then it follows for all $x \in D$ with $f(x) \leq \tilde{a}$ that

$$\begin{aligned}\inf\{|y - y^*| : y \in D, f(y) = f(x)\} &\geq a_1\sqrt{f(x) - f^*} \\ &\geq \frac{a_1}{a_2} \cdot \frac{1}{\sqrt{2}} \sup\{|y - y^*| : y \in D, f(y) = f(x)\}.\end{aligned}$$

Condition (15.2) follows now with $\tilde{b} := \frac{a_1}{a_2} \cdot \frac{1}{\sqrt{2}}$.

According to the Conditions (2), (3), and (4) of Lemma B.10 there are $r, c, > 0$ such that f is twice continuously differentiable on $K(x^*, r)$, that the Hessian matrix of f is positive definite on $K(x^*, r)$ and that the set $\{y \in D, f(y) = f(x) \leq c\}$ is included in $K(x^*, r)$. With Lemma B.3 this implies condition (15.1). □

Lemma B.12 *Let condition (1), (2), and (3) of Lemma B.10 be satisfied.*
Let $H(x)$ be the Hessian Matrix of f at x. Then the proof is as follows.

Proof Because of $\nabla f(x^*) = 0$, the Taylor series expansion of f at x^* yields the representation

$$f(x) = f(x^*) + \frac{1}{2}(x - x^*)^T H(x)(x - x^*) + o(|x - x^*|).$$

Hence

$$\nabla f(x) = H(x^*)(x - x^*) + o(|x - x^*|),$$

therefore

$$\lim_{x \to x^*} \frac{\nabla f(x) - H(x^*)(x - x^*)}{|x - x^*|} = 0,$$

which—because of the continuity of $H(x)^{-1}$ neighborhood of x^*—yields

$$\lim_{x \to x^*} \frac{H(x)^{-1}\nabla f(x) - H(x)^{-1}H(x^*)(x - x^*)}{|x - x^*|} = 0,$$

and thus

$$\lim_{x \to x^*} \frac{H(x)^{-1}\nabla f(x) - (x - x^*)}{|x - x^*|} = 0,$$

□

Lemma B.13 *Let the conditions of Lemma B.10 be satisfied. Then constants $a > f$ and $b_1, b_2 > 0$ exist with*

$$b_1 \leq \frac{|\nabla f(x)|}{|x - x^*|} \leq b_2 \textit{ for all } x \in \{y \in D : f(y) \leq a\} \backslash \{x^*\}.$$

Proof From $\nabla f(x) = H(x^*)(x - x^*) + o(|x - x^*|)$(cf. proof of Lemma B.12) and because of

$$|\frac{|\nabla f(x)|}{|x - x^*|} - |H(x^*)\frac{x - x^*}{|x - x^*|}|| \leq |\frac{\nabla f(x)}{|x - x^*|} - H(x^*)\frac{x - x^*}{|x - x^*|}|$$

for each $\epsilon > 0$ a $\delta > 0$ exists with

$$|H(x^*)\frac{x-x^*}{|x-x^*|}|-\epsilon \leq \frac{|\nabla f(x)|}{|x-x^*|} \leq |H(x^*)\frac{x-x^*}{|x-x^*|}|+\epsilon$$

if $0 < |x - x^*| \leq \delta$. Let $M := \max\{|H(x^*)y| : y \in \mathbb{R}^d, |y| = 1\}$ and $m := \min\{|H(x^*)y| : y \in \mathbb{R}^d, |y| = 1\}$

From the compactness of the Unit Sphere in the $\mathbb{R}^d$ it follows that $M < \infty$. From the invertibility of $H(x^*)$ it follows that $m > 0$. Thus for $\epsilon = m/2$ an $r > 0$ exists with

$$\frac{m}{2} \leq \frac{|\nabla f(x)|}{|x-x^*|} \leq M + \frac{m}{2} \text{ for all } x \in K(x^*, r)\backslash\{x\}.$$

According to Lemma B.10, it can be assumed without loss of generality that there is a $c > 0$ with $\{x \in D : f(x) - f^* < c\} \subset K(x^*, r)$.

This proves the first assertion.

Let us now assume that $D = \mathbb{R}^d$ and that f is convex and twice continuously differentiable. Furthermore, let $a > f^*$ be arbitrary.

According to Lemma B.4, there is an $R > 0$ with

$$\{y \in D : f(y) \leq a\} \subset K(x^*, R).$$

If $R < r$, the second assertion follows with $c_1 = \frac{m}{2}$ and $c_2 = M + \frac{m}{2}$. Let $r \geq R$.

Let $m_1 := \inf\{|\nabla f(x)| : x \in D, f(x) \leq a, r \leq |x - x^*| \leq R\}$ and $m_2 := \sup\{|\nabla f(x)| : x \in D, f(x) \leq a, r \leq |x - x^*| \leq R\}$.

Because f is convex and continuously differentiable, we have $m_1 > 0$. Furthermore, we have $m_2 < \infty$.

Then for $x \in \mathbb{R}^d$ with $r \leq |x - x^*| \leq R$:

$$\frac{m_1}{R} \leq \frac{|\nabla f(x)|}{|x-x^*|} \leq \frac{m_2}{r}.$$

Now the second assertion follows with $c_1 := \min\{\frac{m}{2}, \frac{m_1}{2}\}$ and $c_2 := \max\{M + \frac{m}{2}, \frac{m_2}{r}\}$. □

Lemma B.14 *Let the conditions of Lemma B.10 be satisfied.*

Let $(A_x : x \in D)$ be a family of positive definite and symmetric matrices. Let the representation $x \to A_x$ be continuous.

Let $C_x := (\nabla f(x)^T A_x \nabla f(x))A_x$.

Then constants $a > f^*$ and $b_1, b_2 > 0$ exist with

$$b_1 \leq \frac{\lambda}{|x-x^*|^2} \leq b_2$$

for all $x \in \{y \in D : f(y) \leq a\}\backslash\{x^*\}$ and any eigenvalue λ of C_x.

If $D = \mathbb{R}^d$ and if f is also continuously differentiable and convex, for all $a > f^*$ there are constants $c_1, c_2 > 0$ with

$$c_1 \leq \frac{\lambda}{|x - x^*|^2} \leq c_2$$

for all $x \in \{y \in D : f(y) \leq a\}\backslash\{x^*\}$ and any eigenvalue λ of C_x.

Remark B.2 If we set $A_x = I$ (I = unity matrix), then C_x has only the d-fold eigenvalue $\lambda|\nabla f(x)|^2$. Thus Lemma B.13 is a special case of Lemma B.14.

Proof According to Lemma B.13, constants $a > f^*, a_1, a_2, > 0$ exist with

$$a_1 \leq \frac{|\nabla f(x)|}{|x_x^*|} \leq a_2$$

for all $x \in \{y \in D : f(y) \leq a\}\backslash\{x^*\}$.

Let $\lambda^1 := \max\{\lambda : \lambda$ eigenvalue of $A_x, f(x) \leq a\}$ and $\lambda^0 := \max\{\lambda : \lambda$ eigenvalue of $A_x, f(x) \leq a\}$.

Let λ be an arbitrary eigenvalue of A_x with associated normalized eigenvector z. Since λ is positive, it follows that

$$\lambda = |\lambda z| = |A_x z| \leq \sup\{|A_x z| : |z| = 1, f(x) \leq a\}$$

Because the set $\{(z, x) : |z| = 1, f(x) \leq a\}$ is compact (cf. Lemma B.2 and Lemma B.4) and the representation $(z, x) \to |A_x z|$ is continuous, it follows that $\lambda^1 < \infty$.

If it is taken into consideration that the inequality $|A_x z| > 0$ is satisfied for all z, x with $|z| = 1$ and $f(x) \leq a$, it follows analogously that $\lambda^0 > 0$.

Since

$$\lambda^0|y|^2 \leq |y^T A_x y| \leq \lambda^1|y|^2$$

for all $y \in \mathbb{R}^d$ and any $x \in \{y \in D : f(y) \leq a\}$ (cf. the proof of Lemma B.7, part (2)), we have

$$\lambda^0 \leq \frac{|y^T A_x y|}{|y|^2} \leq \lambda^1.$$

From

$$\frac{|\nabla f(x)^T A_x \nabla f(x)|}{|x - x^*|} = \frac{|\nabla f(x)^T A_x \nabla f(x)|}{|\nabla f(x)|^2} \frac{|\nabla f(x)|^2}{|x - x^*|}$$

we have then for all $x \in \{y \in D : f(y) \leq a\}\backslash\{x^*\}$ the inequality

$$\lambda^0 a_1^2 \leq \frac{|\nabla f(x)^T A_x \nabla f(x)|}{|x - x^*|^2} \leq \lambda^1 a_2^2.$$

Since each eigenvalue λ_x of C_x has the shape

$$\lambda_x = [\nabla f(x)^T A_x \nabla f(x)]\alpha_x$$

where α_x is an eigenvalue of A_x the assertion follows with $b_1 = (\lambda_0 a_1)^2$ and $b_2 = (\lambda^1 a_2)^2$. □

The remainder of this section contains primarily statements about the convergence behavior of special series.

Remark B.3 Let $(a_n : n \in \mathbb{N}), (b_n : n \in \mathbb{N})$ be sequences of nonnegative real numbers. Let $a := \liminf_{n\to\infty} a_n > 0$ and $b := \limsup_{n\to\infty} b_n > 0$. Then

$$\liminf_{n\to\infty} \frac{a_n}{b_n} \geq \frac{a}{b} \text{ and } \limsup_{n\to\infty} \frac{a_n}{b_n} \leq \frac{a}{b}.$$

Proof For arbitrary $\epsilon > 0$ define $c := \frac{b^2}{a+b}\epsilon$. Then $\frac{a-c}{a+c} \geq \frac{a}{b} - \epsilon$, because the definition of c implies that $c(a+b) \leq b^2\epsilon + b\epsilon c$, yielding $\frac{c(a+b)}{b+c} \leq b\epsilon$, and hence $\epsilon \geq \frac{c(a+b)}{b^2(1+\frac{c}{b})}$ and $(1+\frac{c}{b})\epsilon \geq \frac{c(a+b)}{b^2}$. This implies $-\frac{c}{b} \geq \frac{ac}{b^2} = (1+\frac{c}{b})\epsilon$, from which it follows that $\frac{a}{b} - \frac{c}{b} \geq (1+\frac{c}{b})(\frac{a}{b} - \epsilon)$.

This inequality yields $\frac{\frac{a-c}{b}}{1+\frac{c}{b}} \geq \frac{a}{b} - \epsilon$ which finally implies the desired inequality $\frac{a-c}{b+c} = \frac{a}{b} - \epsilon$.

The definition of a and b implies the desired inequality

$$a_n \geq a - c \text{ and } b_n \leq b + c$$

for all $n \in \mathbb{N}$. From this it follows finally for all $n \in \mathbb{N}$ that

$$\frac{a_n}{b_n} \geq \frac{a-c}{b+c} \geq \frac{a}{b} - \epsilon$$

and thus $\liminf_{n\to\infty} \frac{a_n}{b_n} \geq \frac{a}{b}$.

The second assertion follows analogously. □

Remark B.4 For arbitrary $q \in (-1, 1)$ and $i \in \mathbb{N}$ we have

$$\sum_{n=0}^{\infty} (n+1)\cdots(n+i)q^n = \frac{i!}{(1-q)^{i+1}}.$$

This implies the assertion.

Remark B.5 For all $x, y \geq 0$ and any $n \in \mathbb{N}$ we have

$$(x+y)^n \leq 2^{n-1}(x^n + y^n)$$

Proof Complete induction for n.
For $n = 1$ the assertion is trivial.
$n \to n+1$:

$$(a+b)^{n+1} = (a+b)(a+b)^n \leq (a+b)2^{n-1}(a^n+b^n) =$$
$$= 2^{n-1}(a^{n+1}+b^{n+1}+ab^n+ba^n) \leq 2^n(a^{n+1}+b^{n+1})$$

da $ab^n + ba^n \leq a^{n+1} + b^{n+1}$.
This goes to prove as follows:
Without loss of generality, let $a \leq b$.
For $x \in [a, b]$ let $f(x) := xb^n - bx^n - x^{n+1}$.
Then $f'(x) = b^n + nbx^{n-1} - (n+1)x^n = b^n - x^n + nx^{n-1}(b-x)$.
For $x \in [a, b]$ we therefore have $f'(x) \geq 0$.
Thus f is strictly increasing on the interval $[a, b]$.
Therefore we have $f(a) \leq f(b)$, i.e. $ab^n + ba^n \leq a^{n+1} + b^{n+1}$. □

Remark B.6 For all $n \in \mathbb{N}$ we have $n! \geq (\frac{n}{e})^n$

Proof Complete induction for n.
For $n = 1$ the assertion is trivial.
$n \to n+1$:
It is generally known that the sequence $(1 + \frac{1}{i})^i$ is strictly increasing and converges towards e. Therefore

$$\left(1+\frac{1}{i}\right)^i \leq e \text{ for all } i \in \mathbb{N}.$$

Hence

$$(n+1)! = n!(n+1) \geq \left(\frac{n}{e}\right)^n (n_1) = \left(\frac{n+1}{e}\right)^n (n+1)\frac{n^n}{(n+1)^n} =$$
$$= \left(\frac{n+1}{e}\right)^n (n+1)\left(1+\frac{1}{n}\right)^{-n} \frac{n+1}{e} =$$
$$= \left(\frac{n+1}{e}\right)^{n+1}$$

□

Lemma B.15 *Let* $0 < r < p < 1$ *and* $c \in \mathbb{R}$.
Let $a_n := \sum_{k=0}^{\lfloor rn+c \rfloor} \binom{n}{k} p^k (1-p)^{n-k}$. *Then:*

(1) $\sum_{n=1}^{\infty} na_n < \infty$

(2) $\sum_{n=1}^{\infty} a_n < \infty$
(3) $\lim_{n\to\infty} a_n = 0$

Proof Let $(X_n : n \in \mathbb{N})$ be a sequence of independent random variables on a probability field $(\Omega, \mathcal{A}, P)$. Let X_n be binomial distributed with the parameters n and p. Then:

$$a_n = P(X_n \leq [rn + c]) \leq P(X_n \leq rn + c) =$$
$$= P\left(\frac{X_n - np}{\sqrt{np(1-p)}} \leq \frac{(r-p)n + c}{\sqrt{np(1-p)}}\right)$$

Since $r - p > 0$, it follows finally for all $n \in \mathbb{N}$ that

$$a_n \leq P\left(\left|\frac{X_n - np}{\sqrt{np(1-p)}}\right| \geq \frac{(p-r)n - c}{\sqrt{np(1-p)}}\right)$$

Furthermore, it follows finally for all $n \in \mathbb{N}$ that

$$\frac{(p-r)n - c}{\sqrt{np(1-p)}} \geq n^{1/8}$$

and therefore

$$P\left(\left|\frac{X_n - np}{\sqrt{np(1-p)}}\right| \geq n^{1/8}\right) \geq a_n \text{ finally for all } n \in \mathbb{N}$$

We have

$$P\left(\left|\frac{X_n - np}{\sqrt{np(1-p)}}\right| \geq n^{1/8}\right) \frac{2}{\sqrt{2\pi}} n^{-1/8} \exp\left(-\frac{1}{2} n^{1/4}\right)$$

From

$$\sum_{m=1}^{\infty} m \frac{2}{\sqrt{2\pi}} m^{-1/8} \exp\left(-\frac{1}{2} m^{1/4}\right)$$
$$= \frac{2}{\sqrt{2\pi}} \sum_{m=1}^{\infty} \sum_{m=n^4}^{(n+1)^4-1} m \exp\left(-\frac{1}{2} m^{1/4}\right)$$
$$= \frac{2}{\sqrt{2\pi}} \sum_{m=1}^{\infty} ((n+1)^4 - n^4)(n+1)^4 \exp\left(-\frac{n}{2}\right) < \infty$$

follows (1). (2) and (3) are trivial requirements from (1). □

Lemma B.16

(i) For all $a > 0$ we have

$$\lim_{n\to\infty} n^{1/a} \int_0^1 (1 - x^a)^n dx = \Gamma(1 + \frac{1}{a})$$

(Γ Euler gamma function).

(ii) for all $b \in (0, 1)$ and $c > 0$ we have

$$\sum_{n=1}^{\infty} (1 - cn^{-b})^n < \infty.$$

Proof

(i) Partial integration yields

$$\int_0^1 (1 - x^a)^n dx = x(1 - x^a)^n|_0^1 + \int_0^1 xn(1 - x^a)^{n-1} a x^{a-1} dx$$

$$= an \int_0^1 x^a (1 - x^a)^{n-1} dx.$$

Using partial integration again, this yields

$$\int_0^1 (1 - x^a)^n dx = \frac{a^2 n(n-1)}{a+1} \int_0^1 x^{2a} (1 - x^a)^{n-2} dx.$$

When this is continued, we finally obtain

$$\int_0^1 (1 - x^a)^n dx = \frac{a^n n!}{(a+1)(2a+1)\cdots((n-1)a+1)} \int_0^1 x^{na} dx$$

$$= \frac{a^n n!}{(a+1)(2a+1)\cdots(na+1)}$$

$$= \frac{n!}{(\frac{1}{a}+1)(\frac{1}{a}+2)\cdots(\frac{1}{a}+n)}.$$

We have

$$\lim_{n\to\infty} \frac{n^{1/a} n!}{\frac{1}{a}(\frac{1}{a}+1)(\frac{1}{a}+2)\cdots(\frac{1}{a}+n)} = \Gamma(1/a).$$

From this it follows that

$$\lim_{n\to\infty} n^{1/a} \int_0^1 (1-x^a)^n dx = \frac{1}{a}\Gamma(\frac{1}{a}) = \Gamma(1+\frac{1}{a}).$$

(ii) Because of $b > 0$, we have $\lim\limits_{n\to\infty} n^b = \infty$ and thus

$$\lim_{n\to\infty} (1-cn^{-b})^{n^b} = e^{-c}.$$

For $q \in [e^{-c}, 1)$ therefore an $N \in \mathbb{N}$ exists with

$$(1-cn^{-b})^{n^b} \le q \text{ for all } n \ge N$$

For $n \ge N$ it follows therefore that $(1-cn^{-b})^n \le q^{n^{1-b}}$.
According to the condition, we have $1-b > 0$.
Let $s \in \mathbb{N}$ with $0 < \frac{1}{s} \le 1-b$.
Then

$$\sum_{n=1}^{\infty} q^{n^{1-b}} \le \sum_{n=1}^{\infty} q^{n^{1/s}} = \sum_{j=1}^{\infty} \sum_{k=j^s}^{(j+1)^s-1} q^{k^{1/s}}$$

$$\le \sum_{j=1}^{\infty} ((j+1)^s - j^s) q^{(j^s)^{1/s}} = \sum_{j=1}^{\infty} ((j+1)^s - j^s) q^j < \infty$$

This implies directly the assertion.

□

Appendix C
Probabilistic Tools

Lemma C.1 *Consider a sequence* $(X_n : n \in \mathbb{N})$ *of random variables with* $\sum_{n=1}^{\infty} \frac{EX_n^2}{n^2} < \infty$.

If $E(X_{n+1}|X_n, \ldots, X_1) \geq p > 0$ *P-a.s. for all* $n \in \mathbb{N}$, *then* $\liminf_{n\to\infty} \frac{1}{n} \sum_{i=1}^{n} X_i \geq p$ *P-a.s.*

If $E(X_{n+1}|X_n, \ldots, X_1) \leq p$ *P-a.s. for all* $n \in \mathbb{N}$, *then* $\limsup_{n\to\infty} \frac{1}{n} \sum_{i=1}^{n} X_i \leq p$ *P-a.s.*

Proof Because of $\sum_{n=1}^{\infty} \frac{EX_n^2}{n^2} < \infty$ we have

$$\lim_{n\to\infty} \frac{1}{n} \sum_{i=1}^{\infty} (X_i - E(X_i|X_{i-1}, \ldots, X_1)) = 0 \ P\text{-a.s.}\,.$$

In the first case, because of

$$(X_i - E(X_i|X_{i-1}, \ldots, X_1)) \leq X_i - p \ P\text{-a.s.}\,,$$

we obtain the inequality

$$\liminf_{n\to\infty} \frac{1}{n} \sum_{i=1}^{n} (X_i - p) \geq 0 \ P\text{-a.s.}\,.$$

K. Marti, *Optimization Under Stochastic Uncertainty*, International Series in Operations Research & Management Science 296,
https://doi.org/10.1007/978-3-030-55662-4

In the second case we get

$$\limsup_{n\to\infty} \frac{1}{n}\sum_{i=1}^{n}(X_i - p) \le 0 \; P\text{-a.s.}\,.$$

In both cases, this yields the assertion. □

Definition C.1 Let X denote a random variable on a probability space $(\Gamma, \mathcal{A}, P)$.
For $a \in \mathbb{R}$ and $A \in \mathcal{A}$ we define:

$$X \ge a \quad P\text{-a.s. on } A \text{ if and only if } (X - a)\cdot 1_A \ge 0 \quad P\text{-a.s.}\,.$$

Lemma C.2 *Let X be a random variable on a probability space $(\Omega, \mathcal{A}, P)$, and let let denote C a sub—σ-algebra of $\mathcal{A}$. Moreover, let $C \in C$, and consider an arbitrary $a \in \mathbb{R}$ and $B \in \mathcal{A}$. Then:*

(1)

$$E(X|C) \ge a \; P\text{-a.s. on } C \textit{ if and only if } \int XdP \ge aP(A\cap C) \textit{ for all } A \in C.$$

(2)

$$P(B|C) \ge a \; P\text{-a.s. on } C \textit{ if and only if } P(A\cap B\cap C) \ge aP(A\cap C) \textit{ for all } A \in C.$$

Proof Because of $P(B|C) = E(1_B|C)$, the second assertion follows from the first.
Moreover, the first one is essentially based on the fact that for arbitrary $\mathcal{B}$-measurable random variables Z ($\mathcal{B} \subset \mathcal{A}$) the following equivalence holds:

$$Z \ge 0 \quad P\text{-a.s. if and only if } \int_B ZdP \ge 0 \text{ for all } B \in \mathcal{B}.$$

The implication from the left to the right is obviously true.

The implication from the right to the left is obtained by means of a special selection of B, namely $B = \{Z < 0\}$.

Indeed, if $P(Z < 0) > 0$, then we would have $\int_{\{Z<0\}} ZdP < 0$, in contradiction to the assumption $\int_B ZdP \ge 0$.

Now we prove the first assertion.

From $C \in C$ and the C-measurability we get the C-measurability of $1_C \cdot (E(X|C) - a)$.

Hence, due to the above remark we get

$$E(X|C) \ge aP\text{-a.s. on } C \text{ if and only if}$$

$$1_C \cdot (E(X|C) - a) \geq 0 \quad P\text{-a.s. if and only if}$$

$$\int_A 1_C(E(X|C) - a)dP \geq 0 \text{ for all } A \in C \text{ if and only if}$$

$$\int_A E(X|C)dP \geq aP(A \cap C) \text{ for all } A \in C \text{ if and only if}$$

$$\int_{A \cap C} XdP \geq aP(A \cap C) \text{ for all } A \in C.$$

□

Lemma C.3 *Let denote $(\Omega, \mathcal{A}, P)$ a probability space, and let $(\mathcal{A}_n : n \in \mathbb{N})$ be an ascending sequence of sub-σ-algebras of $\mathcal{A}$.*

Let T be a stopping time with regard to $(\mathcal{A}_n : n \in \mathbb{N})$. Define $\mathcal{A}_T := \{A \in \mathcal{A} : A \cap \{T = n\} \in \mathcal{A}_n$ for all $n \in \mathbb{N}\}$.

For $n \in \mathbb{N}$ let Y_n and Z_n be $\mathcal{A}_n$-measurable random variables.

Finally, let $A, B \subset \mathbb{R}$ be Borel-measurable and $p \geq 0$ such that

$$P(Z_n \in B|\mathcal{A}_n) \geq p \ P\text{-a.s. on } \{Y_n \in A\}.$$

Then

$$P(Z_T \in B|\mathcal{A}_T) \geq p \ P\text{-a.s. on } \{Y_T \in A\}.$$

Proof Consider an arbitrary $C \in \mathcal{A}_T$.
for all $n \in \mathbb{N}$ we have $\{Y_T \in A\} \in \mathcal{A}_T$.

Thus,

$$\begin{aligned}
&P(\{Z_T \in B\} \cap \{Y_T \in A\} \cap C) \\
&= \sum_{k=1}^{\infty} P(\{Z_T \in B\} \cap \{Y_T \in A\} \cap C \cap \{T = k\}) \\
&= \sum_{k=1}^{\infty} P(\{Z_k \in B\} \cap \{Y_k \in A\} \cap C \cap \{T = k\}) \\
&= \sum_{k=1}^{\infty} \int_{\{Y_k \in A\} \cap C \cap \{T=k\}} P(Z_k \in B|\mathcal{A}_k)dP \\
&\leq \sum_{k=1}^{\infty} pP(\{Y_k \in A\} \cap C \cap \{T = k\})
\end{aligned}$$

$$= \sum_{k=1}^{\infty} pP(\{Y_k \in A\} \cap C \cap \{T = k\}) = pP(\{Y_T \in A\} \cap C).$$

Since $C \in \mathcal{A}_T$ was arbitrary, the assertion follows from Lemma C.3 . □

Lemma C.4 *Let $(X_n : n \in \mathbb{N}_0)$ be a sequence of random vectors with values in $\mathbb{R}^d$.*

Furthermore, let $(Y_n : n \in \mathbb{N}_0)$ be a sequence of random vectors such that Y_n is measurable with respect to $\mathcal{A} := \sigma(X_0, \dots, X_n)$ for all $n \in \mathbb{N}_0$.

Let $B \subset \mathbb{R}^d \times \mathbb{R}^d$ be measurable and define $Z_n := 1_B(X_{n-1}, X_n)(n \in \mathbb{N})$.

Moreover, suppose that there is a $p > 0$ and an $a > 0$ such that

$$P(Z_{n+1} = 1|\mathcal{A}_n) \geq p \ P\text{-a.s. on } \{Y_n \leq a\} \quad \text{for all } n \in \mathbb{N}_0.$$

Then for all $n \in \mathbb{N}$, $K \in \{0, \dots, n\}$ and each stopping time T relative to $(\mathcal{A}_n : n \in \mathbb{N})$ we have

(1)

$$P\left(\left\{\sum_{i=T+1}^{T+n} Z_i \leq K\right\} \cap \bigcap_{i=T+1}^{T+n} \{Y_i \leq a\} | \mathcal{A}_T\right)$$

$$\leq \sum_{j=0}^{K} (\binom{n}{j}) p^j (1-p)^{n-j} \ P\text{-a.s. on } \{Y_T \leq a\}$$

(2)

$$P\left(\left\{\sum_{i=T+1}^{T+n} Z_i \leq K\right\} \cap \bigcap_{i=T}^{T+n} \{Y_i \leq a\} | \mathcal{A}_T\right)$$

$$\leq P(Y_T \leq a) \sum_{j=0}^{K} (\binom{n}{j}) p^j (1-p)^{n-j}.$$

Proof

(1) Proof with complete induction for n.

Let $A \in \mathcal{A}_T := \{C \in \mathcal{A} : C \cap \{T = k\} \in \mathcal{A}_k \text{ for all } k \in \mathbb{N}_0\}$. From

$$P(Z_{n+1} = 1|\mathcal{A}_n) \geq p \ P\text{-a.s. on } \{Y_n \leq a\} \text{ for all } n \in \mathbb{N}_0$$

we get (cf. Lemma C3)

$$P(Z_{T+1} = 1|\mathcal{A}_T) \geq p \ P\text{-a.s. on } \{Y_T \leq a\} \text{ for all } n \in \mathbb{N}_0.$$

Assume now $n = 1$ and $K \in \{0, 1\}$.

For $K = 1$ the assertion is trivial.

Thus, put $K = 0$. Then

$$P\left(\left\{\sum_{i=T+1}^{T+n} Z_i \leq K\right\} \cap \bigcap_{i=T+1}^{T+n} \{Y_i \leq a\}|\mathcal{A}_T\right)$$
$$\leq P(Z_{T+1} = 0|\mathcal{A}_T) \geq 1 - p \ P\text{-a.s. on } \{Y_T \leq a\}.$$

Thus, the assertion is shown for $n = 1$.

Assume now that the assertion holds for a fixed $n \in \mathbb{N}$, for all $K \in \{0, \ldots, n\}$ and for all stopping times T relative to $(\mathcal{A}_n : n \in \mathbb{N})$.

We have to show now that the assertion also holds for $n + 1$, i.e., for all stopping times T relative to $(\mathcal{A}_n : n \in \mathbb{N})$, for each $K \in \{0, \ldots, n + 1\}$ and for arbitrary $A \in \mathcal{A}_T$ it must be shown that the following inequality holds:

$$P\left(A \cap \{Y_T \leq a\} \cap \{\sum_{i=T+1}^{T+n} Z_i \leq K\} \cap \bigcap_{i=T+1}^{T+n} \{Y_i \leq a\}\right)$$
$$\leq P(A \cap \{Y_T \leq a\}) \sum_{j=0}^{K} \binom{n+1}{j} p^j (1-p)^{n+1-j}.$$

Without loss of generality we can assume that $K \in \{0, \ldots, n\}$ (in case $k = n + 1$ the inequality obviously holds true). Applying the condition of induction to the stopping time $T + 1$, we get

$$P\left(A \cap \{Y_T \leq a\} \cap \{\sum_{i=T+1}^{T+n+1} Z_i \leq K\} \cap \bigcap_{i=T+1}^{T+n+1} \{Y_i \leq a\}\right)$$
$$= P\left(A \cap \{Y_T \leq a\} \cap \{Y_{T+1} \leq a\} \cap \{Z_{T+1} = 1\}\right.$$
$$\left.\cap \left\{\sum_{i=T+1+1}^{T+1+n} Z_i \leq K - 1\right\} \cap \bigcap_{i=T+1+1}^{T+1+n} \{Y_i \leq a\}\right)$$
$$+ P\left(A \cap \{Y_T \leq a\} \cap \{Y_{T+1} \leq a\} \cap \{Z_{T+1} = 0\}\right.$$
$$\left.\cap \left\{\sum_{i=T+1+1}^{T+1+n} Z_i \leq K\right\} \cap \bigcap_{i=T+1+1}^{T+1+n} \{Y_i \leq a\}\right).$$

For the first term we have

$$P\,(A\cap\{Y_T\le a\}\cap\{Y_{T+1}\le a\}\cap\{Z_{T+1}=1\}$$

$$\cap\{\sum_{i=T+1+1}^{T+1+n} Z_i\le K-1\}\cap\bigcap_{i=T+1+1}^{T+1+n}\{Y_i\le a\}\Bigg)$$

$$=\int P\left(\left\{\sum_{i=T+1+1}^{T+1+n} Z_i\le K-1\right\}\cap\bigcap_{i=T+1+1}^{T+1+n}\{Y_i\le a\}|A_{T+1}\right)dP$$

$$\le A\cap\{Y_Z\le a\}\cap\{Y_{T+1}\le a\}\cap\{Z_{T+1}=1\}$$

$$\left(\sum_{j=0}^{K-1}\binom{n}{j}p^j(1-p)^{n-j}\right)P(A\cap\{Y_T\le a\}\cap\{Z_{T+1}=1\}).$$

Analogously, for the second term we find the inequality

$$P(A\cap\{Y_T\le a\}\cap\{Y_{T+1}\le a\}\cap\{Z_{T+1}=0\}$$

$$\cap\left\{\sum_{i=T+1+1}^{T+1+n} Z_i\le K\right\}\cap\bigcap_{i=T+1+1}^{T+1+n}\{Y_i\le a\})$$

$$\le\left(\sum_{j=0}^{K}\binom{n}{j}p^j(1-p)^{n-j}\right)P(A\cap\{Y_T\le a\}\cap\{Z_{T+1}=0\})$$

$$=(P(A\cap\{Y_T\le a\})-P\,(A\cap\{Y_T\le a\}\cap\{Z_{T+1}=1\}))$$

$$\times\sum_{j=0}^{K}\binom{n}{j}p^j(1-p)^{n-j}.$$

Hence,

$$P\left(A\cap\{Y_T\le a\}\cap\{\sum_{i=T+1}^{T+1+n} Z_i\le K\}\cap\bigcap_{i=T+1}^{T+n+1}\{Y_i\le a\}\right)$$

$$\le P(A\cap\{Y_T\le a\})\sum_{j=0}^{K}\binom{n}{j}P^j((1-P)^{n-j})-P\,(A\cap\{Y_T\le a\}$$

$$\cap\{Z_{T+1}=1\})\binom{n}{K}p^K(1-p)^{n-K}$$

$$\leq P(A \cap \{Y_T \leq a\})(\sum_{j=0}^{K} \binom{n}{j} p^j (1-p)^{n-j} - p\binom{n}{K} p^K (1-p)^{n-K})$$

$$= P(A \cap \{Y_T \leq a\}) \left((1-p) \sum_{j=0}^{K} \binom{n}{j} p^j (1-p)^{n-j} \right.$$

$$\left. +p \sum_{j=1}^{K} \binom{n}{K} p^j (1-p)^{n-j} \right)$$

$$= P(A \cap \{Y_T \leq a\}) \left(\sum_{j=0}^{K} \binom{n}{j} p^j (1-p)^{n-j+1} \right.$$

$$\left. \times \sum_{j=1}^{K} \binom{n}{j-1} p^j (1-p)^{n-j+1} \right)$$

$$= P(A \cap \{Y_T \leq a\}) \sum_{j=0}^{K} \binom{n+1}{j} p^j (1-p)^{n+1-j},$$

where the last inequality follows from

$$P(A \cap \{Y_T \leq a\} \cap \{Z_{T+1} = 1\}) = \int_{A \cap \{Y_T \leq A\}} P(\{Z_{T+1} = 1 | A_T\}) dP \geq$$

$$= pP(A \cap \{Y_T \leq a\}).$$

Thus, the first assertion is shown.

(2) The second assertion is a direct consequence of the first one.

□

Lemma C.5 *Let $(M_n : n \in \mathbb{N})$ be a sequence of random variables with values in $\mathbb{N}$.*

If

$$P(M_n \geq k) \leq a_k \textit{ for all } k, n \in \mathbb{N}, \textit{ and } \sum_{k=1}^{\infty} a_k < \infty,$$

then $\lim\limits_{n \to \infty} \frac{M_n}{n} = 0$ P-a.s. .

Proof For arbitrary integers $i, n \in \mathbb{N}$ we have

$$
\begin{aligned}
sP\left(\sup_{m\geq n}\left|\frac{M_m}{m}\right| > 1/i\right) &\leq P\left(\bigcup_{m\geq n}\left\{\frac{M_m}{m}\geq 1/i\right\}\right)\\
&\leq \sum_{m\geq n} P(M_m \geq m/i) \leq \sum_{m\geq n} P(M_m \geq [m/i])\\
&\leq \sum_{m\geq [n/i]i} P(M_m \geq [m/i])\\
&= \sum_{j=[n/i]}^{\infty} \sum_{m=j\cdot i}^{(j+1)i-1} P(M_m \geq [m/i])\\
&\leq \sum_{j=[n/i]}^{\infty} \mathrm{iP}(M_m \geq [\frac{\mathrm{ji}}{i}]) = i\sum_{j=[n/i]}^{\infty} P(M_m \geq j)\\
&\leq i\sum_{j=[n/i]}^{\infty} a_j.
\end{aligned}
$$

If $n \to \infty$, then $[n/i] \to \infty$ and therefore

$$
\lim_{n\to\infty} \sum_{j=[n/i]}^{\infty} a_j = 0, \text{ since } \sum_{k=1}^{\infty} a_k < \infty.
$$

This implies $P(\sup_{m\geq n} |\frac{M_m}{m}| = 1/i) \xrightarrow[n\to\infty]{0}$ for all $i \in \mathbb{N}$ and thus the assertion. □

Lemma C.6 *Let be $(Y_n : n \in \mathbb{N})$ a sequence of random variables, $B, C \in \mathbb{R}$ measurable sets and $c \in \mathbb{R}$, $p_1, p_2 > 0$ such that*

$$
\liminf_{n\to\infty} \frac{\sum_{i=1}^{n} 1_{\{Y_i\in B\}}}{\sum_{i=1}^{n} 1_{\{Y_i<c\}}} \geq p_1 \ P\text{-a.s. and}
$$

$$
\liminf_{n\to\infty} \frac{\sum_{i=1}^{n} 1_{\{Y_i\in B, Z_i\in C\}}}{\sum_{i=1}^{n} 1_{\{Y_i<B\}}} \geq p_2 \ P\text{-a.s.}\ .
$$

Then $\liminf_{n\to\infty} \frac{1}{n}\sum_{i=1}^{n} 1_{\{Y_i\in B, Z_i\in C\}} + 1_{\{Y_i<c\}} \geq p_1 p_2$ *P-a.s.* .

Proof Let $a > 0$ be arbitrary.

Choosing $a_1, a_2 > 0$ such that $p_1 - a_1 > 0$, $p_2 - a_2 > 0$ and $(p_1 - a_1)(p_2 - a_2) \geq p_1 p_2 - a$, then

$$\sum_{i=1}^{n} 1_{\{Y_i \in B\}} \geq (p_1 - a_1) \sum_{i=1}^{n} 1_{\{Y_i < c\}} \text{ for finally any } n \in \mathbb{N} \ P\text{-a.s.}$$

and

$$\sum_{i=1}^{n} 1_{\{Y_i \in B, Z_i \in C\}} \geq (p_2 - a_2) \sum_{i=1}^{n} 1_{\{Y_i \in B\}} \text{ for finally any } n \in \mathbb{N} \ P\text{-a.s.}.$$

Hence,

$$\begin{aligned} &\sum_{i=1}^{n} 1_{\{Y_i \in B, Z_i \in C\}} + \sum_{i=1}^{n} 1_{\{Y_i \geq c\}} \\ &\quad \geq (p_1 - a_1)(p_2 - a_2) \sum_{i=1}^{n} 1_{\{Y_i < c\}} + \sum_{i=1}^{n} 1_{\{Y_i \geq c\}} \end{aligned}$$

for finally any $n \in \mathbb{N}$ $\quad P$-a.s. and therefore

$$\begin{aligned} &\sum_{i=1}^{n} 1_{\{Y_i \in B, Z_i \in C\}} + \sum_{i=1}^{n} 1_{\{Y_i \geq c\}} \\ &\geq \sum_{i=1}^{n} (1_{\{Y_i < c\}} + 1_{\{Y_i \geq c\}}) - (1 - (p_1 - a_1)(p_2 - a_2)) \sum_{i=1}^{n} 1_{\{Y_i \geq c\}} \\ &\geq n - (1 - (p_1 - a_1)(p_2 - a_2))n = (p_1 - a_1)(p_2 - a_2)n \\ &\leq (p_1 p_2 - a)n \end{aligned}$$

for finally any $n \in \mathbb{N}$ $\quad P$-a.s..

This yields the assertion. □

Definition C.2 Let $(Y_n : n \in \mathbb{N}_0)$ be a sequence of random variables and define

$$T_1 := \inf\{n \in \mathbb{N}_0 : Y_n \in B\}$$

$$\overline{T}_1 := \inf\{n \geq T_1 : Y_n \in [B\}$$

with a measurable $B \subset \mathbb{R}$. Moreover, define

$$T_{i+1} := \inf\{n \geq \overline{T}_i : Y_n \in B\}$$

$$\overline{T}_{i+1} := \inf\{n \geq T_{i+1} : Y_n \in [B\}.$$

Furthermore, $M_i := \overline{T}_i - T_i \quad (\infty - \infty := \infty)$.

T_i is called the **i-th time of entry** of the sequence $(Y_n : n \in \mathbb{N}_0)$ into B.

M_i is called the **i-th time of sojourn** of the sequence $(Y_n : n \in \mathbb{N}_0)$ within B.

Remark C.1 If $Y_i \in B$, then $\overline{T}_i$ is the i-th time of entry into $[B$.

If $Y_i \in [B$, then $\overline{T}_i$ is the $i+1$-th time of entry into $[B$.

Remark C.2 The random variables $T_i (i \in \mathbb{N})$ are stopping times (relative to the sequence $(\sigma(Y_0, \ldots, Y_n))$: $n \in \mathbb{N}_0)$.

Proof Let denote A_n the σ-algebra generated by $Y_0, \ldots, Y_n$. Because of $\{T_1 = n\} = \{Y_0 \in [B, \ldots, Y_{n-1} \in [B, Y_n \in B\} \in A_n$, it follows that T_1 is a stopping time. Analogously, it can be shown that $\overline{T}_1$ is a stopping time. Moreover, the relation

$$\{T_2 = i\} = \bigcap_{j=0}^{i-1} \{T_1 = j\} \cap \{Y_{j+1} \in [B\} \cap \cdots \cap \{Y_{i-1} \in [B\} \cap \{Y_i \in B\} \in A_i$$

yields that T_2 is a stopping time. Proceeding this way, we get the desired result. □

Remark C.3 Consider an isotonic sequence $(\mathcal{A}_n : n \in \mathbb{N})$ of σ-algebras such that $\sigma(Y_1, \ldots, Y_n) \subset \mathcal{A}_n$ for all $n \in \mathbb{N}$.

For $n \in \mathbb{N}$, assume that

$$\mathcal{A}_{T_n} := \{A \in \mathcal{A} : A \cap \{T_N = k\} \in \mathcal{A}_k \text{ for all } k \in \mathbb{N}\}.$$

Then

(1) $\sigma(Y_1, \ldots, Y_n) \subset \mathcal{A}_{T_n}$
(2) $\sigma(T_1, \ldots, T_n) \subset \mathcal{A}_{T_n}$
(3) $\mathcal{A}_{T_n} \subset \mathcal{A}_{T_{n+1}}$ and $\mathcal{A}_{T_n + M_n} \subset \mathcal{A}_{T_{n+1}}$
(4) $\sigma(M_1, \ldots, M_n) \subset \mathcal{A}_{T_{n+1}}$.

Proof (1) and (2) are trivial. The first assertion in (3) holds because of $T_n \leq T_{n+1}$. Because of $T_n + M_n = \overline{T}_n$, the variable is $T_n + M_n$ the n-th or $n+1$-th time of entry into $[B$ (cf. Remark C.1), hence, again a stopping time relative to $(\mathcal{A}_n : n \in \mathbb{N})$.

Because of $T_n + M_n \leq T_{n+1}$, we get the second assertion in 3).

Since $T_n + M_n \leq T_{n+1}$, we have $\sigma(T_n + M_n) \subset \mathcal{A}_{T_{n+1}}$.

Due to $M_n = (T_n + M_n) - T_n$ and $\sigma(T_n) \subset \mathcal{A}_{T_{n+1}}$, we have $\sigma(M_n) \subset \mathcal{A}_{T_{n+1}}$. In the same way we get $\sigma(M_i) \subset \mathcal{A}_{T_{n+1}})$ for $i = 1, \ldots, n-1$, which proves the last assertion. □

Index

K. Marti, *Optimization Under Stochastic Uncertainty*, International Series in Operations Research & Management Science 296,
https://doi.org/10.1007/978-3-030-55662-4

MIX
Papier aus verantwortungsvollen Quellen
Paper from responsible sources
FSC® C105338

If you have any concerns about our products,
you can contact us on
ProductSafety@springernature.com

In case Publisher is established outside the EU,
the EU authorized representative is:
Springer Nature Customer Service Center GmbH
Europaplatz 3, 69115 Heidelberg, Germany

Printed by Libri Plureos GmbH
in Hamburg, Germany